Causal Symmetric Spaces
Geometry and Harmonic Analysis

PERSPECTIVES IN MATHEMATICS, Vol. 18
S. Helgason, editor

Causal Symmetric Spaces
Geometry and Harmonic Analysis

Joachim Hilgert
Department of Mathematics
University of Wisconsin

Gestur Ólafsson
Department of Mathematics
Louisiana State University

ACADEMIC PRESS
San Diego London Boston New York Sydney Tokyo Toronto

ACADEMIC PRESS, INC.
525 B Street, Suite 1900, San Diego, CA 92101-4495, USA
1300 Boylston Street, Chestnut Hill, MA 02167, USA
http://www.apnet.com

ACADEMIC PRESS LIMITED
24–28 Oval Road, London NW1 7DX, UK
http://www.hbuk.co.uk/ap/

Library of Congress Cataloging-in-Publication Data

ISBN 0-12-388711-9

Printed in the United States of America
96 97 98 99 00 BB 9 8 7 6 5 4 3 2 1

Contents

Preface

In the late 1970s several mathematicians independently started to study a possible interplay of order and continuous symmetry. The motivation for doing so came from various sources. K.H. Hofmann and J.D. Lawson [67] tried to incorporate ideas from geometric control theory into a systematic Lie theory for semigroups, S. Paneitz [147, 148] built on concepts from cosmology as propagated by his teacher I.E. Segal [157], E.B. Vinberg's [166] starting point was automorphism groups of cones, and G.I. Ol'shanskii [137, 138, 139] was lead to semigroups and orders by his studies of unitary representations of certain infinite-dimensional Lie groups. It was Ol'shanskii who first considered the subject proper of the present book, causal symmetric spaces, and showed how they could play an important role in harmonic analysis. In particular, he exhibited the role of semigroups in the Gelfand-Gindikin program [34], which is designed to realize families of similar unitary representations simultaneously in a unified geometric way. This line of research attracted other researchers such as R.J. Stanton [159], B. Ørsted, and the authors of the present book [159].

This book grew out of the *Habilitationschrift* of G. Ólafsson [129], in which many of the results anounced by Ol'shanskii were proven and a classification of invariant causal structures on symmetric spaces was given. The theory of causal symmetric spaces has seen a rapid development in the last decade, with important contributions in particular by J. Faraut [24, 25, 26, 28] and K.-H. Neeb [114, 115, 116, 117, 120]. Its role in the study of unitarizable highest-weight representations is becoming increasingly clear [16, 60, 61, 88] and harmonic analysis on these spaces turned out to be very rich with interesting applications to the study of integral equations [31, 53, 58] and groupoid C^*-algebras [55]. Present research also deals with the relation to Jordan algebras [59, 86] and convexity properties of gradient flows [49, 125]. Thus it is not possible to write a definitive treatment of ordered symmetric spaces at this point. On the other hand, even results considered "standard" by the specialists in the field so far have either not appeared in print at all or else can be found only in the original literature.

This book is meant to introduce researchers and graduate students with a solid background in Lie theory to the theory of causal symmetric spaces, to make the basic results and their proofs available, and to describe some important lines of research in the field. It has gone through various stages and quite a few people helped us through their comments and corrections, encouragement, and criticism. Many thanks to W. Bertram, F. Betten, J. Faraut, S. Helgason, T. Kobayashi, J. Kockmann, Kh. Koufany, B. Krötz, K.-H. Neeb, and B. Ørsted. We would also like to thank the Mittag-Leffler Institute in Djursholm, Sweden, for the hospitality during our stay there in spring 1996.

Djursholm
Gestur Ólafsson
Joachim Hilgert

Introduction

Symmetric spaces are manifolds with additional structure. In particular they are homogeneous, i.e., they admit a transitive Lie group action. The study of causality in general relativity naturally leads to "orderings" of manifolds [149, 157]. The basic idea is to fix a convex cone (modeled after the light cone in relativity) in each tangent space and to say that a point x in the manifold precedes a point y if x can be connected to y by a curve whose derivative lies in the respective cone wherever it exists (i.e., the derivative is a timelike vector). Various technical problems arise from this concept. First of all, the resulting "order" relation need not be antisymmetric. This describes phenomena such as time traveling and leads to the concept of causal orientation (or quasi-order), in which antisymmetry is not required. Moreover, the relation may not be closed and may depend on the choice of the class of curves admitted. Geometric control theory has developed tools to deal with such questions, but things simplify considerably when one assumes that the field of cones is invariant under the Lie group acting transitively. Then a single cone will completely determine the whole field and it is no longer necessary to consider questions such as continuity or differentiablility of a field of cones. In addition, one now has an algebraic object coming with the relation: If one fixes a base point $\mathbf{o}$, the set of points preceded by $\mathbf{o}$ may be viewed as a positive domain in the symmetric space and the set of group elements mapping the positive domain into itself is a semigroup, which is a very effective tool in studying causal orientations. Since we consider only homogeneous manifolds, we restrict our attention to this simplified approach to causality.

Not every cone in the tangent space at $\mathbf{o}$ of a symmetric space $\mathcal{M}$ leads to a causal orientation. It has to be invariant under the action of the stabilizer group H of $\mathbf{o}$. At the moment one is nowhere near a complete classification of the cones satisfying this condition, but in the case of irreducible semisimple symmetric spaces it is possible to single out the ones which admit such cones. These spaces are then simply called causal symmetric spaces, and it turns out that the existence of a causal orientation puts severe restrictions on the structure of the space. More precisely, associated to each symmetric space $\mathcal{M} = G/H$ there is an involution $\tau: G \to G$ whose infinitesimal version (also denoted by τ) yields an eigenspace decomposition $\mathfrak{g} = \mathfrak{h} + \mathfrak{q}$ of the Lie algebra $\mathfrak{g}$ of G, where $\mathfrak{h}$, the Lie algebra of H, is the eigenspace for the eigenvalue 1 and $\mathfrak{q}$ is the eigenspace for the eigenvalue -1. Then the tangent space of $\mathcal{M}$ at $\mathbf{o}$ can be identified with $\mathfrak{q}$ and causal orientations are in one-to-one correspondence with H-invariant cones in $\mathfrak{q}$.

A heavy use of the available structural information makes it possible to classify all regular (i.e., containing no lines but interior points) H-invariant cones in $\mathfrak{q}$. This classification is done in terms of the intersection with a Cartan subspace $\mathfrak{a}$ of $\mathfrak{q}$. The resulting cones in $\mathfrak{a}$ can then be described explicitly via the machinery of root systems and Weyl groups.

It turns out that causal symmetric spaces come in two families. If $\mathfrak{g} = \mathfrak{k} + \mathfrak{p}$ is a Cartan decomposition compatible with $\mathfrak{g} = \mathfrak{h} + \mathfrak{q}$, i.e., τ and the Cartan involution commute, then a regular H-invariant cone in $\mathfrak{q}$ either has interior points contained in $\mathfrak{k}$ or in $\mathfrak{p}$. Accordingly the resulting causal orientation is called compactly causal or noncompactly causal. Compactly causal and noncompactly causal symmetric spaces show a radically different behavior. So, for instance, noncompactly causal orientations are partial orders with compact order intervals, whereas compactly causal orientations need neither be antisymmetric nor, if they are, have compact order intervals. On the other hand, there is a duality between compactly causal and noncompactly causal symmetric spaces which on the infinitesimal level can be described by the correspondence $\mathfrak{h} + \mathfrak{q} \longleftrightarrow \mathfrak{h} + i\mathfrak{q}$. There are symmetric spaces which admit compactly as well as noncompactly causal orientations. They are called spaces of Cayley type and are in certain respects the spaces most accessible to explicit analysis.

The geometry of causal symmetric spaces is closely related to the geometry of Hermitian symmetric domains. In fact, for compactly causal symmetric spaces the associated Riemannian symmetric space G/K, where K is the analytic subgroup of G corresponding to $\mathfrak{k}$, is a Hermitian symmetric domain and the space $H/(H \cap K)$ can be realized as a bounded *real* domain by a real analog of the Harish–Chandra embedding theorem. This indicates that concepts such as strongly orthogonal roots that can be applied successfully in the context of Hermitian symmetric domains are also important for causal symmetric spaces. Similar things could be said about Euclidian Jordan algebras.

Harmonic analysis on causal symmetric spaces differs from harmonic analysis on Riemannian symmetric spaces in various respects. So, for instance, the stabilizer group of the basepoint is noncompact, which accounts for considerable difficulties in the definition and analysis of spherical functions. Moreover, useful decompositions such as the Iwasawa decomposition do not have global analogs. On the other hand, the specific structural information one has for causal symmetric space makes it possible to create tools that are not available in the context of Riemannian or general semisimple symmetric spaces. Examples of such tools are the order compactification of noncompactly causal symmetric spaces and the various semigroups associated to a causal orientation.

The applications of causal symmetric spaces in analysis, most notably spherical functions, highest-weight representations, and Wiener-Hopf operators, have not yet found a definitive form, so we decided to give a mere outline of the analysis explaining in which way the geometry of causal orientations enters. In addition we provide a guide to the original literature as we know it.

We describe the contents of the book in a little more detail. In Chapter 1 we review some basic structure theory for symmetric spaces. In particular, we introduce the duality constructions which will play an important role in the theory of causal symmetric spaces. The core of the chapter consists of a detailed study of the H- and $\mathfrak{h}$-module structures on $\mathfrak{q}$. The resulting information plays a decisive role in determining symmetric spaces admitting G-invariant causal structures. Two classes of examples are treated in some detail to illustrate the theory: hyperboloids and symmetric spaces obtained via duality from tube domains.

In Chapter 2 we review some basic facts about convex cones and give precise definitions of the objects necessary to study causal orientations. These definitions are illustrated by a series of examples that will be important later on in the text. The central results are a series of theorems due to Kostant, Paneitz, and Vinberg, giving conditions for finite-dimensional representations to contain convex cones invariant under the group action. We also introduce the causal compactification of an ordered homogeneous space, a construction that plays a role in the analysis on such spaces.

Chapter 3 is devoted to the determination of all irreducible symmetric spaces which admit causal structures. The strategy is to characterize the existence of causal structures in terms of the module structure of $\mathfrak{q}$ and then use the results of Chapter 1 to narrow down the scope of the theory to a point where a classification is possible. We give a list of symmetric pairs $(\mathfrak{g}, \tau)$ which come from causal symmetric spaces and give necessary and sufficient conditions for the various covering spaces to be causal.

In Chapter 4 we determine all the cones that lead to causal structures on the symmetric spaces described in Chapter 3. Using duality it suffices to do that for the noncompactly causal symmetric spaces. It turns out that up to sign one always has a minimal and a maximal cone giving rise to a causal structure. The cones are determined by their intersection with a certain (small) abelian subspace $\mathfrak{a}$ of $\mathfrak{q}$ and it is possible to characterize the cones that occur as intersections with $\mathfrak{a}$. Since the cone in $\mathfrak{q}$ can be recovered from the cone in $\mathfrak{a}$, this gives an effective classification of causal structures. A technical result, needed to carry out this program but of independent interest, is the linear convexity theorem, which describes the image of certain coadjoint orbits under the orthogonal proction from $\mathfrak{q}$ to $\mathfrak{a}$.

Chapter 5 is a collection of global geometric results on noncompactly causal symmetric spaces which are frequently used in the harmonic analysis of such spaces. In particular, it is shown that the order on such spaces has compact intervals. Moreover, the causal semigroup associated naturally to the maximal causal structure is characterized as the subsemigroup of G which leaves a certain open domain in a flag manifold of G invariant. The detailed information on the ordering obtained by this characterization allows to prove a nonlinear analog of the convexity theorem which plays an important role in the study of spherical functions. Moreover, one has an Iwasawa-like decomposition for the causal semigroup which makes it possible to describe the positive cone of the symmetric space purely in terms of the "solvable part" of the semigroup, which can be embedded in a semigroup of affine selfmaps. This point of view allows us to compactify the solvable semigroup and in this way makes a better understanding of the causal compactification possible.

In Chapter 6 we pursue the study of the order compactification of noncompactly causal symmetric spaces. The results on the solvable part of the causal semigroup together with realization of the causal semigroup as a compression semigroup acting on a flag manifold yield a new description of the compactification of the positive cone which then can be used to give a very explicit picture of the G-orbit structure of the order compactification.

The last three chapters are devoted to applications of the theory in harmonic analysis.

In Chapter 7 we sketch a few of the connections the theory of causal symmetric spaces has with unitary representation theory. It turns out that unitary highest-weight representations are characterized by the fact that they admit analytic extensions to semigroups of the type considered here. This opens the way to construct Hardy spaces and gives a conceptual interpretation of the holomorphic discrete series for compactly causal symmetric spaces along the lines of the Gelfand–Gindikin program.

Chapter 8 contains a brief description of the spherical Laplace transform for noncompactly causal symmetric spaces. We introduce the corresponding spherical functions, describe their asymptotic behavior, and give an inversion formula.

In Chapter 9 we briefly explain how the causal compactification from Chapter 2 is used in the study of Wiener-Hopf operators on noncompactly causal symmetric spaces. In particular, we show how the results of Chapter 6 yield structural information on the C^*-algebra generated by the Wiener-Hopf operators.

Appendix A consists of background material on reductive Lie groups and their finite-dimensional representations. In particular, there is a collection of our version of the standard semisimple notation in Section A.1. Moreover, in Section A.3 we assemble material on Hermitian Lie groups and Hermitian symmetric spaces which is used throughout the text.

In Appendix B we describe some topological properties of the set of closed subsets of a locally compact space. This material is needed to study compactifications of homogeneous ordered spaces.

Causal Symmetric Spaces
Geometry and Harmonic Analysis

Chapter 1

Symmetric Spaces

In this chapter we present those parts of the theory of symmetric spaces which are essential for the study of causal structures on these spaces. Since in this book we are mainly interested in the group theoretical aspects, we use here the definition in [81] which is given in group theoretical rather than differential geometric terms.

Apart from various well known standard facts we describe several duality constructions which will play an important role in our treatment of causal structures. The central part of this chapter is a detailed discussion of the module structure of the tangent spaces of semisimple non-Riemannian symmetric spaces. The results of this discussion will play a crucial role in our study of causal structures on symmetric spaces. Finally we review some technical decomposition results due to Oshima and Matsuki which will be needed in later chapters.

In order to illustrate the theory with examples that are relevant in the context of causal symmetric spaces we treat the hyperboloids in some detail.

1.1 Basic Structure Theory

Definition 1.1.1 A *symmetric space* is a triple (G, H, τ), where

1) G is a Lie group,

2) τ is a nontrivial involution on G, i.e., $\tau\colon G \to G$ is an automorphism with $\tau^2 = \mathrm{Id}_G$, and

3) H is a closed subgroup of G such that $G^\tau_o \subset H \subset G^\tau$.

Here the subscript $_o$ means *the connected component containing the identity* and G^τ denotes the group of τ-fixed points in G. By abuse of notation we also say that (G, H) as well as G/H is a symmetric space. □

The infinitesimal version of Definition 1.1.1 is

Definition 1.1.2 A pair $(\mathfrak{g}, \tau)$ is called a *symmetric pair* if

1) $\mathfrak{g}$ is a Lie algebra,

2) τ is a nontrivial involution of $\mathfrak{g}$, i.e., $\tau\colon \mathfrak{g} \to \mathfrak{g}$ is an automorphism with $\tau^2 = \mathrm{Id}_{\mathfrak{g}}$. □

Let $(\mathfrak{g}, \tau)$ be a symmetric pair and let G be a connected Lie group with Lie algebra $\mathfrak{g}$. If H is a closed subgroup of G, then (G, H) is called *associated* to $(\mathfrak{g}, \tau)$ if τ integrates to an involution on G, again denoted by τ, such that (G, H, τ) is a symmetric space. In this case the Lie algebra of H is denoted by $\mathfrak{h}$ and it is given by:

$$\mathfrak{h} = \mathfrak{g}(1, \tau) := \{X \in \mathfrak{g} \mid \tau(X) = X\}. \tag{1.1}$$

Note that $\mathfrak{g} = \mathfrak{h} \oplus \mathfrak{q}$, where

$$\mathfrak{q} = \mathfrak{g}(-1, \tau) := \{X \in \mathfrak{g} \mid \tau(X) = -X\}. \tag{1.2}$$

We have the following relations:

$$[\mathfrak{h}, \mathfrak{h}] \subset \mathfrak{h}, \quad [\mathfrak{h}, \mathfrak{q}] \subset \mathfrak{q}, \qquad \text{and} \qquad [\mathfrak{q}, \mathfrak{q}] \subset \mathfrak{h}. \tag{1.3}$$

From (1.3) it follows that $\mathrm{ad}_{\mathfrak{q}} : \mathfrak{h} \to \mathrm{End}(\mathfrak{q})$, $X \mapsto \mathrm{ad}(X)|_{\mathfrak{q}}$, is a representation of $\mathfrak{h}$. In particular, $(\mathfrak{g}, \mathfrak{h})$ is a *reductive pair* in the sense that there exists an $\mathfrak{h}$-stable complement of $\mathfrak{h}$ in $\mathfrak{g}$. On the other hand, if $(\mathfrak{g}, \mathfrak{h})$ is a reductive pair and the commutator relations (1.3) hold, we can define an involution τ of $\mathfrak{g}$ by $\tau|_{\mathfrak{h}} = \mathrm{id}$ and $\tau|_{\mathfrak{q}} = -\mathrm{id}$. Then $(\mathfrak{g}, \tau)$ is a symmetric pair.

Example 1.1.3 (The Group Case I) Let G be a connected Lie group. Define τ by $\tau(a, b) := (b, a)$. Then $(G \times G)^\tau = \Delta(G) := \{(a, a) \mid a \in G\}$ is the diagonal in $G \times G$ and $\mathcal{M} = (G \times G)/\Delta(G)$ is a symmetric space. The map

$$G \times G \ni (a, b) \mapsto ab^{-1} \in G$$

induces a diffeomorphism $\mathcal{M} \simeq G$ intertwining the left action of $G \times G$ on $\mathcal{M}$ and the operation $(a, b) \cdot c = acb^{-1}$ of $G \times G$ on G. The decomposition of $\mathfrak{g} \times \mathfrak{g}$ into τ-eigenspaces is given by

$$\mathfrak{h} = \{(X, X) \mid X \in \mathfrak{g}\} \simeq \mathfrak{g}$$

and

$$\mathfrak{q} = \{(X, -X) \mid X \in \mathfrak{g}\} \simeq \mathfrak{g}. \qquad \square$$

Example 1.1.4 (The Group Case II) Let $\mathfrak{g}$ be a real Lie algebra, $\mathfrak{g}_{\mathbb{C}}$ its complexification, and $\sigma\colon \mathfrak{g}_{\mathbb{C}} \to \mathfrak{g}_{\mathbb{C}}$ the complex conjugation with respect the real form $\mathfrak{g}$ of $\mathfrak{g}_{\mathbb{C}}$, i.e.,

$$\sigma(X + iY) = X - iY, \quad X, Y \in \mathfrak{g}.$$

Further, let $G_{\mathbb{C}}$ be a connected complex Lie group with Lie algebra $\mathfrak{g}_{\mathbb{C}}$ such that σ integrates to a real involution on $G_{\mathbb{C}}$, again denoted by σ. Then $G_{\mathbb{R}}$, the analytic subgroup of $G_{\mathbb{C}}$ with Lie algebra $\mathfrak{g}$, is the connected component of $(G_{\mathbb{C}})^{\sigma}$. In particular, it is closed in $G_{\mathbb{C}}$ and $G_{\mathbb{C}}/G_{\mathbb{R}}$ is a symmetric space. The decomposition of $\mathfrak{g}_{\mathbb{C}}$ into σ-eigenspaces is given by $\mathfrak{g}_{\mathbb{C}} = \mathfrak{h} + \mathfrak{q} = \mathfrak{g} + i\mathfrak{g}$. □

Example 1.1.5 (Bounded Symmetric Domains) Let $\mathfrak{g}$ be a Hermitian Lie algebra and let G be a connected Lie group with Lie algebra $\mathfrak{g}$. Let K be a maximal compact subgroup of G. Then $D = G/K$ is a bounded symmetric domain. Let $H(D)$ be the group of holomorphic isometries $f : D \to D$. Then $H(D)_o$, the connected component containing the identy map, is localy isomorphic to G. Let us assume that $G = H(D)_o$. Let $\sigma_D : D \to D$ be a complex conjugation, i.e., σ_D is an antiholomoprhic involution. We assume that $\sigma(\mathbf{o}) = \mathbf{o}$, where $\mathbf{o} = \{K\}$. Define $\sigma : G \to G$ by

$$\sigma(f) = \sigma_D \circ f \circ \sigma_D.$$

Then σ is a nontrivial involution on G In this case

$$G^{\sigma} = \{f \in G \mid f(D_{\mathbb{R}}) = D_{\mathbb{R}}\}$$

where $D_{\mathbb{R}}$ is the real form of D given by $D_{\mathbb{R}} = \{z \in D \mid \sigma_D(z) = z\}$. □

Example 1.1.6 (Symmetric Pairs and Tube Domains) We have the following construction in the case where G/K is a tube domain and G is contained in the simply connected Lie group $G_{\mathbb{C}}$. Consider the elements $Z^0 = Z_o \in \mathfrak{z}(\mathfrak{k})$, $X_o \in \mathfrak{p}$ and $Y_o \in \mathfrak{p}$, as well as the space $\mathfrak{a} \subset \mathfrak{p}$, from p. 253 and p. 254. We have $\operatorname{spec}(\operatorname{ad} Z^0) = \{0, i, -i\}$ and $\operatorname{spec}(\operatorname{ad} X_o) = \operatorname{spec}(\operatorname{ad} Y_o) = \{0, 1, -1\}$. Let $\mathfrak{h} := \mathfrak{g}(0, Y_o)$, $\mathfrak{q}_+ := \mathfrak{g}(1, Y_o)$, $\mathfrak{q}_- := \mathfrak{g}(-1, Y_o)$ and $\mathfrak{q} = \mathfrak{q}^+ \oplus \mathfrak{q}^-$. Then $\mathfrak{h}$ is a θ-stable subalgebra of $\mathfrak{g}$. As $[\mathfrak{q}^+, \mathfrak{q}^+] \subset \mathfrak{g}(2, Y_o) = \{0\}$ and $[\mathfrak{q}^-, \mathfrak{q}^-] \subset \mathfrak{g}(-2, Y_o) = \{0\}$, it follows that $\mathfrak{q}^+$ and $\mathfrak{q}^-$ are both abelian subalgebras of $\mathfrak{g}$.

Define

$$\tau = \mathrm{Ad}(\exp \pi i Y_o) : \mathfrak{g}_{\mathbb{C}} \to \mathfrak{g}_{\mathbb{C}}\,.$$

Then $\tau|_{\mathfrak{h}_{\mathbb{C}}} = \mathrm{id}$ and $\tau|_{\mathfrak{q}_{\mathbb{C}}} = -\mathrm{id}$. Hence τ is a complex linear involution on $\mathfrak{g}_{\mathbb{C}}$. We notice that $\tau = \mathbf{C}^2$ in the notation of Lemma A.3.2. As usual, we denote the involution on $G_{\mathbb{C}}$ determined via $\tau(\exp X) = \exp(\tau(X))$ for $X \in \mathfrak{g}_{\mathbb{C}}$ by the same letter. We set $H := G^\tau$ and $\mathcal{M} = G/H$. Then (G, H, τ) is a symmetric space. As $\tau = \mathbf{C}^2$, we get from Lemma A.3.2 and Theorem A.3.5 that $\tau(Z_o) = -Z_o$. Hence the induced involution on the tube domain G/K is antiholomorphic. □

Let (G, H, τ) be a symmetric space. We set $\mathcal{M} := G/H$ and $\mathrm{o} := 1/H$. Then the map $\mathfrak{q} \ni X \mapsto X_{\mathrm{o}} \in T_{\mathrm{o}}(\mathcal{M})$, given by

$$X_{\mathrm{o}}(f) := \frac{d}{dt} f(\exp(tX) \cdot \mathrm{o})|_{t=0}, \quad f \in C^\infty(\mathcal{M}), \tag{1.4}$$

is a linear isomorphism of $\mathfrak{q}$ onto the tangent space $T_{\mathrm{o}}(\mathcal{M})$ of $\mathcal{M}$ at o.

Two symmetric pairs $(\mathfrak{g}, \tau)$ and $(\mathfrak{l}, \varphi)$ are called *isomorphic*, $(\mathfrak{g}, \tau) \simeq (\mathfrak{l}, \varphi)$, if there exists an isomorphism of Lie algebras $\lambda : \mathfrak{g} \to \mathfrak{l}$ such that $\lambda \circ \tau = \varphi \circ \lambda$. If λ is only assumed to be injective, then we call $(\mathfrak{g}, \tau)$ a *subsymmetric pair* of $(\mathfrak{l}, \varphi)$. In the same way we can define isomorphisms and homomorphisms for symmetric pairs (G, H) and also for symmetric spaces $\mathcal{M} = G/H$.

The symmetric pair $(\mathfrak{g}, \mathfrak{h})$ is said to be *irreducible* if the only τ-stable ideals in $\mathfrak{g}$ containing the Lie algebra $\mathfrak{g}^{fix} := \bigcap_{g \in G} \mathrm{Ad}(g)\mathfrak{h}$ are $\mathfrak{g}^{fix}$ and $\mathfrak{g}$. Note that it plays no role here which group G with Lie algebra $\mathfrak{g}$ we use in the definition of $\mathfrak{g}^{fix}$.

Let $\mathcal{M} = G/H$ be a symmetric space. For $a \in G$ we denote the diffeomorphism $m \mapsto a \cdot m$ by $\ell_a : \mathcal{M} \to \mathcal{M}$.

Lemma 1.1.7 *Consider the pointwise stabilizer*

$$G^{\mathcal{M}} = \{a \in G \mid \forall m \in \mathcal{M} \ : \ell_a m = m\} = \ker(a \mapsto \ell_a) \subset H$$

of $\mathcal{M}$ in G. Then the Lie algebra of $G^{\mathcal{M}}$ is $\mathfrak{g}^{fix}$.

Proof: $X \in \mathfrak{g}$ is contained in the Lie algebra of $G^{\mathcal{M}}$ if and only if

$$(\exp tX)gH \subset gH$$

for all $g \in G$ and all $t \in \mathbb{R}$. But this is equivalent to $\mathrm{Ad}(g)X \in \mathfrak{h}$ for all $g \in G$ and hence to $X \in \bigcap_{g \in G} \mathrm{Ad}(g)\mathfrak{h} = \mathfrak{g}^{fix}$. □

Denote by $\mathcal{N}(H)$ the set of normal subgroups in G contained in H. Then $\mathcal{N}(H)$ is ordered by inclusion.

Proposition 1.1.8 *1) The group $G^{\mathcal{M}}$ is a normal subgroup of G and contained in H.*

2) If N is a normal subgroup of G contained in H, then $N \subset G^{\mathcal{M}}$. This means that $G^{\mathcal{M}}$ is the unique maximal element of $\mathcal{N}(H)$.

3) Let $\mathrm{Ad}_{\mathfrak{q}}: H \to \mathrm{GL}(\mathfrak{q})$ be defined by $\mathrm{Ad}_{\mathfrak{q}}(h) = \mathrm{Ad}(h)|_{\mathfrak{q}}$. Then $G^{\mathcal{M}} = \ker \mathrm{Ad}_{\mathfrak{q}}$.

Proof: The first part is obvious. If N is a normal subgroup of G contained in H, then for all $a \in G$ and $n \in N$ we have

$$n \cdot (ao) = a(a^{-1}na) \cdot o = a \cdot o$$

as $a^{-1}na \in N$. Hence $N \subset G^{\mathcal{M}}$ and 2) follows. In order to prove 3) we first observe that $G^{\mathcal{M}} \subset \ker \mathrm{Ad}_{\mathfrak{q}}$ since $h \exp(X) h^{-1} H = \exp(X) H$ for all $h \in G^{\mathcal{M}}$ and $X \in \mathfrak{q}$. To show the converse it suffices to prove that $\ker \mathrm{Ad}_{\mathfrak{q}}$ is normal in G. But $\exp \mathfrak{q}$ and H are clearly in the normalizer of $\ker \mathrm{Ad}_{\mathfrak{q}}$ and these two sets generate G. □

Remark 1.1.9 Proposition 1.1.8 shows the assumption $\mathfrak{g}^{fix} = \{0\}$ means that the pair $(\mathfrak{g}, \mathfrak{h})$ is *effective*, i.e., that the representation $\mathrm{ad}_{\mathfrak{q}}$ of $\mathfrak{h}$, $X \mapsto \mathrm{ad}\, X|_{\mathfrak{q}}$, is faithful. If $(\mathfrak{g}, \mathfrak{h})$ is effective, we have $[\mathfrak{q}, \mathfrak{q}] = \mathfrak{h}$ since $[\mathfrak{q}, \mathfrak{q}] + \mathfrak{q}$ clearly is an ideal in $\mathfrak{g}$. □

Remark 1.1.10 Let $(\mathfrak{g}, \tau)$ be an irreducible and effective symmetric pair. If $\mathfrak{r}$ is the radical of $\mathfrak{g}$, then $\mathfrak{r}$ is τ-stable. Therefore $\mathfrak{g}$ is either semisimple or solvable. If $\mathfrak{g}$ is solvable, then $[\mathfrak{g}, \mathfrak{g}]$ is a τ-stable ideal of $\mathfrak{g}$. Hence $\mathfrak{g}$ is abelian. As every τ-stable subspace of an abelian algebra is an τ-stable ideal, it follows that $\mathfrak{g}$ is one dimensional. □

The following theorem is proved in [97], p.171.

Theorem 1.1.11 *Let G be a connected simply connected Lie group and $\tau: G \to G$ an involution. Then the group G^τ of τ-fixed points in G is connected and the quotient space G/G^τ is simply connected.* □

Let G be a connected simply connected Lie group with Lie algebra $\mathfrak{g}$. Then any involution $\tau : \mathfrak{g} \to \mathfrak{g}$ integrates to an involution on G which we also denote by τ. Thus Theorem 1.1.11 yields a canonical way to construct a simply connected symmetric space from a symmetric pair.

Definition 1.1.12 Let $(\mathfrak{g}, \tau)$ be a symmetric pair and $\tilde{G}$ the simply connected Lie group with Lie algebra $\mathfrak{g}$. Then the symmetric space $\tilde{\mathcal{M}} := \tilde{G}/\tilde{G}^\tau$ described by Theorem 1.1.11 is called the ***universal symmetric space*** associated to $(\mathfrak{g}, \tau)$. □

Note that, given a symmetric space $\mathcal{M} = G/H$, the canonical projection $\tilde{G} \to G$ induces a covering map $\tilde{\mathcal{M}} \to \mathcal{M}$, so that $\tilde{\mathcal{M}}$ is indeed the universal covering space of $\mathcal{M}$.

Another canonical way to associate a symmetric space to a symmetric pair is to complexify $\mathfrak{g}$ and then use a simply connected complex Lie group.

Definition 1.1.13 Let $\mathfrak{g}$ be a real Lie algebra and $\mathfrak{g}_{\mathbb{C}}$ the complexification of $\mathfrak{g}$. Consider the following involutions of the *real* Lie algebra $\mathfrak{g}_{\mathbb{C}}$:

1) $\sigma = \sigma_{\mathfrak{g}}$ is the *complex conjugation* of $\mathfrak{g}_{\mathbb{C}}$ with respect to $\mathfrak{g}$,

$$\sigma(X + iY) = X - iY, \qquad X, Y \in \mathfrak{g},$$

2) $\tau : \mathfrak{g}_{\mathbb{C}} \to \mathfrak{g}_{\mathbb{C}}$, the complex linear extension of τ to $\mathfrak{g}_{\mathbb{C}}$.

3) $\eta := \tau \circ \sigma = \sigma \circ \tau$, the antilinear extension of τ to $\gamma_{\mathbb{C}}$. □

The corresponding Lie algebras of fixed points are

$$\mathfrak{g}_{\mathbb{C}}^{\sigma} = \mathfrak{g}, \quad \mathfrak{g}_{\mathbb{C}}^{\tau} = \mathfrak{h}_{\mathbb{C}}, \quad \mathfrak{g}_{\mathbb{C}}^{\eta} = \mathfrak{g}^{c} := \mathfrak{h} + i\mathfrak{q}.$$

Note that η depends on τ as well as on the real form $\mathfrak{g}$. If necessary we will therefore write $\eta(\tau, \mathfrak{g})$.

Given a symmetric pair $(\mathfrak{g}, \tau)$, let $G_{\mathbb{C}}$ be a simply connected complex Lie group with Lie algebra $\mathfrak{g}_{\mathbb{C}}$. Then, according to Theorem 1.1.11, τ, σ, and η can be integrated to involutions on $G_{\mathbb{C}}$ with connected fixed point groups $G_{\mathbb{C}}^{\sigma}, G_{\mathbb{C}}^{\tau}$, and $G_{\mathbb{C}}^{\eta}$, respectively. We set $\check{G} := G_{\mathbb{C}}^{\sigma}, \check{G}^{c} := G_{\mathbb{C}}^{\eta}$ and $\check{H} := (G_{\mathbb{C}}^{\sigma} \cap G_{\mathbb{C}}^{\tau}) = (G_{\mathbb{C}}^{\eta} \cap G_{\mathbb{C}}^{\tau})$. Then

$$\check{\mathcal{M}} := G_{\mathbb{C}}^{\sigma}/(G_{\mathbb{C}}^{\sigma} \cap G_{\mathbb{C}}^{\tau}) = \check{G}/\check{H} \tag{1.5}$$

$$\check{\mathcal{M}}^{c} := G_{\mathbb{C}}^{\eta}/(G_{\mathbb{C}}^{\eta} \cap G_{\mathbb{C}}^{\tau}) = \check{G}^{c}/\check{H} \tag{1.6}$$

are symmetric spaces associated to $(\mathfrak{g}, \tau)$ and $(\mathfrak{g}^{c}, \tau)$, respectively.

Lemma 1.1.14 *Let $(\mathfrak{g}, \tau)$ be a symmetric pair with $\mathfrak{g}$ semisimple and denote the Killing form of $\mathfrak{g}$ by B. Then the following holds:*

1) $\mathfrak{q} = \{X \in \mathfrak{g} \mid B(X, \mathfrak{h}) = 0\}$.

2) The ideal $\mathfrak{g}_1 = [\mathfrak{q}, \mathfrak{q}] \oplus \mathfrak{q}$ τ-stable ideal and $\mathfrak{l} := \{X \in \mathfrak{g} \mid \forall Y \in \mathfrak{g}_1 : B(X, Y) = 0\}$ is an ideal of $\mathfrak{g}$ contained in $\mathfrak{h}$. In particular, if $\mathfrak{g}^{fix} = \{0\}$, then $\mathfrak{g}_1 = \mathfrak{g}$.

Proof: 1) As B is nondegenerate and

$$B(X,Y) = B(\tau(X),Y) = B(X,\tau(Y)) = -B(X,Y)$$

for all $X \in \mathfrak{h}$ and $Y \in \mathfrak{q}$, the claim follows from $\mathfrak{g} = \mathfrak{h} \oplus \mathfrak{q}$.

3) Since $B([X,Y],Z) = -B(Y,[X,Z])$ for all $X,Y,Z \in \mathfrak{g}$, this follows from (1). □

Remark 1.1.15 Let (G,H,τ) be a symmetric space with G a noncompact semisimple Lie group and τ not a Cartan involution. In this case we call $\mathcal{M} := G/H$ a *nonRiemannian semisimple symmetric space.* For such spaces it is possible to find a Cartan involution θ that commutes with τ (cf. [99], p. 337). Consequently we have

$$\mathfrak{g} = \mathfrak{k} \oplus \mathfrak{p} = \mathfrak{h} \oplus \mathfrak{q} = \mathfrak{h}_k \oplus \mathfrak{q}_k \oplus \mathfrak{h}_p \oplus \mathfrak{q}_p, \tag{1.7}$$

where $\mathfrak{k} = \mathfrak{g}(+1,\theta)$, $\mathfrak{p} = \mathfrak{g}(-1,\theta)$ and the subscripts $_k$ and $_p$, respectively denote intersection with $\mathfrak{k}$ and $\mathfrak{p}$, respectively. The above decompositions of $\mathfrak{g}$ are orthogonal w.r.t. the inner product $(\cdot \mid \cdot)_\theta$ on $\mathfrak{g}$ defined in (A.9).

Any subalgebra of $\mathfrak{g}$ invariant under θ is reductive (cf. [168]). In particular, the Lie algebra $\mathfrak{h}$ is reductive. Moreover, the group H is θ-invariant and θ induces a Cartan decomposition

$$H = (H \cap K) \exp \mathfrak{h}_p. \tag{1.8}$$

In fact, let $x = k \exp X \in H$ with $k \in K$ and $X \in \mathfrak{p}$. Then $x = \tau(x) = \tau(k) \exp \tau(X)$. Thus $k = \tau(k)$, $X = \tau(X)$. In particular, $\exp(X) \in H_o \subset H$ whence $k = x \exp(-X) \in H$. □

1.2 Dual Symmetric Spaces

In this section we fix a symmetric pair $(\mathfrak{g},\tau)$ with semisimple $\mathfrak{g}$ and a Cartan decomposition θ of $\mathfrak{g}$ commuting with τ. Denote the complex linear extension of θ to $\mathfrak{g}_\mathbb{C}$ also by θ and recall the involutions σ and η from Definition 1.1.13. The decomposition (1.7) of $\mathfrak{g}$ allows us to construct further symmetric pairs.

1.2.1 The c-Dual Space $\tilde{\mathcal{M}}^c$

The fixed point algebra of η is $\mathfrak{g}^c = \mathfrak{h} + i\mathfrak{q}$. Then we have two natural symmetric spaces associated to the symmetric pair $(\mathfrak{g}^c,\tau)$, the universal symmetric space $\tilde{\mathcal{M}}^c$ (cf. Def. 1.1.12) and the space $\mathcal{M}^c$ (cf. (1.6)). We call $\tilde{\mathcal{M}}^c$ the *c-dual* of $\mathcal{M}$.

The equality

$$\mathfrak{g}^c = (\mathfrak{h}_k \oplus i\mathfrak{q}_p) \oplus (\mathfrak{h}_p \oplus i\mathfrak{q}_k) \tag{1.9}$$

yields a Cartan decomposition of $\mathfrak{g}^c$ corresponding to the Cartan involution $\theta\tau$. We will write $\mathfrak{k}^c := \mathfrak{h}_k \oplus i\mathfrak{q}_p$ and $\mathfrak{p}^c := \mathfrak{h}_p \oplus i\mathfrak{q}_k$. Analogously, we write $\mathfrak{q}^c$ for $i\mathfrak{q}$. If we want to stress which involution we are using, we will write $(\mathfrak{g}, \tau)^c$ or $(\mathfrak{g}, \tau, \theta)^c$.

Comparing the decompositions (1.7) and (1.9), we see that the c-duality "interchanges" the compact and the noncompact parts of the respective (-1)-eigenspaces.

There are cases for which c-dual symmetric pairs are isomorphic. The following lemma describes a way to obtain such isomorphisms.

Lemma 1.2.1 *Let $\mathfrak{g}$ be a semisimple Lie algebra with Cartan involution θ and corresponding Cartan decomposition $\mathfrak{g} = \mathfrak{k} + \mathfrak{p}$.*

1) Let $X \in \mathfrak{g}_\mathbb{C}$ be such that $\mathfrak{g}_\mathbb{C} = \mathfrak{g}_\mathbb{C}(0, X) \oplus \mathfrak{g}_\mathbb{C}(i, X) \oplus \mathfrak{g}_\mathbb{C}(-i, X)$. Further, let

$$\tau_X := \mathrm{Ad}(\exp \pi X) = e^{\pi \,\mathrm{ad}\, X} .$$

Then τ_X is an involution on $\mathfrak{g}_\mathbb{C}$ such that

$$\mathfrak{g}_\mathbb{C}(+1, \tau_X) = \mathfrak{g}_\mathbb{C}(0, X) \quad \text{and} \quad \mathfrak{g}_\mathbb{C}(-1, \tau_X) = \mathfrak{g}_\mathbb{C}(i, X) \oplus \mathfrak{g}_\mathbb{C}(-i, X) .$$

If $X \in \mathfrak{k} \cup i\mathfrak{p}$, then $\mathfrak{g}$ is τ_X-stable and $\tau_X|_\mathfrak{g}$ commutes with θ.

2) If $X \in \mathfrak{k} \cup i\mathfrak{p}$, define

$$\varphi_X = \mathrm{Ad}\left(\exp(\tfrac{\pi}{2} X)\right) = e^{(\pi/2)\,\mathrm{ad}\, X}$$

and

$$(\mathfrak{g}, \mathfrak{h}) := (\mathfrak{g}, \mathfrak{g}(1, \tau_X)).$$

Then $\varphi_X^2 = \tau_X$, $\varphi_X|_\mathfrak{h} = \mathrm{id}$, $\varphi_X|_{\mathfrak{g}(\pm i, X)} = \pm i \,\mathrm{id}$, and in the case that $X \in i\mathfrak{p}$, φ_X defines an isomorphism $(\mathfrak{g}, \mathfrak{h}) \simeq (\mathfrak{g}^c, \mathfrak{h})$.

Proof: All statements except the last part of 1) follow from the fact that

$$\exp\left(z \,\mathrm{ad}(X)\right)|_{\mathfrak{g}_\mathbb{C}(\lambda, X)} = e^{z\lambda}\,\mathrm{id} .$$

Let $X \in \mathfrak{k} \cup i\mathfrak{p}$. Then $\sigma(X) = X$ if $X \in \mathfrak{k}$ and $\sigma(X) = -X$ if $X \in i\mathfrak{p}$. As τ_X is an involution, we have $\tau_X = \tau_{-X}$. Hence

$$\sigma \circ \tau_X = \tau_{\sigma(X)} \circ \sigma = \tau_X \circ \sigma$$

and τ_X commutes with σ. Thus $\mathfrak{g}$ is τ_X-stable. That τ_X commutes with θ follows in the same way. □

1.2.2 The Associated Dual Space $\mathcal{M}^a$

The involution $\tau^a := \tau\theta$ is called the *associated* or *a-dual* involution of τ. We have

$$\mathfrak{h}^a = \mathfrak{g}^{\tau^a} = \mathfrak{h}_k \oplus \mathfrak{q}_p, \quad \mathfrak{q}^a = \mathfrak{g}(-1,\tau^a) = \mathfrak{q}_k \oplus \mathfrak{h}_p$$

and call

$$(\mathfrak{g},\tau,\theta)^a = (\mathfrak{g},\tau^a,\theta) \tag{1.10}$$

the a-dual or associated triple. Since all ideals in $|fg$ are θ-invariant we see that τ-invariant ideals are automatically τ^a-invariant and conversely. In particular, we see that $(\mathfrak{g},\tau)$ is irreducible and effective if and only if $(\mathfrak{g},\tau^a)$ is irreducible and effective. To define a canonical symmetric space associated to $(\mathfrak{g},\tau^a)$, let

$$G_o^{\tau^a} \subset H^a := (K \cap H)\exp(\mathfrak{q}_p) \subset G^{\tau^a}. \tag{1.11}$$

H^a is a group since $K \cap H$ is a group normalizing $\mathfrak{q}_p$. Thus (G^a, H^a, τ^a), where $G^a = G$, is a symmetric space. We denote G/H^a by $\mathcal{M}^a$.

1.2.3 The Riemannian Dual Space $\mathcal{M}^r$

The dual (Riemannian) triple $(\mathfrak{g}^r, \tau^r, \theta^r)$ is defined by

$$(\mathfrak{g}^r,\tau^r,\theta^r) := (\mathfrak{g}^c, \theta|\mathfrak{g}^c, \theta^c)^c. \tag{1.12}$$

As $\sigma|_{\mathfrak{g}^c} = \eta$, it follows that

$$\eta(\mathfrak{g}^c,\theta) = \eta\circ\theta = \sigma\circ\tau^a = \eta(\mathfrak{g},\theta\circ\tau).$$

Thus

$$\mathfrak{g}^r = (\mathfrak{g}^a)^c = (\mathfrak{g}^c)^r. \tag{1.13}$$

Furthermore,

$$\mathfrak{g}^r = (\mathfrak{h}_k \oplus i\mathfrak{h}_p) \oplus (i\mathfrak{q}_k \oplus \mathfrak{q}_p) = \mathfrak{k}^r \oplus \mathfrak{p}^r \tag{1.14}$$

is a Cartan decomposition of $\mathfrak{g}^r$ corresponding to the Cartan involution $\theta^r = \tau|_{\mathfrak{g}^r}$. The symmetric pair $(\mathfrak{g}^r, \tau^r)$ is Riemannian and it is called the *Riemannian dual of* $\mathfrak{g}$ or the *Riemannian pair associated to* $(\mathfrak{g},\tau,\theta)$.

Let G^r be a connected Lie group with Lie algebra $\mathfrak{g}^r$ and K^r the analytic subgroup of G^r with Lie algebra $\mathfrak{k}^r$. Then $\mathcal{M}^r = G^r/K^r$ is said to be *the (dual) Riemannian form of* $\mathcal{M}$. We note that $\mathcal{M}^r$ is automatically simply connected and does not depend on the choice of G^r.

Example 1.2.2 (The Group Case III) Let L be a connected semisimple Lie group and $\mathcal{M} = (L \times L)/\Delta(L)$ as in Example 1.1.3. The associated symmetric pair is $(\mathfrak{l} \times \mathfrak{l}, \tau)$, where $\tau(X, Y) = (Y, X)$ for $X, Y \in \mathfrak{l}$. We define $\mathfrak{g} := \mathfrak{l} \times \mathfrak{l}$, $G := L \times L$ and $H := \Delta(L)$. Then we have

$$\mathfrak{g}^c = \{(X, X) + i(Y, -Y) \mid X, Y \in \mathfrak{l}\},$$

and the projection onto the first component

$$(X, X) + i(Y, -Y) = (X + iY, X - iY) \mapsto X + iY$$

is a linear isomorphism $\mathfrak{g}^c \to \mathfrak{l}_{\mathbb{C}}$ which transforms τ into the complex conjugation $\sigma : \mathfrak{l}_{\mathbb{C}} \to \mathfrak{l}_{\mathbb{C}}$ with respect to $\mathfrak{l}$. Thus the symmetric pair $(\mathfrak{g}^c, \tau)$ is isomorphic to $(\mathfrak{l}_{\mathbb{C}}, \sigma)$ (cf. Example 1.1.4). If now $L_{\mathbb{C}}$ is a simply connected complex Lie group with Lie algebra $\mathfrak{l}_{\mathbb{C}}$ and $L_{\mathbb{R}}$, is the analytic subgroup of $L_{\mathbb{C}}$ with Lie algebra $\mathfrak{l}$, then σ integrates to an involution on $L_{\mathbb{C}}$, again denoted by σ and $L_{\mathbb{R}} = L_{\mathbb{C}}^{\sigma}$ (cf. Theorem 1.1.11). Thus $L_{\mathbb{C}}/L_{\mathbb{R}}$ is simply connected and hence isomorphic to $\tilde{\mathcal{M}}^c$.

Let θ_L be a Cartan involution on L. Let $\mathfrak{l} = \mathfrak{l}_k \oplus \mathfrak{l}_p$ be the corresponding Cartan decomposition. Define a Cartan involution on G by $\theta(a, b) = (\theta_L(a), \theta_L(b))$. Then

$$\tau^a(a, b) = (\theta(b), \theta(a)).$$

In particular,

$$\begin{aligned} H^a &= \{(a, \theta(a)) \mid a \in L\} \simeq L, \\ \mathfrak{h}^a &= \{(X, \theta(X)) \mid X \in \mathfrak{l}\} \simeq \mathfrak{l}, \\ \mathfrak{q}^a &= \{(Y, -\theta(Y)) \mid Y \in \mathfrak{l}\} \simeq \mathfrak{l}, \end{aligned}$$

and

$$\mathcal{M}^a \simeq L,$$

where the isomorphism is now given by $(a, b)H^a \mapsto a\theta(b)^{-1}$. For the Riemannian dual we notice that

$$\begin{aligned} \mathfrak{h}_k &= \{(X, X) \mid X \in \mathfrak{l}_k\} \simeq \mathfrak{l}_k, \\ \mathfrak{h}_p &= \{(X, X) \mid X \in \mathfrak{l}_p\} \simeq \mathfrak{l}_p, \\ \mathfrak{q}_k &= \{(X, -X) \mid X \in \mathfrak{l}_k\} \simeq \mathfrak{l}_k, \\ \mathfrak{q}_p &= \{(X, -X) \mid X \in \mathfrak{l}_p\} \simeq \mathfrak{l}_p. \end{aligned}$$

The projection onto the first factor,

$$(X + iY + iS + T, X + iY - iS - T) \mapsto X + iY + iS + T,$$

defines an isomorphism $\mathfrak{g}^r \simeq \mathfrak{l}_{\mathbb{C}}$ mapping $\mathfrak{k}^r$ into the compact real form $\mathfrak{u} = \mathfrak{l}_k \oplus i\mathfrak{l}_p$. Hence $\mathcal{M}^r \simeq L_{\mathbb{C}}/U$, where U is the maximal compact subgroup corresponding to $\mathfrak{u}$. □

Example 1.2.3 Let $\mathcal{M} = G/H$ be as in Example 1.1.6. We want to determine the dual symmetric spaces of $\mathcal{M}$. To do this we have to invoke Lemma 1.2.1. We recall that

$$\tau = \tau_{iY_o}\,.$$

Hence

$$\varphi_{iY_o} = \mathrm{Ad}\left(\exp\frac{\pi i}{2}Y_o\right) : \mathfrak{g} \to \mathfrak{g}^c$$

is an isomorphism commuting with τ. Furthermore,

$$\varphi_{iY_o} \circ \theta = \theta \circ \varphi_{iY_o}^{-1} = (\theta \circ \tau) \circ \varphi_{iY_o}\,.$$

Hence $\varphi_{iY_o} : (\mathfrak{g}, \tau, \theta) \to (\mathfrak{g}^c, \tau, \theta\tau)$ is an isomorphism. We note that $\varphi_{iY_o} = \mathbf{C}_h$ in the notation in Lemma A.3.2, page 255.

An $\mathfrak{sl}_2$-reduction proves the following lemma.

Lemma 1.2.4 $X_o = \mathrm{Ad}\left(\exp -\frac{\pi}{2}Z_o\right) Y_o = [-Z_o, Y_o]$. □

Let $\varphi = \varphi_{-Z_o}$. As $\theta = \tau_{Z_o}$ we get $\varphi \circ \tau = \tau \circ \varphi^{-1} = \tau^a \circ \varphi$ and $\tau^a = \tau_{iX_o}$. In particular, φ defines an isomorphism $(\mathfrak{g}, \mathfrak{h}^a) \simeq (\mathfrak{g}, \mathfrak{h})$. Finally, we have

$$\varphi_{iX_o} \circ \theta = \theta \circ \varphi_{-iX_o} = \tau \circ \varphi_{iX_o}\,.$$

As $\varphi_{iX_o}|_{\mathfrak{h}_k \oplus \mathfrak{q}_p} = \mathrm{id}$ and φ_{iX_o} is multiplication by $\pm i$ on $\mathfrak{h}_p \oplus \mathfrak{q}_k$, we get that $\varphi_{iX_o} : (\mathfrak{g}, \mathfrak{k}) \to (\mathfrak{g}^r, \mathfrak{k}^r)$ is an isomorphism interchanging the role of τ and θ. We collect this in a lemma.

Lemma 1.2.5 *Let $\tau = \tau_{iY_o}$ and $\theta = \tau_{Z_o}$ as above. Then the following holds:*

1) *$\varphi_{iY_o} : (\mathfrak{g}, \tau, \theta) \to (\mathfrak{g}^c, \tau, \theta\tau)$ is an isomorphism.*

2) *Let $X_o = [-Z_o, Y_o]$. Then $\tau^a = \tau_{iX_o}$ and $\varphi_{-Z_o} : (\mathfrak{g}, \tau, \theta) \to (\mathfrak{g}, \tau^a, \theta)$ is an isomorphism.*

3) *$\varphi_{iX_o} : (\mathfrak{g}, \tau, \theta) \to (\mathfrak{g}^r, \tau^r, \theta^r)$ is an isomorphism.* □

□

1.3 The Module Structure of $T_o(G/H)$

Let $\mathcal{M} = G/H$ be a non-Riemannian semisimple symmetric space as defined in Remark 1.1.15. Let τ be the corresponding involution and θ a Cartan involution of G commuting with τ. Further assume that $(\mathfrak{g}, \mathfrak{h})$ is irreducible and effective. We will use the notation introduced in Remark 1.1.15. In this section we will study the H- and $\mathfrak{h}$-module structures of $\mathfrak{q}$ under the assumptions just spelled out.

The spaces

$$\mathfrak{q}^{H\cap K} := \{X \in \mathfrak{q} \mid \forall k \in H \cap K \ : \ \mathrm{Ad}(k)X = X\} \tag{1.15}$$

and

$$\mathfrak{q}^{H_o\cap K} = \{X \in \mathfrak{q} \mid \forall k \in H_o \cap K \ : \ \mathrm{Ad}(k)X = X\} \tag{1.16}$$

will play a crucial role in the study of causal structures on irreducible non-Riemannian semisimple symmetric spaces. We will explore the special properties of symmetric spaces for which $\mathfrak{q}^{H\cap K}$ or $\mathfrak{q}^{H_o\cap K}$ are nontrivial.

Lemma 1.3.1 *Let $\mathfrak{l}$ be a θ-stable subalgebra of $\mathfrak{g}$. Let $\mathfrak{l}^\perp := \{X \in \mathfrak{g} | \forall Y \in \mathfrak{l} : B(Y,X) = 0\}$. Then $\mathfrak{g} = \mathfrak{l} \oplus \mathfrak{l}^\perp$ and the Lie algebra generated by $\mathfrak{l}^\perp$, $\mathfrak{g}_1 = [\mathfrak{l}^\perp, \mathfrak{l}^\perp] + \mathfrak{l}^\perp$, is a θ-stable ideal in $\mathfrak{g}$.*

Proof: That $\mathfrak{g} = \mathfrak{l} \oplus \mathfrak{l}^\perp$ follows from the fact that $-B(\cdot, \theta(\cdot))$ is positive definite inner product on $\mathfrak{g}$. Let $X, Y \in \mathfrak{l}$ and let $Z \in \mathfrak{l}^\perp$. Then

$$B(X, [Y, Z]) = -B([Y, X], Z) = 0,$$

as $\mathfrak{l}$ is an algebra. By the Jacobi identity it follows now that $\mathfrak{g}_1$ is an ideal. As $\mathfrak{l}$ is θ-stable, the same holds true for $\mathfrak{l}^\perp$ and hence for $\mathfrak{g}_1$. □

Lemma 1.3.2 *The algebra $\mathfrak{h}$ is a maximal θ-stable subalgebra of $\mathfrak{g}$.*

Proof: Let $\mathfrak{l}$ be a θ-stable subalgebra of $\mathfrak{g}$ containing $\mathfrak{h}$ and assume that $\mathfrak{l} \neq \mathfrak{g}$. Then $\mathfrak{l}^\perp \neq \{0\}$. Furthermore, $\mathfrak{l}^\perp \subset \mathfrak{h}^\perp = \mathfrak{q}$. Now the above lemma shows that $[\mathfrak{l}^\perp, \mathfrak{l}^\perp] \oplus \mathfrak{l}^\perp$ is an τ-stable ideal in $\mathfrak{g}$. As $(\mathfrak{g}, \mathfrak{h})$ is assumed to be irreducible, we get $[\mathfrak{l}^\perp, \mathfrak{l}^\perp] \oplus \mathfrak{l}^\perp = \mathfrak{g}$. Thus $\mathfrak{l}^\perp = \mathfrak{q}$ and the lemma follows. □

Remark 1.3.3 Lemma 1.3.2 becomes false if we replace θ-stable by τ-stable as is shown by $\mathfrak{sl}(2, \mathbb{R})$ with the involution τ:

$$\tau \begin{pmatrix} a & b \\ c & -a \end{pmatrix} = \begin{pmatrix} a & -b \\ -c & -a \end{pmatrix}.$$

In this case $\mathfrak{h} = \mathbb{R}\begin{pmatrix} 1 & 0 \\ 0 & -1 \end{pmatrix}$. Furthermore,

$$\mathfrak{h} \subset \mathfrak{p}_{\min} = \left\{ \begin{pmatrix} a & x \\ 0 & -a \end{pmatrix} \middle| a, x \in \mathbb{R} \right\},$$

$\mathfrak{p}_{\min}$ is τ-stable and a proper subalgebra of $\mathfrak{sl}(2,\mathbb{R})$. □

Lemma 1.3.4 *Let $(\mathfrak{g}, \tau)$ be an irreducible effective semisimple symmetric pair. If $\mathfrak{q}$ is not irreducible as an $\mathfrak{h}$-module, then the following holds.*

1) *$\mathfrak{q}$ splits into two irreducible components,*

$$\mathfrak{q}^+ := \mathfrak{g}(1, Y^0) \quad and \quad \mathfrak{q}^- := \mathfrak{g}(-1, Y^0), \tag{1.17}$$

such that $\theta(\mathfrak{q}^+) = \mathfrak{q}^-$ for any Cartan involution θ commuting with τ.

2) *The submodules $\mathfrak{q}^\pm$ are isotropic for the Killing form and abelian subalgebras of $\mathfrak{g}$.*

3) *The subalgebras $\mathfrak{h} + \mathfrak{q}^\pm$ of $\mathfrak{g}$ are maximal parabolic.*

4) *The $\mathfrak{h}$-modules $\mathfrak{q}^+$ and $\mathfrak{q}^-$ are not isomorphic. In particular, $\mathfrak{q}^+$ and $\mathfrak{q}^-$ are the only nontrivial η-submodules of $\mathfrak{q}$.*

Proof: Let $V \subset \mathfrak{q}$ be an $\mathfrak{h}$-submodule and $V^\perp \subset \mathfrak{q}$ its orthogonal complement w.r.t. the Killing form B, which is nondegenerate on $\mathfrak{q}$. But $\mathfrak{b} := V + [V, V]$ and note that this is an $\mathfrak{h}$-invariant subalgebra of $\mathfrak{g}$. Moreover, we have

$$B([V^\perp, V], \mathfrak{h}) = B(V^\perp, [V, \mathfrak{h}]) \subset B(V^\perp, V) = \{0\}.$$

Since the restriction of B to $\mathfrak{h}$ is also nondegenerate, we conclude that $[V^\perp, V] = \{0\}$.

If $V \subset \mathfrak{q}$ is a nontrivial irreducible $\mathfrak{h}$-submodule which is not isotropic, then the restriction of B to V is nondegenerate and so $\mathfrak{q} = V \oplus V^\perp$. Since $[\mathfrak{b}, V^\perp] = \{0\}$ and $[\mathfrak{b}, V] \subset V$, we see that $\mathfrak{b}$ is a τ-stable ideal of $\mathfrak{g}$, i.e., equal to $\mathfrak{g}$. But this is impossible unless $V = \mathfrak{q}$. Thus every proper submodule of $\mathfrak{q}$ must be isotropic. Moreover, if $\mathfrak{q}^+$ is such a submodule, the subspace $\mathfrak{q}^+ + \theta(\mathfrak{q}^+)$ is also an $\mathfrak{h}$-submodule which is not isotropic, hence coincides with $\mathfrak{q}$. To complete the proof of (1) and (2) we only have to note that $(\mathfrak{q}^+)^\perp = \mathfrak{q}^+$, since $\mathfrak{q}^+$ is isotropic and of half the dimension of $\mathfrak{q}$.

To show (3), note first that we now know

$$[\mathfrak{q}^+, \mathfrak{q}^-] \subset \mathfrak{h}, \quad [\mathfrak{q}^+, \mathfrak{h}] \subset \mathfrak{q}^+, \quad [\mathfrak{q}^+, \mathfrak{q}^+] \subset \{0\}.$$

Therefore $\mathfrak{q}^+$ acts on $\mathfrak{g}$ by nilpotent linear maps and the subalgebra $\mathfrak{h}+\mathfrak{q}^+$ is not reductive in $\mathfrak{g}$. On the other hand, it is maximal since any subalgebra of $\mathfrak{g}$ strictly containing $\mathfrak{h}+\mathfrak{q}^+$ contains an $\mathfrak{h}$-submodule of $\mathfrak{q}$ strictly containing $\mathfrak{q}^+$, i.e., all of $\mathfrak{q}$. Now [10], Chap. 8, §10, Cor. 1 of Th. 2, shows that $\mathfrak{h}+\mathfrak{q}^+$, and hence also $\mathfrak{h}+\mathfrak{q}^-$ is parabolic.

Part (3) shows that there is an $0 \neq X \in \mathfrak{h} \cap \mathfrak{p}$ that is central in $\mathfrak{h}$ and for which $\mathrm{ad}(X)|_{\mathfrak{q}^+}$ has only positive eigenvalues. Since $\theta(X) = -X$, it follows that $\mathrm{ad}(X)|_{\mathfrak{q}^-}$ has only negative eigenvalues. Thus $\mathfrak{q}^+$ and $\mathfrak{q}^-$ cannot be isomorphic as $\mathfrak{h}$-modules. □

Let $\mathfrak{l}$ be a Lie algebra and $\mathfrak{a}, \mathfrak{b} \subset \mathfrak{l}$. Then we denote the *centralizer* of $\mathfrak{a}$ in $\mathfrak{b}$ by

$$\mathfrak{z}_{\mathfrak{b}}(\mathfrak{a}) = \{X \in \mathfrak{b} \,|\, \forall Y \in \mathfrak{a} \;:\; [X,Y] = 0\}. \tag{1.18}$$

The center $\mathfrak{z}_{\mathfrak{l}}(\mathfrak{l})$ of $\mathfrak{l}$ will simply be denoted by $\mathfrak{z}(\mathfrak{l})$.

Lemma 1.3.5 *Let $X \in \mathfrak{q}$. Then the following holds.*

1) *Assume $X \in \mathfrak{q}_k$. Then $[X, \mathfrak{h}_k] = 0$ if and only if $[X, \mathfrak{q}_k] = 0$, i.e., $\mathfrak{z}_{\mathfrak{q}_k}(\mathfrak{h}_k) = \mathfrak{z}_{\mathfrak{q}_k}(\mathfrak{q}_k) = \mathfrak{z}(\mathfrak{k}) \cap \mathfrak{q}$.*

2) *Assume $X \in \mathfrak{q}_p$. Then $[X, \mathfrak{h}_k] = 0$ if and only if $[X, \mathfrak{q}_p] = 0$, i.e., $\mathfrak{z}_{\mathfrak{q}_p}(\mathfrak{h}_k) = \mathfrak{z}_{\mathfrak{q}_p}(\mathfrak{q}_p) = \mathfrak{z}(\mathfrak{h}^a) \cap \mathfrak{q}$.*

3) *$\mathfrak{q}^{H\cap K}$ and $\mathfrak{q}^{H_o\cap K}$ are θ-stable, i.e.,*

$$\mathfrak{q}^{H\cap K} = \mathfrak{q}_k^{H\cap K} \oplus \mathfrak{q}_p^{H\cap K} \quad \text{and} \quad \mathfrak{q}^{H_o\cap K} = \mathfrak{q}_k^{H_o\cap K} \oplus \mathfrak{q}_p^{H_o\cap K}.$$

4) *$\mathfrak{q}_k^{H_o\cap K} \subset \mathfrak{z}(\mathfrak{k})$ and $\mathfrak{q}_p^{H_o\cap K} \subset \mathfrak{z}(\mathfrak{h}^a)$.*

Proof: It is obvious that $\mathfrak{q}^{H\cap K}$ is θ-stable. Let $B(\cdot,\cdot)$ be the Killing form of $\mathfrak{g}$. Then $B(\cdot,\cdot)$ is negative definite on $\mathfrak{k}$. Let $X \in \mathfrak{q}_k$ be such that $[X, \mathfrak{h}_k] = 0$. If $Y \in \mathfrak{q}_k$, then $[X,Y] \in \mathfrak{h}_k$. Thus

$$0 \geq B([X,Y],[X,Y]) = -B(Y,[X,[X,Y]]) = 0.$$

Thus $[X,Y] = 0$. The other claims are proved similarly. □

From Lemma 1.3.5.4) we immediately obtain the following corollary.

Corollary 1.3.6 *If $\mathfrak{q}_k^{H_o\cap K} \neq 0$, then $\mathfrak{z}(\mathfrak{k}) \neq 0$.* □

Lemma 1.3.7 *Suppose that $\mathfrak{q}^{H_o\cap K} \neq \{0\}$. Then $H \neq K$ and one of the following cases occurs:*

1) *$\mathfrak{g}$ is noncompact, simple with no complex structure, and τ is not a Cartan involution.*

2) $\mathfrak{g} = \mathfrak{g}_1 \times \mathfrak{g}_1$, *where* $\mathfrak{g}_1$ *is noncompact, simple with no complex structure, and* τ *is the involution* $(X,Y) \mapsto (Y,X)$.

3) $\mathfrak{g}$ *is simple with a complex structure and* $\mathfrak{h}$ *is a noncompact real form of* $\mathfrak{g}$.

Proof. If $H = K$, then $\mathcal{M} = G/K$ is Riemannian. Then $\mathfrak{g}$ is simple since $(\mathfrak{g}, \mathfrak{h})$ is irreducible and effective. . Therefore $\mathfrak{q}^{H\cap K} \subset \mathfrak{p}$ and $\mathfrak{q}^{H\cap K}$ commutes with $\mathfrak{k}$. But this is impossible, as $\mathfrak{k}$ is a maximal subalgebra of $\mathfrak{g}$ (apply Lemma 1.3.2 to θ).

In order to prove the lemma, according to [33], p. 6, we have to exclude two further possibilities:

a) Suppose that $\mathfrak{g}$ is complex and τ is complex linear. Then it follows from Lemma 1.3.5 that $\mathfrak{q}^{H_o\cap K}$ is a complex abelian algebra that commutes with $\mathfrak{h}$ since $\mathfrak{h} = \mathfrak{h}_k \oplus i\mathfrak{h}_k$. But this contradicts Lemma 1.3.2.

b) Suppose that $\mathfrak{g} = \mathfrak{g}_1 \times \mathfrak{g}_1$, where $\mathfrak{g}_1$ is noncompact, simple with complex structure, and τ is the involution $(X,Y) \mapsto (Y,X)$. This case is c-dual to Case a), as can be seen from Example 1.2.2 (we have to use the complex structure on $\mathfrak{g}_\mathbb{C}$ given by the identification with $(\mathfrak{g}_1 \times \mathfrak{g}_1) \times (\mathfrak{g}_1 \times \mathfrak{g}_1)$), so it also cannot occur if $\mathfrak{q}^{H_o\cap L} \neq \{0\}$. □

For information and notation concerning bounded symmetric domains, refer to Appendix A.3. As τ commutes with θ, we can define an involution on G/K, also denoted by τ, via $\tau(aK) = \tau(a)K$.

Theorem 1.3.8 *Suppose that* $\mathcal{M} = G/H$ *is an irreducible effective non-Riemannian semisimple symmetric space such that* $\mathfrak{q}_k^{H_o\cap K} \neq \{0\}$. *Then we have:*

1) G/K *is a bounded symmetric domain, and the complex structure can be chosen such that* $\tau : G/K \to G/K$ *is antiholomorphic.*

2) $\mathfrak{g}$ *is either simple Hermitian or of the form* $\mathfrak{g}_1 \times \mathfrak{g}_1$ *with* $\mathfrak{g}_1$ *simple Hermitian and* $\tau(X,Y) = (Y,X)$, $X, Y \in \mathfrak{g}_1$.

3) $\mathfrak{q}_k^{H\cap K} \neq \{0\}$.

Proof. Lemma 1.3.5 shows that $\{0\} \neq \mathfrak{q}_k^{H_o\cap K} \subset \mathfrak{z}(\mathfrak{k})$. Therefore, according to Lemma 1.3.7, $\mathfrak{g}$ is either simple, of the form $\mathfrak{g} = \mathfrak{h}_\mathbb{C}$ with $\mathfrak{h}$ simple, or of the form $\mathfrak{g} = \mathfrak{g}_1 \times \mathfrak{g}_1$ with $\mathfrak{g}_1$ simple.

Suppose that Case 3) of Lemma 1.3.7 holds, i.e., τ is the complex conjugation of $\mathfrak{g}$ with respect to $\mathfrak{h}$. Then $i\mathfrak{q}_k = \mathfrak{h}_p$ so that $\mathfrak{h}_p$ has a $(H_o\cap K)$-fixed point which contradicts the assumption that $\mathfrak{h}$ is simple.

Suppose that Case 1) of Lemma 1.3.7 holds. Then $\mathfrak{q}_k^{H_o\cap K} \subset \mathfrak{z}(\mathfrak{k})$ so that $\mathfrak{g}$ is Hermitian. In fact, we even see that $\mathfrak{q}_k^{H_o\cap K} = \mathfrak{z}(\mathfrak{k}) = \mathbb{R}Z^0$ in the notation of Appendix A.3, and obviously $Z^0 \in \mathfrak{q}_k^{H\cap K}$. Note that $\mathrm{ad}(Z^0)$ induces a complex structure on $\mathfrak{p}$ and then, by K-invariance, also on G/K. As $\tau(Z^0) = -Z^0$, it follows that $\tau : G/K \to G/K$ is antiholomorphic w.r.t. this complex structure.

Finally we suppose that Case 2) of Lemma 1.3.7 holds. Then, similarly as for Case 1), we find

$$\mathfrak{q}_k^{H_o\cap K} = \{(X, -X) \mid X \in \mathfrak{z}(\mathfrak{k}_1)\}.$$

Thus $\mathfrak{z}(\mathfrak{k}_1) \neq \{0\}$ and $\mathfrak{g}_1$ is Hermitian. Let Z define a complex structure on G_1/K_1. Then $Z^0 = (Z, -Z) \in \mathfrak{q}_k^{H\cap K}$ defines a complex structure on G/K anticommuting with τ. This implies the claim. □

Remark 1.3.9 There is a converse of Theorem 1.3.8, as Example 1.1.5 shows: Every antiholomorphic involution of a bounded symmetric domain $D = G/K$ fixing the origin $\mathbf{o} = \{K\}$ gives rise to an involution τ with $\mathfrak{q}_k^{H_o\cap K} \neq \{0\}$. This follows from the fact that an involution on D is antiholomorphic if and only if it anticommutes with $\mathrm{ad}(Z^0)|_{\mathfrak{g}/\mathfrak{k}}$. But then $[\tau(Z^0) + Z^0, \mathfrak{g}] \subset \mathfrak{k}$, which implies $\tau(Z^0) = -Z^0$. Therefore τ and θ commute.

Consider the special case $G = G_1 \times G_1$. Then we have

$$\tau(z, w) = (w, z)$$

for $z, w \in G_1/K_1$, where the complex structure on the second factor is the opposite of the complex structure on the first factor. □

Lemma 1.3.10 *Suppose that $\mathcal{M} = G/H$ is an irreducible non-Riemannian semisimple symmetric space with $\mathfrak{q}^{H_o\cap K} \neq \{0\}$. Then we have:*

1) *$Z(H) \cap K$ is discrete. In particular, the center of $\mathfrak{h}$ is contained in $\mathfrak{p}$. Furthermore, $\dim \mathfrak{z}(\mathfrak{h}) \leq 1$.*

2) *$H_o \cap K$ is connected with Lie algebra $\mathfrak{h}_k$.*

3) *$\mathfrak{h}_k \subset [\mathfrak{h}, \mathfrak{h}]$.*

4) *Let H_o' be the semisimple analytic subgroup of G with Lie algebra $[\mathfrak{h}, \mathfrak{h}]$. Then $H_o \cap K = H_o' \cap K$.*

Proof: If $\mathfrak{z}(\mathfrak{h}) \neq \{0\}$, then $\mathfrak{g}$ is simple without complex structure (Lemma 1.3.7). Then [44], p. 443, implies that $\mathfrak{g}_\mathbb{C}$ is simple. Thus $\mathfrak{g}^r$ is also simple. Since $\mathfrak{z}(\mathfrak{h})$ is θ-invariant, we have

$$\mathfrak{z}(\mathfrak{k}^r) = \mathfrak{z}(\mathfrak{h}) \cap \mathfrak{k} + i(\mathfrak{z}(\mathfrak{h}) \cap \mathfrak{p}) \neq \{0\}.$$

Therefore $\mathfrak{g}^r$ is Hermitian and $\mathfrak{z}(\mathfrak{k}^r)$ is one-dimensional. Thus $\mathfrak{z}(\mathfrak{h}) \subset \mathfrak{h}_k$ or $\mathfrak{z}(\mathfrak{h}) \subset \mathfrak{h}_p$. Moreover, $\mathrm{ad}\, Z|_{\mathfrak{p}^r}$ is nonsingular for every nonzero $Z \in \mathfrak{z}(\mathfrak{k}^r)$.

1) Let $Z \in \mathfrak{z}(\mathfrak{h})$ be nonzero. If $Z \in \mathfrak{h}_k$, then $\mathrm{ad}(Z)|_{\mathfrak{q}^{H_o \cap K}} = 0$, which contradicts the regularity of $\mathrm{ad}(Z)|_{\mathfrak{p}^r}$. Thus $\mathfrak{z}(\mathfrak{h}) \subset \mathfrak{h}_p$, and 1) follows.

2) Since H_o is θ-invariant, the Cartan involution of G restricts to a Cartan involution on H_o as was noted in (1.8). Therefore $K \cap H_o$, being the group of θ-fixed points in the connected group H_o is connected.

3) This follows from the θ-invariance of $[\mathfrak{h}, \mathfrak{h}]$ and $\mathfrak{z}(\mathfrak{h}) \subset \mathfrak{p}$.

4) In view of 1) and 2), this is an immediate consequence of 3). □

Theorem 1.3.11 *Let $\mathcal{M} = G/H$ be an irreducible non-Riemannian semisimple symmetric space with $\mathfrak{q}^{H_o \cap K} \neq \{0\}$. Then the following statements are equivalent:*

1) $\dim \mathfrak{z}(\mathfrak{h}) = 1$.

2) *$\mathfrak{q}$ is reducible as an $\mathfrak{h}$-module.*

3) $\dim \mathfrak{q}^{H_o \cap K} > 1$.

4) $\dim \mathfrak{q}^{H_o \cap K} = 2$.

5) *G/K is a tube domain and up to conjugation by an element of K, we have $\mathfrak{h} = \mathfrak{g}(0, Y_o)$ (cf.* Appendix A.3 *for the notation).*

If these conditions are satisfied, then in addition the following properties hold:

a) *There exists an, up to sign unique, element $Y^0 \in \mathfrak{z}(\mathfrak{h}) \cap \mathfrak{p}$ such that*

$$\mathfrak{h} = \mathfrak{g}(0, Y^0), \quad \text{and} \quad \mathfrak{q} = \mathfrak{g}(+1, Y^0) \oplus \mathfrak{g}(-1, Y^0) \tag{1.19}$$

is the decomposition of $\mathfrak{q}$ into irreducible $\mathfrak{h}$-modules.

b) *The spaces $\mathfrak{g}(\pm 1, Y^0)$ are not equivalent as $\mathfrak{h}$-modules and*

$$\theta(\mathfrak{g}(+1, Y^0)) = \mathfrak{g}(-1, Y^0).$$

c) $\mathfrak{g}(+1, Y^0)^{H_o \cap K} \neq \{0\} \neq \mathfrak{g}(-1, Y^0)^{H_o \cap K}$ *and*

$$\mathfrak{q}^{H_o \cap K} = \mathfrak{g}(+1, Y^0)^{H_o \cap K} \oplus \mathfrak{g}(-1, Y^0)^{H_o \cap K}. \tag{1.20}$$

d) $\dim \mathfrak{q}_k^{H_o \cap K} = \dim \mathfrak{q}_p^{H_o \cap K} = 1$

e) $\mathfrak{q}_k^{H_o \cap K} = \mathfrak{q}_k^{H \cap K}$.

Proof: (1)⇒(2): Suppose that $\mathfrak{z}(\mathfrak{h}) \neq \{0\}$. Then up to sign there is a unique $Z_r^0 \in \mathfrak{z}(\mathfrak{k}^r)$ such that $\mathrm{ad}(Z_r^0)$ has the eigenvalues $\pm i$ on $\mathfrak{p}_{\mathbb{C}}^r$ since $\mathfrak{g}^r$ is Hermitian (cf. Lemma 1.3.10). Then $Y^0 := -iZ_r^0 \in \mathfrak{z}(\mathfrak{h}) \cap \mathfrak{p}$ and $\mathrm{ad}_{\mathfrak{q}_{\mathbb{C}}}(Y^0)$ has eigenvalues 1 and -1. Let $\sigma : \mathfrak{g}_{\mathbb{C}} \to \mathfrak{g}_{\mathbb{C}}$ be the conjugation with respect to $\mathfrak{g}$. Then $\sigma(Y^0) = Y^0$. Hence the eigenspaces $\mathfrak{g}(j, Y^0)$, $j = -1, 0, 1$, are σ-stable. It follows that $\mathfrak{g}_{\mathbb{C}}(j, Y^0) = \mathfrak{g}(j, Y^0)_{\mathbb{C}}$. Therefore $\theta(Y^0) = -Y^0$ implies $\mathfrak{g}(\pm 1, Y^0) \neq \{0\}$, $\mathfrak{q} = \mathfrak{g}(+1, Y^0) \oplus \mathfrak{g}(-1, Y^0)$ and, $\theta(\mathfrak{g}(+1, Y^0)) = \mathfrak{g}(-1, Y^0)$. As Y^0 is central in $\mathfrak{h}$, we get that $\mathfrak{g}(+1, Y^0)$ and $\mathfrak{g}(-1, Y^0)$ are $\mathfrak{h}$-invariant. This shows that $\mathfrak{q}$ is reducible as $\mathfrak{h}$-module, i.e., (2).

(2)⇒(1): Conversely, suppose that $\mathfrak{q}$ is reducible. Then $\mathfrak{q}$ is the direct sum of two irreducible $\mathfrak{h}$-modules $\mathfrak{q}^{\pm}$ with $\theta(\mathfrak{q}^+) = \mathfrak{q}^-$ (Lemma 1.3.4).

We get

$$\{0\} \neq \mathfrak{q}^{H_o \cap K} = (\mathfrak{q}^+)^{H_o \cap K} \oplus (\mathfrak{q}^-)^{H_o \cap K}$$

and both of the spaces on the right-hand side are nonzero. Let $0 \neq X \in (\mathfrak{q}^+)^{H_o \cap K}$ and define

$$Z := [X, \theta(X)] \in \mathfrak{p} \cap [\mathfrak{q}, \mathfrak{q}] \subset \mathfrak{h}_p.$$

Then Z commutes with $\mathfrak{h}_k$. Apply Lemma 1.3.5 to the involution $\tau \circ \theta$ to see that Z is central in $\mathfrak{h}$. Thus we have to show that Z is nonzero. To do that define $Y := X + \theta(X) \in \mathfrak{q}^{H_o \cap K} \cap \mathfrak{k}$. Then $Y \in \mathfrak{z}(\mathfrak{k})$ by Lemma 1.3.5 and $Y \neq 0$. From Lemma 1.3.7 we see that $\mathfrak{g}$ is either simple Hermitian or the direct sum of two copies of a simple Hermitian algebra. Calculating in each simple factor separately, we obtain

$$0 \neq [X + \theta(X), X - \theta(X)] = -2Z.$$

(1),(2)⇒ a),b),c): Assertion a) follows from Lemma 1.3.4 and the fact that Y^0 has different eigenvalues on the two irreducible pieces. To prove b) and c), consider $0 \neq X \in \mathfrak{q}^{H_o \cap K}$. Let X_+ and X_- be the projections of X onto $\mathfrak{g}(+1, Y^0)$ and $\mathfrak{g}(-1, Y^0)$, respectively. Then $X_\pm \in \mathfrak{g}(\pm 1, Y^0)^{H_o \cap K}$, and either X_+ or X_- is nonzero. Obviously,

$$\theta\left(\mathfrak{g}(+1, Y^0)^{H_o \cap K}\right) = \mathfrak{g}(-1, Y^0)^{H_o \cap K},$$

and $\theta \circ \mathrm{Ad}(k) = \mathrm{Ad}(k) \circ \theta$ for all $k \in H \cap K$. Thus (1.20) follows.

(3)⇒(1): Suppose that $\dim \mathfrak{q}^{H_o \cap K} > 1$. As the commutator algebra $[\mathfrak{h}, \mathfrak{h}]$ is semisimple, it follows from Lemma A.2.5 that $\mathfrak{q}$ is reducible as an $[\mathfrak{h}, \mathfrak{h}]$-module. If $\mathfrak{z}(\mathfrak{h})$ were zero we would have (2) and a contradiction to the equivalence of (1) and (2).

(1)⇒(4): In view of Theorem 1.3.11.4), we see that Lemma A.2.5, applied to H_o', proves $\dim(\mathfrak{q}^{H_o \cap K}) = 2$, since $\mathfrak{q}$ contains precisely two irreducible $[\mathfrak{h}, \mathfrak{h}]$-submodules by a).

(4)⇒(3): This is obvious.

(1)⇒ d): Since the spaces $\mathfrak{g}(\pm, Y^0)^{H_o \cap K}$ get interchanged by θ, we see that $\mathfrak{q}^{H_o \cap K}$ contains nonzero elements in $\mathfrak{k}$ and in $\mathfrak{p}$. This proves $\dim(\mathfrak{q}^+)^{H_o \cap K} = \dim(\mathfrak{q}^-)^{H_o \cap K} = 1$.

(1)⇒ e): Apply Theorem 1.3.8 to obtain $\mathfrak{q}_k^{H \cap K} = \mathfrak{q}_k^{H_o \cap K} \neq \{0\}$ from d).

(1)⇒(5): Since $\mathfrak{q}_k^{H_o \cap K} \neq \{0\}$, we can apply Theorem 1.3.8, which shows that $\mathfrak{g}$ is Hermitian since $\mathfrak{h}$ is not semisimple. Choose a maximal abelian subspace $\mathfrak{a}^0$ of $\mathfrak{p}$ containing Y^0. Since $\mathfrak{h} = \mathfrak{g}(0, Y^0)$, we have $\mathfrak{a}^0 \subset \mathfrak{h}$. The restricted roots vanishing on Y^0 are precisely the restricted roots of the pair $(\mathfrak{h}, \mathfrak{a}^0)$. In particular, there is at least one restricted root not vanishing on Y^0. But the spectrum of $\mathrm{ad}(Y^0)$ is $\{-1, 0, 1\}$ so that Moore's Theorem A.3.4 shows that G/K is of tube type (cf. also Theorem A.3.5).

There exists a $k \in K$ such that $\mathrm{Ad}(k)\mathfrak{a}^0 = \mathfrak{a}$ is the maximal abelian subspace of $\mathfrak{p}$ described in Appendix A.3. Thus we may assume that $Y^0 \in \mathfrak{a}$. Conjugating with a suitable Weyl group element, we may even assume that Y^0 in the closure of the positive Weyl chamber. But then all positive restricted roots which do not belong to the pair $(\mathfrak{h}, \mathfrak{a})$ take the value 1 on Y^0, which proves that $Y^0 = Y_o$.

(5)⇒(1): Since $\mathfrak{h}$ is the centralizer of an element in $\mathfrak{g}$ it clearly has a nontrivial center. □

In the following we will choose an element $Y^0 \in \mathfrak{z}(\mathfrak{h}) \cap \mathfrak{p}$ in the way Theorem 1.3.11 describes it, whenever it is possible.

Example 1.3.12 Consider the conjugation

$$\tau\left(\begin{pmatrix} a & b \\ c & d \end{pmatrix}\right) = \begin{pmatrix} 0 & 1 \\ 1 & 0 \end{pmatrix}\begin{pmatrix} a & b \\ c & d \end{pmatrix}\begin{pmatrix} 0 & 1 \\ 1 & 0 \end{pmatrix} = \begin{pmatrix} d & c \\ b & a \end{pmatrix} \tag{1.21}$$

on $G = \mathrm{SL}(2, \mathbb{R})$. It commutes with the Cartan involution $\theta(g) = {}^t g^{-1}$ and satisfies

$$G^\tau = \pm\left\{ h(t) := \begin{pmatrix} \cosh t & \sinh t \\ \sinh t & \cosh t \end{pmatrix} \middle| \, t \in \mathbb{R} \right\}.$$

Moreover, we have

$$K = G^\theta = \left\{ k(s) := \begin{pmatrix} \cos s & -\sin s \\ \sin s & \cos s \end{pmatrix} \middle| \, s \in \mathbb{R} \right\}.$$

Note that G is contained in its simply connected complexification $G_\mathbb{C} = \mathrm{SL}(2, \mathbb{C})$. The above formula defines an involution τ on $G_\mathbb{C}$ which on $\mathfrak{g}_\mathbb{C}$ is the complex linear extension of τ on $\mathfrak{g}$. The involution η on $G_\mathbb{C}$ is given by

$$\eta\begin{pmatrix} a & b \\ c & d \end{pmatrix} = \begin{pmatrix} \overline{d} & \overline{c} \\ \overline{b} & \overline{a} \end{pmatrix},$$

which implies $\check{G}^c = \mathrm{SU}_o(1,1)$ and $\check{H} = \mathrm{SO}(1,1)$. In particular, we see that $\check{H}$ is not connected. The corresponding space $\check{\mathcal{M}} = G/\check{H}$ can be realized as the $\mathrm{Ad}(G)$-orbit of the element $\begin{pmatrix} 1 & 0 \\ 0 & -1 \end{pmatrix} \in \mathfrak{g}$, and this orbit is

$$\left\{ \begin{pmatrix} a & b+c \\ b-c & -a \end{pmatrix} \mid a^2 + b^2 - c^2 = 1 \right\},$$

i.e., a one-sheeted hyperboloid.

On the Lie algebra level, τ is given by the same conjugation and we find

$$\begin{aligned} \mathfrak{h} &= \mathbb{R} \begin{pmatrix} 0 & 1 \\ 1 & 0 \end{pmatrix} \subset \mathfrak{p}, \\ \mathfrak{q}_k &= \mathbb{R} \begin{pmatrix} 0 & 1 \\ -1 & 0 \end{pmatrix}, \\ \mathfrak{q}_p &= \mathbb{R} \begin{pmatrix} 1 & 0 \\ 0 & -1 \end{pmatrix} = \mathfrak{h}^a . \end{aligned}$$

Moreover, we see that $\mathfrak{g}^c = \mathfrak{su}(1,1)$ and $\mathfrak{k}^c = i\mathfrak{q}_p$. Set

$$\begin{aligned} Y^0 &= \frac{1}{2}\begin{pmatrix} 0 & 1 \\ 1 & 0 \end{pmatrix} \in \mathfrak{h}_p, \\ X^0 &= \frac{1}{2}\begin{pmatrix} 1 & 0 \\ 0 & -1 \end{pmatrix} \in \mathfrak{q}_p, \\ Z^0 &= \frac{1}{2}\begin{pmatrix} 0 & -1 \\ 1 & 0 \end{pmatrix} \in \mathfrak{q}_k, \\ Y_+ &= \begin{pmatrix} 0 & 1 \\ 0 & 0 \end{pmatrix} = Y^0 - Z^0, \\ Y_- &= \begin{pmatrix} 0 & 0 \\ 1 & 0 \end{pmatrix} = Y^0 + Z^0 = -\theta(Y_+) = \tau(Y_+), \\ X_+ &= \frac{1}{2}\begin{pmatrix} 1 & -1 \\ 1 & -1 \end{pmatrix} = X^0 + Z^0, \\ X_- &= \frac{1}{2}\begin{pmatrix} 1 & 1 \\ -1 & -1 \end{pmatrix} = X^0 - Z^0 = -\theta(X_+). \end{aligned}$$

Then we have

$$\mathfrak{g}(\pm 1, Y^0) = \mathbb{R}\, X_\pm, \quad \mathfrak{g}(0, Y^0) = \mathbb{R}\, Y^0 = \mathfrak{h}\,.$$

The spaces $\mathfrak{g}(\pm 1, Y^0)$ are the irreducible components of $\mathfrak{q}$ as G^τ- and $\mathfrak{h}$-modules. More precisely. we have

$$\mathrm{Ad}\, h(t)(rX_+ + sX_-) = e^{2t} rX_+ + e^{-2t} sX_- .$$

Further, we have

$$\mathfrak{q}_k^{G^\tau\cap K} = \mathbb{R}\, Z^0, \quad \mathfrak{q}_p^{G^\tau\cap K} = \mathbb{R}\, X^0,$$

$$\mathrm{spec}(\mathrm{ad}\, X^0) = \mathrm{spec}(\mathrm{ad}\, Y^0) = \{-1, 0, 1\},$$

and

$$\mathrm{spec}(\mathrm{ad}\, Z^0) = \{-i, 0, i\}. \qquad \square$$

Corollary 1.3.13 *Let $\mathcal{M}$ be an irreducible non-Riemannian semisimple symmetric space with $\mathfrak{q}^{H_o\cap K} \neq \{0\}$. Then $\mathfrak{q}$ is reducible as an $\mathfrak{h}$-module if and only if $(\mathfrak{g}^c, \mathfrak{h})$ is isomorphic to $(\mathfrak{g}, \mathfrak{h})$. In that case there exists an $Y^0 \in \mathfrak{z}(\mathfrak{h}) \cap \mathfrak{p}$ such that*

$$\tau = \exp(i\pi\, \mathrm{ad}\, Y^0) \tag{1.22}$$

and

$$H_o = H_o' \times \exp \mathbb{R} Y^0 . \tag{1.23}$$

Proof: Suppose that $\mathfrak{q}$ is reducible as an $\mathfrak{h}$-module. Then Theorem 1.3.11 shows that we can find an element $Y^0 \in \mathfrak{z}(\mathfrak{h}) \cap \mathfrak{p}$ such that

$$\mathfrak{q} = \mathfrak{g}(+1, Y^0) \oplus \mathfrak{g}(-1, Y^0).$$

Now apply Lemma 1.2.1 to iY^0 to see that $\tau = \tau_{iY^0}$ and that $\varphi_{iX^0} : (\mathfrak{g}, \mathfrak{h}) \to (\mathfrak{g}^c, \mathfrak{h})$ is an isomorphism.

Conversely, suppose that $(\mathfrak{g}, \mathfrak{h})$ and $(\mathfrak{g}^c, \mathfrak{h})$ are isomorphic. Choose an isomorphism $\lambda: \mathfrak{g}^c \to \mathfrak{g}$. We may assume that $\mathfrak{q}_k^{H_o\cap K} \neq \{0\}$. In fact, otherwise we replace λ by $\lambda^{-1}: \mathfrak{g} \to \mathfrak{g}^c$. As $i\mathfrak{q}_k = \mathfrak{q}_p^c$, we get $(\mathfrak{q}^c)_p^{H_o\cap K} \neq \{0\}$. But $\mathfrak{q}_k \cap \lambda\left(\mathfrak{q}_p^c\right) = \{0\}$, so $\dim \mathfrak{q}^{H_o\cap K} = 2$. Thus Theorem 1.3.11 shows that $\mathfrak{q}$ is reducible as an $\mathfrak{h}$ module and $\mathfrak{h} = [\mathfrak{h}, \mathfrak{h}] + \mathbb{R}Y^0$. Considering the Cartan decomposition of H_o, we see that we also have the global version of this fact, which is (1.23). $\square$

Lemma 1.3.14 *Let $\mathcal{M}$ be an irreducible non-Riemannian semisimple symmetric space. Suppose that $\mathfrak{q}$ is reducible as an $\mathfrak{h}$-module but irreducible as an H-module. Fix Y^0 as in* Theorem 1.3.11 *and set*

$$H_1 = \{h \in H \mid \mathrm{Ad}(h)(\mathfrak{g}(+1, Y^0)) = \mathfrak{g}(+1, Y^0)\}.$$

Then H_1 is a θ-stable normal subgroup of H. Moreover, there exists an element $k \in K \cap H$ such that

$$\mathrm{Ad}(k)\mathfrak{g}(+1, Y^0) = \mathfrak{g}(-1, Y^0)$$

and

$$H = H_1 \dot{\cup} kH_1 .$$

Proof: We write $\mathfrak{q}^{\pm} := \mathfrak{g}(\pm 1, Y^0)$ for the $\mathfrak{h}$-irreducible submodules of $\mathfrak{q}$. Since $\mathfrak{q}$ is irreducible as an H-module, and $H = (H \cap K)\exp\mathfrak{h}_p$ by (1.8), we can find a $k \in K \cap H$ such that

$$\mathrm{Ad}(k)\mathfrak{q}^+ \neq \mathfrak{q}^+.$$

As $\mathrm{Ad}(k)\mathfrak{q}^+ \cap \mathfrak{q}^+$ is H_o-stable and $\mathfrak{q}^+$ is irreducible as an H_o-module, we get $\mathrm{Ad}(k)\mathfrak{q}^+ \cap \mathfrak{q}^+ = \{0\}$. Thus

$$\mathfrak{q} = \mathfrak{q}^+ \oplus \mathrm{Ad}(k)\mathfrak{q}^+ = \mathfrak{q}^+ \oplus \mathfrak{q}^-.$$

The H_o-representations $\mathfrak{q}^+$ and $\mathfrak{q}^-$ are inequivalent by Theorem 1.3.11. Thus $\mathrm{Ad}(k)\mathfrak{q}^+ = \mathfrak{q}^-$. Fix an $h \in H \setminus H_1$. Then it follows as above that $\mathrm{Ad}(h)\mathfrak{q}^+ = \mathfrak{q}^- = \mathrm{Ad}(k)\mathfrak{q}^+$. Hence $\mathrm{Ad}(k^{-1}h)\mathfrak{q}^+ = \mathfrak{q}^+$ and H_1 is subgroup of index 2 in H. Therefore H_1 is normal in H. To prove the θ-stability let $h \in H_1$ and recall that $\exp\mathfrak{p} \subset H_o \subset H_1$. Then (1.8) shows that there is a $\tilde{k} \in H \cap K$ and an $X \in \mathfrak{h}_p$ with $h = \tilde{k}\exp(X)$. Thus $\tilde{k} = h\exp(-X) \in H_1$ and H_1 is θ-stable. □

The following example shows that the situation of Lemma 1.3.14 actually occurs.

Example 1.3.15 Let $\mathcal{M} = G/H$ be as in Example 1.1.6 and consider the space $\mathrm{Ad}(G)/\mathrm{Ad}(G)^\tau$, where $\mathrm{Ad}(G)^\tau = \{\varphi \in \mathrm{Ad}(G) \mid \varphi\tau = \tau\varphi\}$. Then $\mathrm{Ad}(G)/\mathrm{Ad}(G)^\tau$ is a non-Riemannian semisimple symmetric space which is locally isomorphic to $\mathcal{M}$. Moreover, $\theta \in \mathrm{Ad}(G)^\tau$, so that $\mathfrak{q}$ is irreducible as an $\mathrm{Ad}(G)^\tau$-module. □

1.4 A-Subspaces

Let $\mathcal{M} = G/H$ be a symmetric space with G connected semisimple. In this section we recall some results from [143] on the orbit decomposition of $\mathcal{M}$ with respect to H. It can be skipped at first reading because there are no proofs in this section and the results will be used only much later.

A maximal abelian subspace $\mathfrak{a}_q$ of $\mathfrak{q}$ is called an *A-subspace* if it consists of semisimple elements of $\mathfrak{g}$. For $X \in \mathfrak{q}$ consider the polynomial

$$\det(t - \mathrm{ad}(X)) = \sum_{j=1}^{n} d_j(X)t^j.$$

Let k be the least integer such that d_k does not vanish. Then the elements

$$\mathfrak{q}' := \{X \in \mathfrak{q} \mid d_k(X) \neq 0\}$$

are called $\mathfrak{q}$-*regular elements* . The set of $\mathfrak{q}$-regular elements is open and dense in $\mathfrak{q}$.

Let $\phi : G \to G$ be the map defined by

$$\phi(g) = g\tau(g^{-1}).$$

This map factors to G/G^τ and defines a homeomorphism between G/G^τ and the closed submanifold $\phi(G)$ of G ([143], p.402). We have the following maps:

$$\begin{array}{ccccc} G & \longrightarrow & G/H & & \\ & & \downarrow & & \\ & & G/G^\tau & \longrightarrow & \phi(G). \end{array}$$

For $x \in \phi(G)$ consider the polynomial

$$\det(1 - t - \mathrm{Ad}(x)) = \sum_{j=1}^{n} D_j(x)t^j.$$

Then the elements of

$$\phi(G)' := \{x \in \phi(G) \mid D_k(x) \neq 0\}$$

are called the $\phi(G)$-*regular elements* . The centralizer

$$A_q := Z_{\phi(G)}(\mathfrak{a}_q)$$

of an A-subspace $\mathfrak{a}_q$ is called an A-*subset.* We set

$$A'_q := A_q \cap \phi(G)'.$$

Oshima and Matsuki, in [143], prove the following results.

Theorem 1.4.1 *Let the notation be as above. Then the following holds:*

1) *Each A-subspace is conjugate under $(G^\tau)_o$ to a θ-invariant one.*

2) *The number of H-conjugacy classes of A-subspaces is finite. This number can be described in terms of root systems.*

3) *The decomposition of an A-subset into connected components has the form*

$$A_q = \bigcup_{j \in J} k_j \exp \mathfrak{a}_q,$$

where $k_j \in \phi(K)$ and J is finite if the center of G is finite (which is the case if $G \subset G_{\mathbb{C}}$).

4) If $\mathfrak{a}_q$ is an A-subspace, then

$$Z_{\phi(G_{\mathbb{C}})}((\mathfrak{a}_q)_{\mathbb{C}}) = \exp(\mathfrak{a}_q)_{\mathbb{C}}.$$

5) $\phi(G)'$ is open dense in $\phi(G)$. □

Theorem 1.4.2 *Let $(\mathfrak{a}_{q,1}, \ldots, \mathfrak{a}_{q,r})$ be a set of θ-invariant A-subspaces representing the H-conjugacy classes. Then the set*

$$\bigcup_{j=1}^{r} H\phi^{-1}\left(A'_{q,j}\right)$$

is open dense in G. □

1.5 The Hyperboloids

Let $p, q \in \mathbb{Z}^+$, $n = p+q$, and $\mathbf{V} = \mathbb{R}^n$. We write elements of $\mathbf{V}$ as $v = \binom{x}{y}$ with $x \in \mathbb{R}^p$ and $y \in \mathbb{R}^q$. Note that for $p = 1$, x is a real number. We write $\mathrm{pr}_1(v) := x$ and $\mathrm{pr}_2(v) = y$. Define the bilinear form $Q_{p,q}$ on $\mathbb{R}^n$ by

$$\begin{aligned} Q_{p,q}(v,w) &= (\mathrm{pr}_1(v)|\,\mathrm{pr}_1(w)) - (\mathrm{pr}_2(v)|\,\mathrm{pr}_2(w)) && (1.24)\\ &= v_1w_1 + \ldots + v_pw_p - v_{p+1}w_{p+1} - \ldots - v_nw_n. && (1.25)\end{aligned}$$

Here $(\cdot|\cdot)$ is the usual inner product on $\mathbb{R}^n$.

For $r \in \mathbb{R}^+$, $p, q \in \mathbb{N}$, $n = p+q \geq 1$ define

$$Q^{p,q}_{-r} = Q_{-r} := \{v \in \mathbb{R}^{n+1} \mid Q_{p,q+1}(v,v) = -r^2\} \qquad (1.26)$$

and

$$Q^{p,q}_{+r} = Q_{+r} := \{x \in \mathbb{R}^{n+1} \mid Q_{p+1,q}(x,x) = +r^2\}. \qquad (1.27)$$

Then $Q_{\pm r}$ has dimension n and

$$T_m(Q_{+r}) \simeq \{v \in \mathbb{R}^{n+1}_{p+1} \mid Q_{p+1,q}(v,m) = 0\} \simeq \mathbb{R}^n_p.$$

The linear isomorphism

$$L_n : \mathbb{R}^n \ni {}^t(x_1, \ldots, x_n) \mapsto {}^t(x_n, \ldots, x_1) \in \mathbb{R}^n$$

satisfies $Q_{q,p} \circ (L_n, L_n) = -Q_{p,q}$. Hence L_{n+1} maps $Q^{p,q}_{+r}$ bijectively onto $Q^{q,p}_{-r}$. Furthermore,

$$S^q \times \mathbb{R}^p \ni (v,w) \mapsto F(v,w) := {}^t\left(w, \sqrt{r^2 + \|w\|^2}\, v\right) \in \mathbb{R}^{n+1}$$

satisfies

$$Q_{p,q+1}(F(v,w),F(v,w)) = \|w\|^2 - (r^2 + \|w\|^2) = -r^2$$

and induces a diffeomorphism

$$Q_{-r} \simeq S^q \times \mathbb{R}^p, \tag{1.28}$$

where the inverse is given by

$$\begin{pmatrix} v \\ w \end{pmatrix} \mapsto \begin{pmatrix} \frac{w}{\|w\|}, v \end{pmatrix}.$$

Using that on Q_{-r}, we have $\|w\|^2 = r^2 + \|v\|^2$. In the same way, or by using the map L_{n+1}, it follows that $S^p \times \mathbb{R}^{n-p} \simeq Q_{+r}$. In particular, Q_{-r}, respectively Q_{+r}, is connected except for $p = n$ (respectively $p = 0$), where it has two components.

Let $\mathrm{O}(p,q)$ be the group

$$\mathrm{O}(p,q) := \{a \in \mathrm{GL}(n,\mathbb{R}) \mid \forall v,w \in \mathbb{R}^n : Q_{p,q}(av,aw) = Q_{p,q}(v,w)\}.$$

Consider $I_{p,q} := \begin{pmatrix} I_p & 0 \\ 0 & -I_q \end{pmatrix}$, where I_m is the identity matrix of size $m \times m$. With this notation we have

$$\mathrm{O}(p,q) = \{a \in \mathrm{GL}(n,\mathbb{R}) \mid {}^t a I_{p,q} a = I_{p,q}\}.$$

In block form $\mathrm{O}(p,q)$ can thus be written as the group of block matrices $\left(\begin{smallmatrix} A & B \\ C & D \end{smallmatrix}\right) \in \mathrm{GL}(n,\mathbb{R})$ satisfying the following relations:

$$A \in M(p,\mathbb{R}), \quad B \in M(p \times q,\mathbb{R}), \quad D \in M(q,\mathbb{R}), \quad C \in M(q \times p,\mathbb{R}),$$

and

$$\begin{aligned} {}^tAA - {}^tCC &= I_p, \\ {}^tAB - {}^tCD &= 0, \\ {}^tBB - {}^tDD &= -I_q. \end{aligned} \tag{1.29}$$

Here $M(l \times m, \mathbb{K})$ denotes the $m \times l$ matrices with entries in $\mathbb{K}$ and $M(m,\mathbb{K}) := M(m \times m, \mathbb{K})$.

The group $\mathrm{O}(p,q)$, $pq \neq 0$ has four connected components, described in the following way, writing $\left(\begin{smallmatrix} A & B \\ C & D \end{smallmatrix}\right) \in \mathrm{O}(p,q)$.

1) $\det A \geq 1$ and $\det D \geq 1$ (the identity component), representative I_n,

2) $\det A \geq 1$ and $\det D \leq -1$, representative $\begin{pmatrix} I_p & 0 \\ 0 & I_{q-1,1} \end{pmatrix}$,

3) $\det A \leq -1$ and $\det D \geq 1$, representative $\begin{pmatrix} I_{p-1,1} & 0 \\ 0 & I_q \end{pmatrix}$,

4) $\det A \leq -1$ and $\det D \leq -1$, representative $\begin{pmatrix} I_{p-1,1} & 0 \\ 0 & I_{q-1,1} \end{pmatrix}$.

Let $\mathrm{SO}(p,q) := \mathrm{O}(p,q) \cap \mathrm{SL}(n,\mathbb{R})$. Then

$$\mathrm{O}(p,q)_o = \mathrm{SO}_o(p,q) := \mathrm{SO}(p,q)_o .$$

For the Lie algebras we have

$$\mathfrak{o}(p,q) = \mathfrak{so}(p,q) = \{X \in \mathfrak{gl}(n,\mathbb{R}) \mid I_{p,q}X + {}^tXI_{p,q} = 0\}$$

or, in block form,

$$\mathfrak{so}(p,q) = \left\{ \begin{pmatrix} A & B \\ C & D \end{pmatrix} \middle| \ {}^tA = -A, {}^tD = -D, {}^tB = C \right\}. \tag{1.30}$$

It is clear that $\mathrm{O}(p+1,q)$ acts on Q_{+r}. Let $\{e_1, \ldots, e_n\}$ be the standard basis for $\mathbb{R}^n$.

Lemma 1.5.1 *For $p, q > 0$ the group $\mathrm{SO}_o(p+1,q)$ acts transitively on Q_{+r}. The isotropy subgroup at re_1 is isomorphic to $\mathrm{SO}_o(p,q)$. Whence, as a manifold,*

$$Q_{+r} \simeq \mathrm{SO}_o(p+1,q)/\mathrm{SO}_o(p,q).$$

Proof: We may and will assume that $r = 1$. Let $v = \binom{x}{y} \in Q_1$. Then, using Witt's theorem, we can find $A \in \mathrm{SO}_o(p+1)$ and $D \in \mathrm{SO}_o(q)$ such that

$$Ax = \|x\|e_1 \qquad \text{and} \qquad Dy = \|y\|e_{p+2}.$$

As $\begin{pmatrix} A & 0 \\ 0 & D \end{pmatrix} \in \mathrm{SO}_o(p+1,q)$, we may assume $x = \lambda e_1$ and $y = \mu e_{p+2}$ with $\lambda, \mu > 0$ and $\lambda^2 - \mu^2 = 1$. But then $a(e_1) = v$ with

$$a = \begin{pmatrix} \lambda & 0 & \mu & 0 \\ 0 & I_p & 0 & 0 \\ \mu & 0 & \lambda & 0 \\ 0 & 0 & 0 & I_{q-1} \end{pmatrix} \in \mathrm{SO}_o(p+1,q),$$

where the block structure of a is according to the partition $(1, p, 1, q-1)$ of $n+1$. The last statement is a direct calculation and is left to the reader.□

Now $L_{n+1}(r,0,\dots,0) = (0,\dots,0,r)$, and on the group level the conjugation

$$\mathrm{Ad}(L_{n+1}) : a \mapsto L_{n+1} a L_{n+1}^{-1}$$

sets up an isomorphism of groups $\mathrm{O}(q+1,p) \simeq \mathrm{O}(p,q+1)$ mapping the stabilizer of $(r,0,\dots,0)$ onto the stabilizer $\mathrm{O}(p,q)$ of $(0,\dots,0,r)$. Thus

$$Q_{-r} \simeq \mathrm{O}(p,q+1)/\mathrm{O}(p,q) = \mathrm{SO}_o(p,q+1)/\,\mathrm{SO}_o(p,q).$$

Next we show that the hyperboloids $Q_{\pm r}$ are symmetric spaces. We will only treat the case $\mathrm{O}(p+1,q)/\mathrm{O}(p,q)$, as the other case follows by conjugation with L_{n+1}. The involution τ on $\mathrm{O}(p+1,q)$ is conjugation by $I_{1,n}$. Then

$$\mathrm{O}(p+1,q)_o^{\tau} = \mathrm{SO}_o(p,q) \subset \mathrm{O}(p+1,q)^{re_1} \subset \mathrm{O}(p+1,q)^{\tau}.$$

The involution τ on the Lie algebra $\mathfrak{g} = \mathfrak{so}(p+1,q)$ is also conjugation by $I_{1,n}$. Then $\mathfrak{h} = \mathfrak{so}(p,q) = \mathfrak{g}^{\tau}$ and defining

$$q(v) := \begin{pmatrix} 0 & -{}^t(vI_{p,q}) \\ v & 0 \end{pmatrix} \tag{1.31}$$

for $v \in \mathbb{R}^n$ we find a linear isomorphism

$$\mathbb{R}^n \ni v \mapsto q(v) \in \mathfrak{q} \tag{1.32}$$

which satisfies

$$\begin{aligned} q(av) &= aq(v)a^{-1},\ a \in \mathrm{SO}_o(p,q), \\ Q_{p,q}(v,w) &= -\frac{1}{2}\operatorname{Tr} q(v)q(w). \end{aligned}$$

The c-dual of a hyperboloid is again a hyperboloid or at least a covering of a hyperboloid. More precisely,

$$(\mathfrak{so}(2,n-1),\mathfrak{so}(1,n-1))^c = (\mathfrak{so}(1,n),\mathfrak{so}(1,n-1))$$

which, by abuse of notation, can be written $Q_{+1}^c \simeq Q_{-1}$. In fact, let q and q_1 denote the map from (1.32) for the case of $(\mathfrak{so}(2,n-1),\mathfrak{so}(1,n-1))$ and $(\mathfrak{so}(1,n),\mathfrak{so}(1,n-1))$, respectively. For $X \in \mathfrak{so}(1,n-1) \subset \mathfrak{so}(2,n-1)$ we define $\lambda(X) \in \mathfrak{so}(1,n-1) \subset \mathfrak{so}(1,n)$ by

$$\lambda\left(\begin{pmatrix} 0 & \\ & X \end{pmatrix}\right) = \begin{pmatrix} X & \\ & 0 \end{pmatrix}.$$

Then

$$\mathfrak{so}(2,n-1)^c \ni X + iq(v) \mapsto \lambda(X) + q_1(v) \in \mathfrak{so}(1,n)$$

is a Lie algebra isomorphism.

Notes for Chapter 1

The material of the first section in this chapter is mostly standard and can be found in [81, 97]. The classification of semisimple symmetric spaces is due to M. Berger [5]. The dual constructions presented here and the relations between them can be found in [146]. The importance of the c-dual for causal spaces was pointed out in [129, 130], where one can also find most of the material on the $\mathfrak{h}$-module structure of $\mathfrak{q}$. The version of Lemma 1.3.4 presented here was communicated to us by K.-H. Neeb.

Chapter 2

Causal Orientations

In this chapter we recall some basic facts about convex cones, their duality, and linear automorphism groups which will be used throughout the book. Then we define causal orientations for homogeneous manifolds and show how they are determined by a single closed convex cone in the tangent space of a base point invariant under the stabilizer group of this point. Finally, we describe various causal orientations for the examples treated in Chapter 1.

2.1 Convex Cones and Their Automorphisms

Let $\mathbf{V}$ be a finite-dimensional real Euclidean vector space with inner product $(\cdot|\cdot)$. Let $\mathbb{R}^+ := \{\lambda \in \mathbb{R} \mid \lambda > 0\}$ and $\mathbb{R}_0^+ = \mathbb{R}^+ \cap \{0\}$. A subset $C \subset \mathbf{V}$ is a *cone* if $\mathbb{R}^+ C \subset C$ and a *convex cone* if C in addition is a convex subset of $\mathbf{V}$, i.e., $u, v \in C$ and $\lambda \in [0,1]$ imply $\lambda u + (1-\lambda)v \in C$. Thus C is *a convex cone* if and only if for all $u, v \in C$ and $\lambda, \mu \in \mathbb{R}^+$, $\lambda u + \mu v \in C$. The cone C is called *nontrivial* if $C \neq -C$. Note that $C \neq \{0\}$, and $C \neq \mathbf{V}$ if C is nontrivial. We set

1) $\mathbf{V}^C := C \cap -C$,

2) $< C > := C - C = \{u - v \mid u, v \in C\}$,

3) $C^* := \{u \in \mathbf{V} \mid \forall v \in C : (v \mid u) \geq 0\}$.

Then $\mathbf{V}^C$ and $< C >$ are vector spaces called the *edge* and the *span* of C. The set C^* is a closed convex cone called the *dual cone* of C. Note that this definition agrees with the usual one under the identification of the dual space $\mathbf{V}^*$ with $\mathbf{V}$ by use of the inner product $(\cdot|\cdot)$. If C is a closed convex cone we have $C^{**} = C$ and

$$(C^* \cap -C^*) = < C >^{\perp}, \tag{2.1}$$

where for a subset U in $\mathbf{V}$ we set $U^{\perp} = \{v \in \mathbf{V} \mid \forall u \in U \ : \ (u \mid v) = 0\}$.

Definition 2.1.1 Let C be a convex cone in $\mathbf{V}$. Then C is called *generating* if $< C >= \mathbf{V}$ and *pointed* if there exists a $v \in \mathbf{V}$ such that for all $u \in C\backslash\{0\}$ we have $(u \mid v) > 0$. If C is closed, it is called *proper* if $\mathbf{V}^C = \{0\}$, *regular* if it is generating and proper, and and *self-dual* if $C^* = C$.

The set of interior points of C is denoted by C^o or $\operatorname{int}(C)$. The interior of C in its linear span $< C >$ is called the *algebraic interior* of C and denoted $\operatorname{algint}(C)$.

Let $S \subset \mathbf{V}$. Then the closed convex cone generated by S is denoted by $\operatorname{cone}(S)$:

$$\operatorname{cone}(S) := \overline{\left\{ \sum_{\text{finite}} r_s s \,\middle|\, s \in S,\ r_s \geq 0 \right\}}. \tag{2.2}$$

The set of closed regular convex cones in $\mathbf{V}$ is denoted by $\operatorname{Cone}(\mathbf{V})$. □

If C is a closed convex cone, then its interior C^o is an open convex cone. If Ω is an open convex cone, then its closure $\overline{\Omega} := cl(\Omega)$ is a closed convex cone. For an open convex cone we define the dual cone by

$$\Omega^* := \{u \in \mathbf{V} \mid \forall v \in \overline{\Omega} \setminus \{0\} : (u \mid v) > 0\} = \operatorname{int}(\overline{\Omega}^*).$$

If $\overline{\Omega}$ is *proper* we have $\Omega^{**} = \Omega$. With this modified version of duality for open convex cones it also makes sense to talk about open self-dual cones.

Example 2.1.2 (The Forward Light Cone) For $n \geq 2$, $q = n - 1$ and $p = 1$ we define the (semialgebraic) cone C in $\mathbb{R}^n$ by

$$C := \{v \in \mathbb{R}^n \mid Q_{1,q}(v, v) \geq 0,\ x \geq 0\}$$

and set

$$\Omega := C^o = \{v \in \mathbb{R}^n \mid Q_{1,q}(v, v) > 0,\ x > 0\}.$$

C is called the *forward light cone* in $\mathbb{R}^n$. We have $v = \binom{x}{y} \in C$ if and only if $x \geq \|y\|$. If $v \in C \cap -C$, then $0 \leq x \leq 0$ and thus $x = 0$. We get $\|y\| = 0$ and hence $y = 0$. Thus $v = 0$ and C is proper. For $v, v' \in C$ we calculate

$$(v' \mid v) = x'x + (y' \mid y) \geq \|y'\|\,\|y\| + (y' \mid y) \geq 0$$

so that $C \subset C^*$. For the converse let $v = \binom{x}{y} \in C^*$. Then it follows by testing against $e_1 \in \Omega$ that $x \geq 0$. We may assume, that $y \neq 0$. Define w by $\mathrm{pr}_1(w) = \|y\|$ and $\mathrm{pr}_2(w) = -y$. Then $w \in C$ and

$$0 \leq (w|v) = x\|y\| - \|y\|^2 = (x - \|y\|)\|y\|.$$

If follows that $x \geq \|y\|$. Therefore $y \in C$ so that $C^* \subset C$ and hence C is self-dual. In the same way one can show that also Ω is self-dual. □

Remark 2.1.3 Let C be a closed convex cone in $\mathbf{V}$. Then the following are equivalent:

1) C^o is nonempty.

2) C contains a basis of $\mathbf{V}$.

3) $< C >= \mathbf{V}$. □

Proposition 2.1.4 *Let C be a nonempty closed convex cone in $\mathbf{V}$. Then the following properties are equivalent:*

1) C is pointed,

2) C is proper.

3) $\mathrm{int}(C^*) \neq \emptyset$.

Proof: The implications "(1)⇒(2)" and "(3)⇒(1)" are immediate. Assume now that C is proper. Then by (2.1) we have

$$< C^* >= (\mathbf{V}^C)^{\perp} = \mathbf{V},$$

so C^* is generating. Now Remark 2.1.3 shows $\mathrm{int}(C^*) \neq \emptyset$ and (3) follows. □

Corollary 2.1.5 *Let C be a closed convex cone. Then C is proper if and only if C^* is generating.* □

Corollary 2.1.6 *Let C be a convex cone in $\mathbf{V}$. Then $C \in \mathrm{Cone}(\mathbf{V})$ if and only if $C^* \in \mathrm{Cone}(\mathbf{V})$.* □

A *face* F of a closed convex cone $C \subset \mathbf{V}$ is subset of C such that $v, v' \in C \setminus F$ with $v, v' \in C$ implies $v \in C \setminus F$ or $v' \in C \setminus F$. We denote the set of faces of a cone C by $\mathrm{Fa}(C)$. Note that $\mathrm{Fa}(C)$ is a lattice with respect to the inclusion order.

A cone C in a finite-dimensional vector space $\mathbf{V}$ is said to be *polyhedral* if it is an intersection of finitely many half-spaces. For the following result see [55].

Remark 2.1.7 Let V be a finite-dimensional vector space and $W \subset V$. Then 1)–3) are equivalent and imply 4), and 5).

1) There exists a finite subset $E \subset V$ such that $W = \text{cone}(E) = \sum_{v \in E} \mathbb{R}^+ v$.

2) W is a polyhedral cone.

3) The dual wedge $W^* \subset V^*$ is polyhedral.

4) For every face $F \in \text{Fa}(W)$ there exists a finite subset $D \subset E$ such that the following assertions hold:

   a) $F = \text{cone}(D)$.
   b) $W - F = \text{cone}(E \cup -D)$ is a cone.
   c) $\mathbf{V}^{W-F} = F - F = \text{cone}(D \cup -D)$.
   d) $< F > \cap W = F$.

5) The mapping

$$\text{op} : \text{Fa}(W) \to \text{Fa}(W^*), \quad F \mapsto F^\perp \cap W^* \tag{2.3}$$

defines an antiisomorphism of finite lattices. Moreover,

$$< \text{op}(F) >= F^\perp \qquad \forall F \in \text{Fa}(W).$$ □

We introduce some notation that will be used throughout the book. Let $\mathbf{W}$ be a Euclidean vector space and let $\mathbf{V}$ be a subspace. Denote by pr_V or simply pr the orthogonal projection $\text{pr}_V : \mathbf{W} \to \mathbf{V}$. If C is a cone in $\mathbf{W}$, then we define $P_W^V(C), I_W^V(C) \subset \mathbf{V}$ by

$$P_V^W(C) = \text{pr}_V(C) \quad \text{and} \quad I_V^W(C) = C \cap \mathbf{V}. \tag{2.4}$$

If the role of $\mathbf{W}$ and $\mathbf{V}$ is clear from the context, we simply write $P(C)$ and $I(C)$.

Lemma 2.1.8 *Let* $\mathbf{W}$ *be Euclidean vector space and let* $\mathbf{V}$ *be a subspace of* $\mathbf{W}$. *Denote by* pr: $\mathbf{W} \to \mathbf{V}$ *the orthogonal projection onto* $\mathbf{V}$. *Given* $C \in \text{Cone}(\mathbf{W})$, *we have* $I(C^*) = P(C)^*$.

Proof: Let $Z \in C$ and let $X = \text{pr}(Z) \in P(C)$. If $Y \in I(C^*)$, then $(Y|X) = (Y|Z) \geq 0$. Thus $Y \in P(C)^*$. Conversely, assume that $X \in P(C)^*$ and $Y \in C$. Then $Y = \text{pr}\, Y + Y^\perp$ with $Y^\perp \perp \mathbf{V}$. In particular, $\text{pr}(Y) \in P(C)$. Therefore we have

$$(Y|X) = (\text{pr}(Y)|X) \geq 0$$

and hence $X \in I(C^*)$. □

Lemma 2.1.9 *Suppose that $\tau: \mathbf{W} \to \mathbf{W}$ is an involutive isometry with fixed point set $\mathbf{V}$. Let $\tilde{C} \in \mathrm{Cone}(\mathbf{W})$ with $-\tau(\tilde{C}) = \tilde{C}$. Then $I_V^W(\tilde{C}) = P_V^W(\tilde{C})$ and $I_V^W(\tilde{C}^*) = I_V^W(\tilde{C})^*$.*

Proof: Let $X = X_+ + X_- \in \tilde{C}$, where the subscripts denote the projections onto the (± 1) eigenspaces of τ. As $-\tau(\tilde{C}) = \tilde{C}$, it follows that $-\tau(X) = -X_+ + X_- \in \tilde{C}$. Hence $X_- = \frac{1}{2}(X - \tau(X)) \in \tilde{C}$. Thus $P(\tilde{C}) \subset I(\tilde{C})$. But we always have $I(\tilde{C}) \subset P(\tilde{C})$. Hence $P(\tilde{C}) = I(\tilde{C})$. Let $Y = Y_+ + Y_- \in \tilde{C}^*$. Since $-\tau(\tilde{C}) = \tilde{C}$ implies $-\tau(\tilde{C}^*) = \tilde{C}^*$, we find $Y_- \in \tilde{C}^*$. Lemma 2.1.8 shows that $P(\tilde{C}^*) = I(\tilde{C})^*$ and thus $I(\tilde{C}^*) = I(\tilde{C})^*$. □

Next we turn to linear automorphism groups of convex cones. For a convex cone C we denote the *automorphism group* of C by

$$\mathrm{Aut}(C) := \{a \in \mathrm{GL}(\mathbf{V}) \mid a(C) = C\}. \tag{2.5}$$

If C is open or closed, then $\mathrm{Aut}(C)$ is closed in $\mathrm{GL}(\mathbf{V})$. In particular, $\mathrm{Aut}(C)$ is a linear Lie group. If we denote the transpose of a linear operator a by ${}^t a$, we obtain

$$\mathrm{Aut}(C^*) = {}^t\,\mathrm{Aut}(C) \tag{2.6}$$

whenever C is an open or closed convex cone.

Remark 2.1.10 Equation (2.6) shows that if C is an open or closed self-dual cone in $\mathbf{V}$, then $\mathrm{Aut}(C)$ is a reductive subgroup of $\mathrm{GL}(\mathbf{V})$ invariant under the involution $a \mapsto \theta(a) := {}^t a^{-1}$. The restriction of θ to the commutator subgroup of the connected component $\mathrm{Aut}(C)_o$ of $\mathrm{Aut}(C)_o$ is a Cartan involution. □

Definition 2.1.11 Let G be a group acting (linearly) on $\mathbf{V}$. Then a cone $C \subset \mathbf{V}$ is called *G-invariant* if $G \cdot C = C$. We denote the set of invariant regular cones in $\mathbf{V}$ by $\mathrm{Cone}_G(\mathbf{V})$. A convex cone C is called *homogeneous* if $\mathrm{Aut}(C)$ acts transitively on C. □

For $C \in \mathrm{Cone}(\mathbf{V})$ we have $\mathrm{Aut}(C) = \mathrm{Aut}(C^o)$ and $C = \partial C \cup C^o = (C \setminus C^o) \cup C^o$ is a decomposition of C into $\mathrm{Aut}(C)$-invariant subsets. In particular, a nontrivial closed regular cone can never be homogeneous.

Remark 2.1.12 Let $\mathbf{V} \subset \mathbf{W}$ be Euclidian vector spaces with orthogonal projection $\mathrm{pr}_V: \mathbf{W} \to \mathbf{V}$. Suppose that L is a group acting on $\mathbf{W}$ and N is a subgroup of L leaving $\mathbf{V}$ invariant. Let $C \subset \mathbf{W}$ be an L-invariant convex cone. Then $I(C)$ is N-stable. If furthermore ${}^t N = N$, then $P(C)$ is N-invariant, too. In fact, the first claim is trivial. For the second, note that ${}^t N = N$ implies that $\mathbf{V}^\perp$ is N-invariant. Hence $\mathrm{pr}(n \cdot \mathbf{w}) = n \cdot \mathrm{pr}(\mathbf{w})$ for all $\mathbf{w} \in \mathbf{W}$ and $P(C)$ is N-invariant.

Now let c be an N-invariant cone in $\mathbf{V}$. We define the *extension* $E_{\mathbf{W},L}^{\mathbf{V},N}(c)$ of c to $\mathbf{W}$ by

$$E_{\mathbf{V},N}^{\mathbf{W},L}(C) = \overline{\mathrm{conv}\,(L \cdot C)}\,. \tag{2.7}$$

If the roles of $L, N, \mathbf{V}$, and $\mathbf{W}$ are clear, we will write $E(C)$, $E_N^L(C)$ or $E_{\mathbf{V}}^{\mathbf{W}}(C)$ instead of $E_{\mathbf{V},N}^{\mathbf{W},L}(C)$. If N_1 is a subgroup of N acting trivially on $\mathbf{V}$, then the group N/N_1 acts on $\mathbf{V}$. By abuse of notation we replace N by N/N_1 in that case. □

Theorem 2.1.13 *Let G be a Lie group acting linearly on the Euclidean vector space $\mathbf{V}$ and $C \in \mathrm{Cone}_G(\mathbf{V})$. Then the stabilizer in G of a point in C^o is compact.*

Proof: Let $\Omega := C^o$. First we note that for every $v \in \Omega$ the set $U = \Omega \cap (v - \Omega)$ is open, nonempty ($\frac{1}{2}v \in U$), and bounded. Thus we can find closed balls $\overline{B_r(\frac{1}{2}v)} \subset U \subset \overline{B_R(\frac{1}{2}v)}$. Let $a \in \mathrm{Aut}(\Omega)^v = \{b \in \mathrm{Aut}(\Omega) \mid b \cdot v = v\}$. From $a \cdot \Omega \subset \Omega$ and $a \cdot v = v$ we obtain $a(U) \subset U$. Therefore $a(\overline{B_r(\frac{1}{2}v)}) \subset \overline{B_R(\frac{1}{2}v)}$ and $a(\frac{1}{2}v) = \frac{1}{2}v$ implies $\|a\| \leq R/r$. Thus $\mathrm{Aut}(\Omega)^v$ is closed and bounded, i.e., compact. □

Example 2.1.14 ($\mathrm{H}^+(m, \mathbb{K})$) For $\mathbb{K} = \mathbb{R}$ or $\mathbb{C}$ we let $\mathbf{V}$ be the real vector space $\mathrm{H}(m, \mathbb{K})$ of Hermitian matrices over $\mathbb{K}$,

$$\mathbf{V} = \mathrm{H}(m, \mathbb{K}) := \{X \in M(m, \mathbb{K}) \mid X^* = X\},$$

where $M(m, \mathbb{K})$ denotes the set of $m \times m$ matrices with entries in $\mathbb{K}$ and $X^* := {}^t\overline{X}$. Then $n = \dim_{\mathbb{R}} \mathbf{V}$ is given by

1) $n = m(m+1)/2$ for $\mathbb{K} = \mathbb{R}$

2) $n = m^2$ for $\mathbb{K} = \mathbb{C}$.

Define an inner product on $\mathbf{V}$ by

$$(X \mid Y) := \mathrm{Re}\,\mathrm{Tr}\, XY^*, \quad X, Y \in \mathbf{V}.$$

Then the set

$$\Omega = \mathrm{H}^+(m, \mathbb{K}) := \{X \in \mathrm{H}(m, \mathbb{K}) \mid X > 0\}, \tag{2.8}$$

where $>$ means *positive definite*, is an open convex cone in $\mathbf{V}$. The closure C of Ω is the closed convex cone of all positive semidefinite matrices in $\mathrm{H}(m, \mathbb{K})$. Let $Y = I_m$ denote the $m \times m$ identity matrix. Then for $X \in C$, $X \neq 0$, $(X \mid Y) > 0$ as all the eigenvalues of X are non-negative and $X \neq 0$. Thus C is proper.

We claim that C and Ω are self-dual. To prove that, let $e_1, \ldots, e_m$ denote the standard basis for $\mathbb{K}^m$, $e_j = {}^t(\delta_{1,j}, \ldots, \delta_{m,j})$ and define $E_{ij} \in M(m, \mathbb{K})$ by

$$E_{ij}e_k = \delta_{j,k}e_i.$$

Then $E_{11}, \ldots, E_{mm}$ is a basis for $M(m, \mathbb{K})$.

Now suppose that $X, Y \in \Omega$. Then $Y = aa^*$ for some matrix a and hence $\mathrm{Tr}(XY) = \mathrm{Tr}(a^*Xa) \geq 0$ so that $\Omega \subset \Omega^*$. Conversely, let $Y \in \Omega^*$. As Y is Hermitian, we may assume that Y is diagonal: $Y = \sum_{i=1}^m \lambda_i E_{ii}$. Let $X = E_{jj} \in \overline{\Omega} \setminus \{0\}$. Then

$$0 < (X \mid Y) = \lambda_j.$$

Thus Y is positive definite and hence in Ω. This proves that Ω is self-dual and hence also the self-duality of C. As a consequence, we see

$$\{0\} = C \cap -C = C^* \cap -C^* = < C >^{\perp}$$

and $C \in \mathrm{Cone}(\mathbf{V})$.

The group $G = \mathrm{GL}(m, \mathbb{K})$ acts on $M(m, \mathbb{K})$ by

$$a \cdot X := aXa^*, \; a \in \mathrm{GL}(m, \mathbb{K}), \; X \in M(m, \mathbb{K})$$

and Ω and C are $\mathrm{GL}(m, \mathbb{K})$-invariant. As every positive definite matrix can be written in the form $aa^* = a \cdot I$ for some $a \in \mathrm{GL}(m, \mathbb{K})$, it follows that Ω is homogeneous. The stabilizer of I in G is just $K = U(m, \mathbb{K})$. We have

1) $K \cap \mathrm{SL}(m, \mathbb{K}) = \mathrm{SO}(m)$ for $\mathbb{K} = \mathbb{R}$, and

2) $K \cap \mathrm{SL}(m, \mathbb{K}) = \mathrm{SU}(m)$ for $\mathbb{K} = \mathbb{C}$.

We can see that Theorem 2.1.13 does not hold in general for an element in the boundary of a convex cone. In fact, let $X = \left(\begin{smallmatrix} I_k & 0 \\ 0 & 0 \end{smallmatrix}\right)$. Let $g = \left(\begin{smallmatrix} A & B \\ C & D \end{smallmatrix}\right) \in \mathrm{GL}(m, \mathbb{K})^X$. Here $A \in M(k, \mathbb{K})$, $B \in M(k \times (m-k), \mathbb{K})$, $C \in M((m-k) \times k, \mathbb{K})$ and $D \in M(m-k, \mathbb{K})$, where $M(r \times s, \mathbb{K})$ denotes the $r \times s$ matrices with entries in $\mathbb{K}$. Then

$$X = gXg^* = \begin{pmatrix} AA^* & AC^* \\ CA^* & CC^* \end{pmatrix}.$$

Thus

$$\mathrm{GL}_m(\mathbb{K})^X = \left\{ \begin{pmatrix} A & B \\ 0 & D \end{pmatrix} \middle| A \in U(k), D \in \mathrm{GL}_{m-k}(\mathbb{K}), B \in M_{k \times (m-k)}(\mathbb{K}) \right\}$$

and this group is noncompact. □

Consider a subset $L \subset \mathbf{V}$. The convex hull of L is the smallest convex subset of $\mathbf{V}$ containing L. We denote this set by $\operatorname{conv}(L)$. Then

$$\operatorname{conv}(L) = \left\{ \sum_J \lambda_j v_j \,\middle|\, J \text{ finite, } \lambda_j \geq 0, \sum_J \lambda_j = 1, v_j \in L \right\}.$$

Caratheodory's theorem ([153], p. 73) says that one can always choose λ_j and v_j such that the cardinality of J is less than or equal to $\dim \mathbf{V} + 1$. If L is a cone, then the convex hull of L can also be described as

$$\operatorname{conv}(L) = \left\{ \sum_J \lambda_j v_j \,\middle|\, J \text{ finite, } \lambda_j \geq 0, v_j \in L \right\}.$$

Lemma 2.1.15 *Let G be a Lie group acting linearly on $\mathbf{V}$ and let $K \subset G$ be a compact subgroup. If $C \subset \mathbf{V}$ is a nontrivial G-invariant proper cone in $\mathbf{V}$, then there exists a K-fixed vector $u \in C \setminus \{0\}$. If C is also generating, then u may be chosen in C^o.*

Proof: Choose $v \in C^*$ such that $(u|v) > 0$ for all $u \in C$, $u \neq 0$. Fix a $u \in C \setminus \{0\}$. Then $(k \cdot u|v) > 0$ for all $k \in K$. It follows that

$$u_K := \int_K (k \cdot u)\, dk \in \operatorname{conv}(K \cdot u) \subset C$$

is K-fixed and

$$(u_K|v) = \int_K (k \cdot u|v)\, dk > 0.$$

Thus $u_K \neq 0$. As K is compact, it follows that $K \cdot u$ is also compact and thus $\overline{\operatorname{conv}(K \cdot u)} = \operatorname{conv}(K \cdot u)$ is compact, too. If $u \in C^o$, then $u_K \in \overline{\operatorname{conv} K \cdot u} = \operatorname{conv} K \cdot u \subset C^o$ since C is convex. □

Example 2.1.16 (The Forward Light Cone Continued) Recall the notation from Example 2.1.2 and the groups $\mathrm{SO}_o(p,q)$ described in Section 1.5.

Obviously, the forward light cone C is invariant under the usual operation of $\mathrm{SO}_o(1,q)$ and under all dilations λI_n, $\lambda > 0$. We claim that the group $\mathrm{SO}_o(1,q)\mathbb{R}^+ I_{q+1}$ acts transitively on $\Omega = C^o$ if $q \geq 2$. In particular, this says that Ω is homogeneous.

To prove the claim, assume that $q \geq 2$. We will show that

$$\Omega = \mathrm{SO}_o(1,q)\mathbb{R}^+ \begin{pmatrix} 1 \\ 0 \end{pmatrix}.$$

In fact, using

$$a_t := \begin{pmatrix} \cosh(t) & \sinh(t) & 0 \\ \sinh(t) & \cosh(t) & 0 \\ 0 & 0 & I_{n-2} \end{pmatrix} \in \mathrm{SO}_o(1,q)$$

we get

$$a_t \cdot \begin{pmatrix} \lambda \\ 0 \end{pmatrix} = \lambda^t\,(\cosh(t), \sinh(t), 0, \ldots, 0) \in \Omega$$

for all $t \in \mathbb{R}$. As $\mathrm{SO}(q)$ acts transitively on S^{q-1} and $\begin{pmatrix} 1 & 0 \\ 0 & A \end{pmatrix} \in \mathrm{SO}_o(1,q)$ for all $A \in \mathrm{SO}(q)$, the claim now follows in view of the fact that $\coth(t)$ runs through $]1, \infty[$ as t varies in $]0, \infty[$.

We remark that the homomorphism $\mathrm{SO}(q) \hookrightarrow \mathrm{SO}_o(1,q)$ realizes $\mathrm{SO}(q)$ as a maximal compact subgroup of $\mathrm{SO}_o(1,q)$, leaving the nonzero vector $e_1 \in \Omega$ invariant, and a straightforward calculation shows that $\mathrm{SO}_o(1,q)^{e_1} = \mathrm{SO}(q)$. According to Lemma 2.1.15, any $\mathrm{SO}_o(1,q)$-invariant regular cone in $\mathbb{R}^n$ contains an $\mathrm{SO}(q)$-fixed vector, i.e., a multiple of e_1. Therefore homogeneity of Ω implies that

$$\mathrm{Cone}_{\mathrm{SO}_o(1,q)}(\mathbb{R}^n) = \{C, -C\}. \tag{2.9}$$

for $q > 1$. Note that for $q = 1$ the equality (2.9) no longer holds. In fact, the four connected components of $|x| \neq |y|$ in $\mathbb{R}^2$ are all $\mathrm{SO}_o(1,1)$-invariant cones. □

We now prove two fundamental theorems in the theory of invariant cones. The first result is due to Kostant [157], whereas the second theorem is due to Vinberg [166]. For the notation, refer to Appendix A.2.

Theorem 2.1.17 (Kostant) *Suppose that* $\mathbf{V}$ *is a finite-dimensional real vector space. Let* L *be a connected reductive subgroup of* $\mathrm{GL}(\mathbf{V})$ *acting irreducibly, and* $G = L'$ *the commutator subgroup of* L. *Further, let* K *be a maximal compact subgroup of* G. *Then the following properties are equivalent:*

1) There exists a regular L*-invariant closed convex cone in* $\mathbf{V}$.

2) The G*-module* $\mathbf{V}$ *is spherical.*

Proof: Note first that G is closed by Remark A.2.7, so K is a compact subgroup of $\mathrm{GL}(\mathbf{V})$. We may assume that $\mathbf{V}$ is Euclidian and that $K \subset \mathrm{SO}(\mathbf{V})$. Then the implication (1)⇒(2) follows from Lemma 2.1.15.

Lemma A.2.8 shows that the connected component of $Z(L)$ acts on $\mathbf{V}$ by positive real numbers. Thus it only remains to show the existence of a G-invariant proper cone. To do that we note that $\mathbf{V}_{\mathbb{C}}$ is irreducible by Lemma A.2.5 and consider a highest-weight vector u of $\mathbf{V}_{\mathbb{C}}$. Let $v_K \in \mathbf{W}^K$ be a nonzero K-fixed vector. We can choose v_K such that $(u|v_K) > 0$. In fact, if u and v_K were orthogonal, then $G = KAN$ would imply that all of $\mathbf{V}$ is orthogonal to v_K. Now $(G \cdot u|v_K) = (A \cdot u|v_K) = \mathbb{R}^+$. This shows that $\text{cone}(G \cdot u)) \subseteq (\mathbb{R}^+u)^*$ is a nontrivial G-invariant convex cone in $\mathbf{V}$. It has to be regular by irreducibility. □

Remark 2.1.18 1) In the situation of Theorem 2.1.17 we see that $K \cdot u$ is a compact subset of C bounded away from zero. This shows that for $v \in \overline{\text{conv}(G \cdot u)} \setminus \{0\}$ there exists a compact interval $J \subset \mathbb{R}^+$ such that $v \in \text{conv}\, \pi(K)Ju$. In particular, $(\text{conv}\, G \cdot u) \cup \{0\}$ is closed.

2) The assumption in the next theorem that $\mathbf{V}$ is irreducible is not needed for the implication (1) ⇒ (2). □

Theorem 2.1.19 (Vinberg) *Let L be a connected reductive Lie group and $\mathbf{V}$ a finite-dimensional irreducible real L-module. Then the following properties are equivalent:*

1) $\text{Cone}_L(\mathbf{V}) \neq \emptyset$.

2) *The G-module $\mathbf{V}$ is spherical, where $G = L'$ is the commutator subgroup of L.*

3) *There exists a ray in $\mathbf{V}$ through 0 which is invariant with respect to some minimal parabolic subgroup P of G.*

If these conditions hold, every invariant pointed cone in $\mathbf{V}$ is regular.

Proof: Note first that any nontrivial L-invariant cone C is automatically generating, as $< C >$ is an L-invariant subspace of $\mathbf{V}$ and $\mathbf{V}$ is assumed irreducible.

Denote the representation of L on $\mathbf{V}$ by π. Then $\pi(G)$ is closed in $\text{GL}(\mathbf{V})$ and $\pi(K)$ is compact, as linear semisimple groups always have compact fixed groups for the Cartan involution. In fact, this shows that $\pi(K)$ is maximal compact in $\pi(G)$. Thus the equivalence of (1) and (2) follows from Kostant's Theorem 2.1.17 applied to $\pi(L)$.

Now suppose that $\mathbf{V}$ is K-spherical. Then Lemma A.2.6 yields the desired ray.

Finally, assume that the ray $\mathbb{R}^+ \cdot u$ is MAN-invariant. Since N is nilpotent and $\pi(M)$ compact, we see that $\mathfrak{n}$ and $\mathfrak{m}$ act trivially on u. Thus u is a highest-weight vector and the corresponding highest weight satisfies the conditions of Theorem A.2.2, i.e., $\mathbf{V}$ is K-spherical. □

Remark 2.1.20 In the situation of Vinberg's theorem with a spherical G-module $\mathbf{V}$, Theorem A.2.2 shows that for any highest-weight vector $v \in \mathbf{V}_{\mathbb{C}}$ also $\overline{v}$ is a highest-weight vector. This implies that $\mathbf{V}$ contains a highest weight vector u. In other words, $\mathbf{V}$ is a *real highest-weight module.* □

Another well-known fact about invariant cones that we will often use is the following description of the minimal and maximal invariant cones due to Vinberg and Paneitz.

Theorem 2.1.21 (Paneitz,Vinberg) *Assume that L is a connected reductive Lie group. Let $\mathbf{V}$ be a finite-dimensional real irreducible L-module with* $\mathrm{Cone}_L(\mathbf{V}) \neq \emptyset$. *Equip $\mathbf{V}$ with an inner product as in* Lemma A.2.3. *Then there exists a-unique up to multiplication by (-1)-minimal invariant cone $C_{\min} \in \mathrm{Cone}_L(\mathbf{V})$ given by*

$$C_{\min} = \mathrm{conv}(G \cdot u) \cup \{0\} = \overline{\mathrm{conv}\, G\, (\mathbb{R}^+ \cdot v_K)}, \tag{2.10}$$

where u is a highest-weight vector, v_K is a nonzero K-fixed vector unique up to scalar multiple, and $(u|v_K) > 0$. The unique (up to multiplication by -1) maximal cone is then given by $C_{\max} = C^*_{\min}$.

Proof: We know by now that there is a pointed invariant cone in $\mathbf{V}$ if and only if $\dim \mathbf{V}^K = 1$ and that every pointed invariant cone contains either v_K or $-v_K$. Furthermore, we know that every pointed invariant cone is regular. Thus Remark 2.1.18 implies that $C := \mathrm{conv}(G \cdot u) \cup \{0\}$ is a regular cone and $v_K \in C^o$. In particular, we get that the closed G-invariant cone C_1 generated by v_K is contained in C. But Lemma A.2.6 shows that $u \in C_1$ and thus $C_{\min} = C = C_1$ is in fact minimal. The equality $C_{\max} = C^*_{\min}$ follows from ${}^t\pi(g) = \pi\big(\theta(g)\big)^{-1}$. Now the claim follows from Lemma A.2.8. □

Remark 2.1.22 We have seen in Remark 2.1.20 that in the situation of Vinberg's theorem 2.1.19 the G-module $\mathbf{V}$ is a real highest-weight module with highest-weight vector $u \in \mathbf{V}$. It follows from Lemma II.4.18 in [46] that the stabilizer G^u of u in G contains the group MN. This implies that $u \in \partial C$. In fact, according to Lemma 2.1.10, the stabilizer group of a point in the interior of an invariant cone acts as a *compact* group. □

2.2 Causal Orientations

Let $\mathcal{M}$ be a C^∞-manifold. For $m \in \mathcal{M}$ we denote the tangent space of $\mathcal{M}$ at m by $T_m(\mathcal{M})$ or $T_m\mathcal{M}$ and the tangent bundle of $\mathcal{M}$ by $T(\mathcal{M})$.

The derivative of a differentiable map $f: \mathcal{M} \to \mathcal{N}$ at m will be denoted by $d_m f: T_m(\mathcal{M}) \to T_{f(m)}(\mathcal{N})$.

A smooth *causal structure* on $\mathcal{M}$ is a map which assigns to each point m in $\mathcal{M}$ a nontrivial closed convex cone $C(m)$ in $T_m\mathcal{M}$ and which is smooth in the following sense: One can find an open covering $\{U_i\}_{i\in I}$ of $\mathcal{M}$, smooth maps

$$\varphi_i : U_i \times \mathbb{R}^n \to T(\mathcal{M})$$

with $\varphi_i(m, v) \in T_m(\mathcal{M})$, and a cone C in $\mathbb{R}^n$ such that

$$C(m) = \varphi_i(m, C).$$

The causal structure is called generating (proper, regular) if $C(m)$ is generating (proper, regular) for all m. A map $f: \mathcal{M} \to \mathcal{M}$ is called *causal* if $d_m f\,(C(m)) \subset C\,(f(m))$ for all $m \in \mathcal{M}$. If a Lie group G acts smoothly on $\mathcal{M}$ via $(g, m) \mapsto g \cdot m$, we denote the diffeomorphism $m \mapsto g \cdot m$ by ℓ_g.

Definition 2.2.1 Let $\mathcal{M}$ be a manifold with a causal structure and G a Lie group acting on $\mathcal{M}$. Then the causal structure is called *G-invariant* if all ℓ_g, $g \in G$, are causal. □

If $\mathcal{M} = G/H$ is homogeneous, then a G-invariant causal structure is determined completely by the cone $C := C(\mathbf{o}) \subset T_{\mathbf{o}}(\mathcal{M})$, where $\mathbf{o} := \{H\} \in G/H$. Furthermore, C is proper, generating, etc., if and only if this holds for the causal structure. We also note that C is invariant under the action of H on $T_{\mathbf{o}}(\mathcal{M})$ given by $h \mapsto d_{\mathbf{o}}\ell_h$. On the other hand, if $C \in \mathrm{Cone}_H(T_{\mathbf{o}}(\mathcal{M}))$, then we can define a field of cones by

$$\mathcal{M} \ni aH \mapsto C(aH) := d_{\mathbf{o}}\ell_a(C) \subset T_{a\cdot\mathbf{o}}(\mathcal{M}),$$

and this cone field is clearly G-invariant, regular, and satisfies $C(\mathbf{o}) = C$. What is not so immediate is the fact that $m \mapsto C(m)$ is smooth in the sense described above. This can be seen using a smooth local section of the quotient map $\mathcal{M} = G/H \to G$ and then one obtains the following theorem, which we can also use as a definition since we will exclusively deal with G-homogeneous regular causal structures.

Theorem 2.2.2 *Let $\mathcal{M} = G/H$ be homogeneous. Then*

$$C \mapsto (aH \mapsto d_{\mathbf{o}}\ell_a(C)) \tag{2.11}$$

defines a bijection between $\mathrm{Cone}_H(T_{\mathbf{o}}(\mathcal{M}))$ and the set of G-invariant, regular causal structures on $\mathcal{M}$. □

Let $\mathcal{M} = G/H$ and $C \in \mathrm{Cone}_G(T_o\mathcal{M})$. An absolutely continuous curve $\gamma: [a,b] \to \mathcal{M}$ is called *C-causal* (also called *conal*, cf. [52]) if $\gamma'(t) \in C(\gamma(t))$ whenever the derivative exists. Here a continuous mapping $\gamma: [a,b] \to \mathcal{M}$ is called *absolutely continuous* if for any coordinate chart $\phi: U \to \mathbb{R}^n$ the curve $\eta = \phi \circ \gamma: \gamma^{-1}(U) \to \mathbb{R}^n$ has absolutely continuous coordinate functions and the derivatives of these functions are locally bounded.

We define a relation $\preceq_s$ (s for strict) on $\mathcal{M}$ by saying that $m \preceq_s n$ if there exists a C-causal curve γ connecting m with n. The relation $\preceq_s$ clearly is reflexive and transitive. We call such relations *causal orientations.* Elsewhere they are also called *quasiorders* .

Example 2.2.3 (Vector Spaces) Let $\mathbf{V}$ be a finite-dimensional vector space and $C \subset \mathbf{V}$ a closed convex cone in $\mathbf{V}$. Then we define a causal $\mathrm{Aut}(C)$-invariant orientation on $\mathbf{V}$ by

$$u \preceq v \iff v - u \in C.$$

Then $\preceq$ is antisymmtric if and only if C is proper. In particular, $\mathrm{H}^+(n, \mathbb{K})$ defines a $\mathrm{GL}(n, \mathbb{K})$-invariant global ordering in $\mathrm{H}(n, \mathbb{K})$. Also, the light cone $C \subset \mathbb{R}^{n+1}$ defines a $\mathrm{SO}_o(1,n)$-invariant ordering in $\mathbb{R}^{n+1}$. The space $\mathbb{R}^{n+1}$ together with this global ordering is called *the* $(n+1)$*-dimensional Minkowski space* . □

In general, the *graph* $\mathcal{M}_{\preceq_s} := \{(m,n) \in \mathcal{M} \times \mathcal{M} \mid m \preceq_s n\}$ of $\preceq_s$ will not be closed in $\mathcal{M} \times \mathcal{M}$. This makes $\preceq_s$ difficult to work with and we will mostly use its closure $\preceq$, defined via

$$m \preceq n :\iff (m,n) \in \overline{\mathcal{M}_{\preceq_s}},$$

instead. It turns out that $\preceq$ is again a causal orientation. The only point that is not evident is transitivity. So suppose that $m \preceq n \preceq p$ and let m_k, n_k, n'_k, p_k be sequences with

$$m_k \preceq_s n_k, n'_k \preceq_s p_k, \quad m_k \to m, n_k \to n, n'_k \to n, p_k \to p.$$

Then we can find a sequence g_k in G converging to the identity such that $n'_k = g_k \cdot n_k$. Thus $g_k m_k \to m$ and $g_k m_k \preceq_s p_k$ implies $m \preceq p$.

Given any causal orientation $\leq$ on $\mathcal{M}$, we write for $A \subset \mathcal{M}$

$$\uparrow A := \{y \in Y \mid \exists a \in A : a \leq y\} \tag{2.12}$$

and

$$\downarrow A := \{y \in Y \mid \exists a \in A : y \leq a\}. \tag{2.13}$$

We also write simply $\uparrow x := \uparrow\{x\}$ and $\downarrow x := \downarrow\{x\}$. Then the *intervals* with respect to this causal orientation are

$$[m,n]_{\leq} := \{z \in \mathcal{M} \mid m \leq z \leq n\} = \uparrow m \cap \downarrow n$$

and

$$[m,\infty[_{\leq} := \uparrow m, \quad]-\infty,m]_{\leq} := \downarrow m.$$

The following proposition shows that replacing $\preceq_s$ by $\preceq$ does not change intervals too much.

Proposition 2.2.4 *Let $\mathcal{M} = G/H$ be a homogeneous space with a causal structure determined by $C \in \mathrm{Cone}_G(T_o\mathcal{M})$ and $\preceq_s, \preceq$ the associated causal orientations. Then $[m,\infty[_{\preceq} = \overline{[m,\infty[_{\preceq_s}}$*

Proof: Let $m \preceq n$. Then there exists a sequence $(m_k, n_k) \in \mathcal{M}_{\preceq_s}$ converging to (m,n). Let U be a neighborhood of m in $\mathcal{M}$ such that there exists a continuous section σ of the quotient map $\pi: G \to \mathcal{M} = G/H$. Then $\sigma(m_k) \to \sigma(m)$ and hence $g_k = \sigma(m)\sigma(m_k)^{-1}$ converges to $1 \in G$. Therefore

$$m = \sigma(m)\cdot \mathrm{o} = g_k \cdot m_k \preceq_s g_k \cdot n_k$$

and $g_k \cdot n_k \to n$ so that $n \in \overline{[m,\infty[_{\preceq_s}}$ proving the first claim. For the second, suppose that $m \preceq n \preceq l$. Then, according to the first part we can find sequences (n_k) and (l_k) converging to n and l, respectively, such that $m \preceq_s n_k$ and $n \preceq_s l_k$. As above, we find $g_k \in G$ converging to 1 such that $g_k \cdot n_k = n$. Thus $g_k \cdot m \preceq_s g_k \cdot n_k = n \preceq_s l_k$ implies $(g_k \cdot m, l_k) \in \mathcal{M}_{\preceq_s}$ and hence $(m,l) \in \mathcal{M}_{\preceq}$. □

For easy reference we introduce some more definitions.

Definition 2.2.5 Let $\mathcal{M}$ be manifold.

1) A causal orientation $\leq$ on $\mathcal{M}$ is called *topological* if its graph $\mathcal{M}_{\leq}$ in $\mathcal{M} \times \mathcal{M}$ is closed.

2) A space $(\mathcal{M}, \leq)$ with a topological causal orientation is called a *causal space*. If $\leq$ is in addition antisymmetric, i.e., a partial order, then $(\mathcal{M}, \leq)$ is called *globally ordered* or simply *ordered*.

3) Let $(\mathcal{M}, \leq)$ and $(\mathcal{N}, \leq)$ be two causal spaces and let $f : \mathcal{M} \longrightarrow \mathcal{N}$ be continuous. Then f is called *order preserving* or *monotone* if

$$m_1 \leq m_2 \Longrightarrow f(m_1) \leq f(m_2).$$

4) Let G be a group acting on $\mathcal{M}$. Then a causal orientation $\leq$ is called *G-invariant* if

$$m \leq n \Longrightarrow \forall a \in G : a \cdot m \leq a \cdot n.$$

5) A triple $(\mathcal{M}, \leq, G)$ is called a *causal G-manifold* or simply *causal* if $\leq$ is a topological G-invariant causal orientation. □

2.3 Semigroups

Invariant causal orientations on homogeneous spaces are closely related to semigroups. We assume that $\mathcal{M} = G/H$ carries a causal orientation $\leq$ such that $(\mathcal{M}, \leq, G)$ is causal. Then we define the semigroup $S_\leq$ by

$$S_\leq := \{a \in G \mid \mathbf{o} \leq a \cdot \mathbf{o}\},$$

called the *causal semigroup* of $(\mathcal{M}, \leq, G)$. If m is another point in $\mathcal{M}$, then we can find an $a \in G$ such that $m = a \cdot \mathbf{o}$. Thus for the corresponding semigroup $S'_\leq := \{a \in G \mid m \leq a \cdot m\}$ we have $S' = aSa^{-1}$.

Lemma 2.3.1 *1) For all $m, n \in \mathcal{M}$ the intervals $[m, n]$ and $[m, \infty[$ are closed.*

2) The semigroup $S_\leq$ is closed.

3) $G_{S_\leq} := S_\leq \cap S_\leq^{-1}$ is the closed subgroup of G given by

$$G_{S_\leq} = \{a \in G \mid \mathbf{o} \leq a \cdot \mathbf{o} \leq \mathbf{o}\}.$$

$G_{S_\leq}$ contains the stabilizer H of $\mathbf{o}$ and normalizes $S_\leq$.

4) $G_{S_\leq} = H$ if and only if $\leq$ is a partial order.

Proof: Let $\{z_j\}$ be a sequence in $[m, n]$ converging to $z \in \mathcal{M}$. Then $\mathcal{M}_\leq \ni (m, z_j) \to (m, z)$. As $\mathcal{M}_\leq$ is closed, it follows that $(m, z) \in \mathcal{M}_\leq$, i.e., $m \leq z$ so that $[m, \infty]$ is closed. Similarly, we find $z \leq n$ whence $[m, n[$ is also closed.

Let $\{a_j\}$ be a sequence in S, $\lim a_j = a \in G$. Then again $\mathcal{M}_\leq \ni (\mathbf{o}, a_j \cdot \mathbf{o}) \to (\mathbf{o}, a \cdot \mathbf{o}) \in \mathcal{M}_\leq$ as $\mathcal{M}_\leq$ is closed. It follows that $S_\leq$ is closed and so is $G_{S_\leq}$. Now $G_{S_\leq}^{-1} = G_{S_\leq}$, and the G-invariance of $\leq$ and the transitivity of $\leq$ imply (3). Using (3) and the G-invariance of $\leq$, we see that $\leq$ is a partial order if and only if $\mathbf{o} \leq g \cdot \mathbf{o} \leq \mathbf{o}$ is equivalent to $g \cdot \mathbf{o} = \mathbf{o}$, i.e., if and only if $G_{S_\leq} = H$. □

Remark 2.3.2 If $(\mathcal{M} = G/H, \leq, G)$ is causal and $H = G_{S_\leq}$, then $\leq$ can be recovered from $S_\leq$ via $a \cdot \mathbf{o} \leq b \cdot \mathbf{o} \Longleftrightarrow a^{-1}b \in S_\leq$. Conversely, given a closed subsemigroup S of G, one obtains a causal G-manifold $(G/H, \leq_S, G)$ via $H = S \cap S^{-1}$ and $a \cdot \mathbf{o} \leq_S b \cdot \mathbf{o} :\Longleftrightarrow a^{-1}b \in S$. □

Let $\mathcal{M} = G/H$ be a homogeneous space with a causal structure determined by a cone $C \in \mathrm{Cone}_G(T_\mathbf{o}\mathcal{M})$ and $\preceq$ the associated topological causal orientation. Then $S_\preceq = \{g \in G \mid \mathbf{o} \preceq g \cdot \mathbf{o}\}$ is a closed subsemigroup of G. But there is another semigroup canonically associated to C: Let W be the preimage of C under $T_1 G = \mathfrak{g} \to T_\mathbf{o}\mathcal{M} = \mathfrak{g}/\mathfrak{h}$ and S_W the closed subsemigroup of G generated by $\exp W$. For the following theorems, refer to [52], Proposition 4.16, and Theorem 4.21.

Theorem 2.3.3 $S_\preceq = \overline{S_W H}$. □

Theorem 2.3.4 *The following statements are equivalent:*

1) $\preceq$ *is a partial order.*

2) $S_\preceq \cap S_\preceq^{-1} = H$.

3) $\mathbf{L}(S_\preceq) := \{X \in \mathfrak{g} \mid \exp \mathbb{R}^+ X \subset S_\preceq\} = W$. □

Theorem 2.3.4 shows in particular that one can recover the causal structure from $\preceq$ provided $\preceq$ is a partial order. To do that one has to calculate the tangent cone $\mathbf{L}(S_\preceq)$ of the semigroup $S_\preceq$.

We can also build causal orientations starting with a closed subsemigroup S of G. Write $H := S \cap S^{-1}$ for the *group of units* of S. If $\mathcal{M} = G/H$ is the associated homogeneous space, let $\pi : G \to \mathcal{M}, g \mapsto gH$ be the canonical projection, and $\mathbf{o} := \pi(1)$ the base point. We define a left invariant causal orientation on G by the prescription

$$g \leq_S g' \quad \text{if} \quad g' \in gS. \tag{2.14}$$

Then $g \leq_S g' \leq_S g$ is equivalent to $gH = g'H$ and the prescription

$$\pi(g) \leq \pi(g') \quad \text{if} \quad g \leq_S g' \tag{2.15}$$

defines a partial order on $\mathcal{M}$ such that $\pi : (G, \leq_S) \to (\mathcal{M}, \leq)$ is monotone. We say that a function $f : G \to \mathbb{R}$ is *S-monotone* if

$$f : (G, \leq_S) \to (\mathbb{R}, \leq)$$

is a monotone mapping. We write $\mathrm{Mon}(S)$ for the set of all S-monotone continuous functions on G.

The construction of a causal orientation from a semigroup is of particular interest if the semigroup S can be recovered from its tangent cone $\mathbf{L}(S) = \{X \in \mathfrak{g} \mid \exp(\mathbb{R}^+ X) \subset S\}$.

Definition 2.3.5 1) S is called a *Lie semigroup* if $\overline{\langle \exp \mathbf{L}(S) \rangle} = S$, i.e., if the subsemigroup of G generated by $\exp \mathbf{L}(S)$ is dense in S.

2) S is called an *extended Lie semigroup* if $\overline{G_S \langle \exp \mathbf{L}(S) \rangle} = S$.

3) S is called *generating* if $\mathfrak{g}$ is the smallest subalgebra of $\mathfrak{g}$ containing $\mathbf{L}(S)$. □

Remark 2.3.6 For every generating extended Lie semigroup $S \subset G$ we know that S^o is a dense semigroup ideal in S and $S^o \subset \langle \exp \mathbf{L}(S) \rangle H$ (cf.[52], Lemma 3.7). Moreover, we have $\uparrow g = gS$ and $\downarrow g = gS^{-1}$ for $g \in G$ and these sets have dense interior. Similarly, $\uparrow x = \pi(gS)$ and $\downarrow x = \pi(gS^{-1})$ for $x = gH \in G/H$ and these sets also have dense interior. □

2.4 The Order Compactification of an Ordered Homogeneous Space

Compactifications are an indispensible tool whenever one wants to describe the behavior of mathematical objects at infinity in a quantitative manner. Which type of compactification is suitable depends very much on the specific situation given. For an ordered homogeneous space $\mathcal{M} = G/H$ one can define a compactification that takes the order into account and therefore turns out to be particularly useful. The basic idea is to identify an element $gH \in \mathcal{M}$ with the set of elements $g'H$ smaller than gH. One has the Vietoris topology (cf. Appendix B) on the set $\mathcal{F}(\mathcal{M})$ of closed subsets of $\mathcal{M}$ which makes $\mathcal{F}(\mathcal{M})$ a compact space. Then one can close up $\mathcal{M}$ to obtain a compactification.

In this section we assume that G is a connected Lie group and $S \subset G$ an extended Lie semigroup with unit group H; cf. page 45. We describe a compactification of $\mathcal{M} := G/H$ which is particularly suited for analytic questions taking into account the order structure $\leq$ on $\mathcal{M}$ induced from $\leq_S$. Recall the notation from Appendix B and note that both G and G/H are metrizable and σ-compact. Therefore the results of Appendix B apply in particular to $\mathcal{F}(G)$ and $\mathcal{F}(G/H)$.

Lemma 2.4.1 *1) The set $\mathcal{F}_\downarrow(G) := \{F \in \mathcal{F}(G) \mid \downarrow F = F\} \subset \mathcal{F}(G)^H$ is closed.*

2) The set $\mathcal{F}_\downarrow(G/H) := \{F \in \mathcal{F}(G/H) \mid \downarrow F = F\}$ is closed.

Proof: 1) The condition $\downarrow F = F$ is equivalent to $Fs \subset F$ for all $s \in S^{-1}$. For every $s \in S^{-1}$ the set

$$\mathcal{F}_s := \{F \in \mathcal{F}(G) : Fs \subset F\}$$

is closed in view of Lemma B.1.2 because $\mathcal{F}(G)$ is a pospace. Therefore

$$\mathcal{F}_{\downarrow}(G) = \bigcap_{s \in S} \mathcal{F}_s$$

is closed.

2) This follows from Proposition B.1.4 and 1). □

Lemma 2.4.2 *The mapping $\eta : G \to \mathcal{F}_{\downarrow}(G), g \mapsto \downarrow g$ factors to a continuous order-preserving injective mapping*

$$\overline{\eta} : G/H \to \mathcal{F}_{\downarrow}(G), \quad gH \mapsto \downarrow(gH)$$

of locally compact G-spaces.

Proof: The continuity of the mapping η follows from Lemma B.1.2 and the fact that

$$\eta(g) = g \cdot \eta(1) \qquad \forall g \in G.$$

This mapping is constant on the cosets gH of H in G. Therefore it factors to a continuous mapping $\overline{\eta}$. To see that $\overline{\eta}$ is injective, let $a, b \in G$ with $\eta(a) = \eta(b)$. Then $\downarrow a = \downarrow b$ and therefore $a \leq_S b \leq_S a$. Hence $aH = bH$. This proves the injectivity of $\overline{\eta}$. Finally, suppose $g \leq_S g'$. Then $g' \in gS$ and therefore $\downarrow g = gS^{-1} \subset g'S^{-1} = \downarrow g'$. This shows that $\overline{\eta}$ preserves the order. □

We write $\mathcal{M}_+ := [\mathbf{o}, \infty) = S \cdot \mathbf{o}$ for the *positive cone* in $\mathcal{M} = G/H$ and set

$$\mathcal{M}^{cpt} := \overline{\overline{\eta}(\mathcal{M})} = \overline{\eta(G)} \subset \mathcal{F}(G) \quad \text{and} \quad \mathcal{M}_+^{cpt} := \overline{\overline{\eta}(\mathcal{M}_+)} = \overline{\eta(S)}. \quad (2.16)$$

Then $\mathcal{M}^{cpt}$ is called the *order compactification* of the ordered space $(\mathcal{M}, \leq)$. We refer to $\overline{\eta}$ as to the *causal compactification map* .

Lemma 2.4.3 *Let $F \in \mathcal{M}^{cpt}$. Then the following assertions hold:*

1) $F \in \mathcal{M}_+^{cpt}$ is equivalent to $1 \in F$.

2) $F = \{g \in G \mid g^{-1}F \in \mathcal{M}_+^{cpt}\}$.

3) If $F \neq \emptyset$, then there exists $g \in G$ with $gF \in \mathcal{M}_+^{cpt}$, i.e., $\mathcal{M}^{cpt} \subset G \cdot \mathcal{M}_+^{cpt} \cup \{\emptyset\}$.

Proof: 1) For $s \in S$ we clearly have that $1 \in \eta(s) = \downarrow s$. Therefore $1 \in F$ for all $F \in \mathcal{M}_+^{cpt}$. If, conversely, $1 \in F$, and $F = \lim \eta(g_n)$, $g_n \in S$, then there exists a sequence $a_n \in \eta(g_n) = \downarrow g_n$ such that $a_n \to 1$ (Lemma B.1.1). We conclude with Lemma B.1.2 that

$$1 \in \lim \downarrow(a_n^{-1} g_n) = \lim a_n^{-1} \downarrow g_n = F.$$

2) In view of 1), this follows from the equivalence of $1 \in g^{-1}F$ and $g \in F$.

3) If $F \neq \emptyset$, then there exists $g \in F$, and therefore $g^{-1}F \in \mathcal{M}_+^{cpt}$. □

Note that Lemma 2.4.3 implies that either $\mathcal{M}^{cpt} = G \cdot \mathcal{M}_+^{cpt}$ or $\mathcal{M}^{cpt} = G \cdot \mathcal{M}_+^{cpt} \cup \{\emptyset\}$.

Proposition 2.4.4 *1)* $\mathcal{M}_+^{cpt} = \{A \in \mathcal{M}^{cpt} \mid 1 \in A\}$.

2) $(\mathcal{M}_+^{cpt})^o = \{A \in \mathcal{M}^{cpt} \mid 1 \in A^o\}$.

3) $S = \{g \in G \mid g \cdot (\mathcal{M}_+^{cpt})^o \subset (\mathcal{M}_+^{cpt})^o\}$.

4) $S^o = \{g \in G \mid g \cdot \mathcal{M}_+^{cpt} \subset (\mathcal{M}_+^{cpt})^o\}$.

Proof: 1) This was proved in Lemma 2.4.3.

2) Let $A \in (\mathcal{M}_+^{cpt})^o$. Then $1 \in A$. Moreover, there exists a symmetric neighborhood U of 1 in G such that $U \cdot A \subset \mathcal{M}_+^{cpt}$. Hence $U \subset A$, which proves that $1 \in A^o$. Conversely, if $1 \in A^o$, then there exists an $s \in A^o \cap S^o$ and therefore also a neighborhood V of s contained in $A \cap S$. Then

$$\tilde{V} := \{F \in \mathcal{F}(G) \mid F \cap V \neq \emptyset\}$$

is a neighborhood of A in $\mathcal{F}(G)$. Since each element $A \in M^{cpt}$ satisfies

$$A = \downarrow A = \{g \in G \mid gS \cap A \neq \emptyset\},$$

$F \in \tilde{V} \cap M^{cpt}$ entails $1 \in \downarrow F = F$, hence $F \in \mathcal{M}_+^{cpt}$.

3) Let $s \in S$ and $A \in (\mathcal{M}_+^{cpt})^o$. According to 2), we have

$$1 \leq_S s \in (sA)^o,$$

whence $1 \in (sA)^o$ and therefore $s \cdot A \in (\mathcal{M}_+^{cpt})^o$. If $g \in G$ with $g \cdot (\mathcal{M}_+^{cpt})^o \subset (\mathcal{M}_+^{cpt})^o$, then $g \cdot \mathcal{M}_+^{cpt} \subset \mathcal{M}_+^{cpt}$ and $gS^{-1} \in \mathcal{M}_+^{cpt}$ because $\mathcal{M}_+^{cpt} = \overline{(\mathcal{M}_+^{cpt})^o}$. This implies that $g \in (S^{-1})^{-1} = S$

4) If $s \in S^o$ and $A \in \mathcal{M}_+^{cpt}$ then $1 \in A$ and we find a neighborhood U of 1 in G such that $Us \cdot A \subset \mathcal{M}_+^{cpt}$. Hence $1 \in gs \cdot A$ for every $g \in U$. This leads to $U^{-1} \subset s \cdot A$ and hence to $1 \in (s \cdot A)^o$, i.e., $s \cdot A \in (\mathcal{M}_+^{cpt})^o$, according to 1). To show the converse, we assume that $g \in G$ and $g \cdot \mathcal{M}_+^{cpt} \subset (\mathcal{M}_+^{cpt})^o$. Then, as in 2), we obtain that $gS^{-1} \in (\mathcal{M}_+^{cpt})^o$ and therefore $1 \in (\downarrow g)^o$ or $g \in S^o$ again by 1). □

Lemma 2.4.5 *1) Let $\mathcal{F}_{\downarrow}^{\infty}(G/H)$ denote the set of all closed subsets $F \subset G/H$ with $\downarrow F = F$ such that for every $a \in F$ the connected component of a in $\uparrow a \cap F$ is noncompact. Then $\mathcal{F}_{\downarrow}^{\infty}(G/H)$ is closed in $\mathcal{F}_{\downarrow}(G/H)$.*

2) Let $F \in \overline{\overline{\eta}(G/H)} \backslash \overline{\eta}(G/H)$ and $a \in F$. Then the connected component of a in $\uparrow a \cap F$ is noncompact.

Proof: 1) Let $F = \lim_{n\to\infty} F_n$, $F_n \in \mathcal{F}_{\downarrow}^{\infty}(G/H)$, and $a \in F$. Let U_n denote the $\frac{1}{n}$ ball around a. Then there exists $n_m \in \mathbb{N}$ such that $F_n \cap U_m \neq \emptyset$ for all $n \geq n_m$. We clearly may assume that the sequence $(n_m)_{m\in\mathbb{N}}$ is increasing and that $n_m \geq m$. Let $a_m \in U_m \cap F_{n_m}$. Then our assumption implies that the connected component C_m of a_m in $\uparrow a_m \cap F_{n_m}$ is noncompact. Passing to a subsequence, we even may assume that $C_m \to C$ in $\mathcal{F}(G/H)$. Then $C_m \subset \uparrow a_m$ and the closedness of $\leq$ show that

$$C = \lim C_m \subset \uparrow(\lim a_m) = \uparrow a.$$

In addition, we know that

$$C = \lim C_m \subset \lim_{m\to\infty} F_{n_m} = F.$$

Hence $C \subset \uparrow a \cap F$. Next, Lemma B.1.1 entails that the connected component of a in C is noncompact. Therefore $A \in \mathcal{F}_{\downarrow}^{\infty}(G/H)$.

2) Let $a_n \in (\downarrow a)^o \subset F$ with $a_n \to a$ and $F_m = \downarrow x_m \to F$ with $x_m \in G/H$. For $n \in \mathbb{N}$ there exists $m_n \geq n$ such that $F_m \cap (\uparrow a_n)^o \neq \emptyset$ for all $m \geq m_n$. This clearly implies that $a_n \in F_m$, i.e., $a_n \leq x_m$. Then we find monotone curves $\gamma_n : \mathbb{R}^+ \to G/H$ and $T_n \in \mathbb{R}^+$ such that $\gamma_n(0) = a_n$ and $\gamma_n(T_n) = x_{m_n}$ ([114], 1.19, 1.31). Passing to a subsequence, we may assume that the sequence $C_n := \gamma_n([0, T_n])$ converges to C in $\mathcal{F}(\mathcal{M}_{+}^{cpt})$. Then $a = \lim a_n \in \lim C_n = C$, the sets C_n are connected chains, i.e., totally ordered subsets, in the partially ordered set $(G/H, \preceq)$, and $\cup_{n\geq n_0} C_n$ is not relatively compact for any $n_0 \in \mathbb{N}$ because $x_{m_n} \in C_n$ and $x_{m_n} \to \omega$ since

$$\lim \overline{\eta}(x_{m_n}) \notin \overline{\eta}(G/H).$$

Now, Lemma B.1.1 entails that the connected component of a in C is noncompact. Moreover,

$$C = \lim C_n \subset \uparrow \lim a_n = \uparrow a \quad \text{and} \quad C = \lim C_n \subset \lim F_{m_n} = F.$$

Finally, this proves that the connected component of a in $\uparrow a \cap F$ is noncompact. □

Theorem 2.4.6 *The image $\overline{\eta}(\mathcal{M})$ of $\mathcal{M}$ in $\mathcal{F}_{\downarrow}(G)$ is open in its closure.*

Proof: Let $F \in \overline{\eta(G/H)}$. We note first that the following two conditions are equivalent:

1) For every $a \in F$, the connected component of $\uparrow a \cap F$ is noncompact.

2) $F \notin \overline{\eta}(G/H)$.

In fact, 2) implies 1) by Lemma 2.4.5. If $F = \overline{\eta}(x) = \downarrow x$ with $x \in G/H$, then $x \in F$ and $\uparrow x \cap \downarrow x = \{x\}$ is compact. Therefore 1) implies 2). Now the theorem is a consequence of Lemma 2.4.5. □

Proposition 2.4.7 *Let $A \subset G$ be a closed subset with $\downarrow A = A$. Then*

$$\mu(\partial A) = 0 \quad \textit{and} \quad A = \overline{A^o}.$$

Proof: Suppose that this is false. Then we can find $x \in \partial A$, the boundary of A, and a compact neighborhood V of x with $\mu(\partial A \cap V) > 0$. Choose a sequence $s_n \in S^o$ with $s_{n+1} \in (\downarrow s_n)^o$ and $\lim_{n\to\infty} s_n = 1$. Then $As_n^{-1} \subset A^o$ and therefore $A = \overline{A^o}$. Let $\tilde{\mu}$ denote a right Haar measure on G. It suffices to prove that $\tilde{\mu}(\partial A \cap V) = 0$. If not, we have $\tilde{\mu}((\partial A \cap V)s_n) = \tilde{\mu}(\partial A \cap V) > 0$ for every $n \in \mathbb{N}$ and

$$(\partial A \cap V)s_n \cap (\partial A \cap V)s_m = \emptyset \quad \text{for} \quad m < n$$

because $(\partial A)s_n \subset A^o s_m$. This shows that

$$\begin{aligned}\textstyle\sum_{n=1}^{\infty} \tilde{\mu}(\partial A \cap V) &= \textstyle\sum_{n=1}^{\infty} \tilde{\mu}((\partial A \cap V)s_n) \\ &= \tilde{\mu}\left(\bigcup_{n\in\mathbb{N}}(\partial A \cap V)s_n\right) \leq \tilde{\mu}\left(\bigcup_{n\in\mathbb{N}} Vs_n\right) < \infty.\end{aligned}$$

Whence $\tilde{\mu}(\partial A \cap V) = 0$. □

We will see later on that for specific $\mathcal{M}$ the order compactification can be described in much more concrete terms than has been done in this section. In particular, it will turn out that the space $\mathcal{M}^{cpt}$ is in some sense the smallest compact G-space $\mathcal{X}$ such that there exists an open subset $\mathcal{O} \subset \mathcal{X}$ with the property that

$$S = \{g \in G \mid g \cdot \mathcal{O} \subset \mathcal{O}\} \quad \text{and} \quad S^o = \{g \in G \mid g \cdot \overline{\mathcal{O}} \subset \mathcal{O}\}. \tag{2.17}$$

This will be used to get information about the structure of $\mathcal{M}_+^{cpt}$ and the G-space $\mathcal{M}^{cpt}$ because in special cases there are very natural compact G-spaces with the above property.

2.5 Examples

2.5.1 The Group Case

Recall the way of viewing a group G as a symmetric space from Example 1.1.3. A cone $D \subset \mathfrak{q}$ belongs to $\mathrm{Cone}_{G\times G}(\mathfrak{q})$ if and only if it is of the form

$$D = \{(X, -X) \mid X \in C\},$$

where $C \in \mathrm{Cone}_G(\mathfrak{g})$. There is extensive literature on the classification of $C \in \mathrm{Cone}_G(\mathfrak{g})$ for arbitrary connected Lie groups G (cf. [50, 52, 118]). Here we only recall some basic facts for the case of simple Lie groups. It turns out that only Hermitian simple Lie algebras (cf. Appendix A.3) admit regular invariant cones.

Lemma 2.5.1 *Let G be a simple real Lie group with Lie algebra $\mathfrak{g}$. Then $\mathrm{Cone}_G(\mathfrak{g})$ is nonempty if and only if $\mathfrak{g}$ is Hermitian.*

Proof: Note first that any nontrivial G-invariant cone in $\mathfrak{g}$ is automatically regular since it spans an ideal. Suppose that $G = K\exp(\mathfrak{p})$ is a Cartan decomposition and $\mathfrak{k} \subset \mathfrak{g}$ is the Lie algebra of K. The group $\mathrm{Ad}(K)$ is compact, so Kostant's Theorem 2.1.17 shows that there is a nontrivial G-invariant cone in $\mathfrak{g}$ if and only if there exists $Z \in \mathfrak{g}$, $Z \neq 0$, such that $\mathrm{Ad}(k)Z = Z$ for all $k \in K$. As $\mathfrak{g}$ is irreducible as a G-module, it follows from Lemma A.2.5 that $\dim \mathfrak{g}^K = 1$ and $\mathfrak{g}_\mathbb{C}$ is simple. Let θ be the Cartan involution on $\mathfrak{g}$ corresponding to the Cartan decomposition $\mathfrak{g} = \mathfrak{k} \oplus \mathfrak{p}$. Then the K-invariance of $\theta(Z)$ shows that $\theta(Z) = \pm Z$. If $\theta(Z) = -Z$, then $\mathbb{R}Z$ is a $\mathfrak{k}$-submodule of $\mathfrak{p}$, which is impossible by Lemma 1.3.4 since the Killing form is positive definite on $\mathfrak{p}$. □

Now suppose that $\mathfrak{g}$ is Hermitian and $Z \in \mathfrak{z}(\mathfrak{k})$, $Z \neq 0$. Then the minimal cone (cf. Theorem 2.1.21) is given by

$$C_{\min} = \overline{\mathrm{conv}\, G \cdot \mathbb{R}^+ Z}$$

and the corresponding maximal cone is

$$C_{\max} = \{X \in \mathfrak{g} \mid \forall Y \in C_{\min} : (X|Y)_\theta \geq 0\}.$$

Later on we will describe these cones in more detail.

Lemma 2.5.1 shows that each Hermitian Lie group is a causal manifold with causal structures in bijective correspondence with the G-invariant cones in $\mathfrak{g}$. Moreover, the lemma shows that every nontrivial G-invariant cone contains a nonzero element $Z \in \mathfrak{z}(\mathfrak{k})$. Therefore the causal structure cannot be antisymmetric if K is compact. This follows from the fact that the curve $t \mapsto \exp tZ$ is periodic and causal.

Multiplication by i maps $\mathrm{Cone}_H(\mathfrak{q})$ into $\mathrm{Cone}_H(i\mathfrak{q})$. In this way causal structures of c-dual spaces are in a canonical bijective correspondence. We make this explicit in the group case: Assume that G is contained in a complex group $G_{\mathbb{C}}$ with Lie algebra $\mathfrak{g}_{\mathbb{C}}$. Let $\mathcal{M} := G_{\mathbb{C}}/G$. Then the tangent space at the origin $o = 1G$ is $i\mathfrak{g}$. If C is a G-invariant cone in $\mathfrak{g}$, then $iC \subset i\mathfrak{g}$ is also G-invariant. Therefore Lemma 2.5.1 shows that $\mathcal{M}$ carries a causal structure if and only if G is Hermitian. We will show later that $\mathcal{M}$ is ordered for each of the topological orientations associated to elements of $\mathrm{Cone}_G(i\mathfrak{g})$ and that the intervals $[m, n]$ are compact. Furthermore, it will turn out that the semigroup $S_{\preceq}$ is given by $G \exp iC$ and $\mathbf{L}(S) = \mathfrak{g} \oplus iC$, where $C \in \mathrm{Cone}_G(i\mathfrak{g})$ is the cone inducing $\preceq$.

2.5.2 The Hyperboloids

Recall the hyperboloid $Q_r = Q_r^{1,n} \subseteq \mathbb{R}^{n+1}$ and the map $q: \mathbb{R}^{n+1} \to \mathfrak{q} \subseteq \mathfrak{so}(2, n)$, defined by

$$q(v) := \begin{pmatrix} 0 & -{}^t(vI_{1,n}) \\ v & 0 \end{pmatrix}, \tag{2.18}$$

from Section 1.5. Let $C_1 \subset \mathbb{R}^{n+1}$ be the forward light cone and

$$C := \{q(v) \in \mathfrak{q} \mid v \in C_1\}.$$

Then C is a regular cone in $\mathfrak{q}$ invariant under the group $\mathrm{SO}_o(1, n)$ and $\mathrm{Cone}_{\mathrm{SO}_o(1,n)}(\mathfrak{q}) = \{C, -C\}$ according to Example 2.1.16. In particular, $Q_r \simeq \mathrm{SO}_o(2, n)/\mathrm{SO}_o(1, n)$ carries a causal structure. Let

$$\alpha(t) := \exp(tq(e_1)) \cdot o = \cos te_1 + \sin te_2 .$$

Then

$$\dot{\alpha}(t) = d_{\alpha(0)}\ell_{\alpha(t)}(\dot{\alpha}(0)) = d_{\alpha(0)}\ell_{\alpha(t)}(q(e_1)) \in C(\alpha(t)).$$

Thus α is a closed causal curve. Therefore the causal orientation is not a partial order.

The case Q_{-1} can now be treated in the same way by using the map

$$q^c: \mathbb{R}^{n+1} \ni w \mapsto \begin{pmatrix} 0 & w \\ {}^t(wI_{1,n}) & 0 \end{pmatrix} \in \mathfrak{q} \subset \mathfrak{so}(1, n+1) \tag{2.19}$$

and the $\mathrm{SO}_o(1, n)$-invariant form

$$(q^c(v), q_1(w)) := \mathrm{Tr}\left(q^c(v)q^c(w)\right). \tag{2.20}$$

Let

$$C := \{q^c(v) \in \mathfrak{q} \mid v \in C_1\}$$

Then the cone C is $\mathrm{SO}_o(1,n)$-invariant and regular. Furthermore,

$$\mathrm{Cone}_{\mathrm{SO}_o(1,n)}(\mathfrak{q}) = \{C, -C\}$$

(cf. Example 2.1.16 again). In particular, $Q_{-r} \simeq \mathrm{SO}_o(1,n+1)/\,\mathrm{SO}_o(1,n)$ is causal.

The causal curve $\alpha(t) = \exp tq^c(e_1)\cdot e_{n+1}$ is given by $\alpha(t) = \sinh(t)e_1 + \cosh(t)e_{n+1}$, and this curve is an embedding of $\mathbb{R}$. We will show later that this space is actually ***globally hyperbolic***, i.e., the causal orientation is antisymmetric and all the order intervals $[m,n]$ are compact.

If n is odd, we consider $a = I_{1,n+1} \in \mathrm{SO}_o(1,n+1)$. Then $\tau(a) = a$, a commutes with $\mathrm{SO}_o(1,q)$ but $a \notin \mathrm{SO}_o(1,n)$. On the other hand, $a^2 = 1$ so that $H_1 := \mathrm{SO}_o(1,n) \cup a\,\mathrm{SO}_o(1,n)$ is a group and the group $\{1,a\}$ normalizes $\mathrm{SO}_o(1,n)$ so it acts on Q_{-r}. Then Q_{-r} is a double covering of the quotient space $Q_{-r}/\{1,a\} \cong \mathrm{SO}_o(1,n+1)/H_1$. In particular, *locally* $Q_{-r}/\{1,a\}$ admits a causal structure. But $aq(e_1)a^{-1} = -q(e_1)$. Thus the light cone is not H_1-invariant and does not define a causal structure on $Q_{-r}/\{1,a\}$ globally. Thus the existence of a causal structure may depend on the fundamental group of the space in question.

2.6 Symmetric Spaces Related to Tube Domains

Let $(\mathfrak{g},\mathfrak{h})$ be a symmetric pair associated to a tube domain. In the notation of Example 1.1.6 this means in particular that we have an element $Z^0 \in \mathfrak{z}(\mathfrak{k}) \cap \mathfrak{q}$ and elements $X_o, Y_o \in \mathfrak{p}$ such that $\mathfrak{h} = \mathfrak{g}(0, Y_o)$, $\mathfrak{q} = \mathfrak{q}^+ + \mathfrak{q}^-$ with $\mathfrak{q}^\pm = \mathfrak{g}(\pm 1, Y_o)$ and $X_o = [Y_o, Z^0] \in \mathfrak{q}_p$. For the last relation see Lemma A.3.2. The same lemma shows that the involution $\tau\colon \mathfrak{g} \to \mathfrak{g}$ coincides with $\mathbf{C}_h^2$. So $Z^0 \in \mathfrak{q}$ is a K-fixed point, whereas $Y_o \in \mathfrak{h}$ is an H_o-fixed point. Therefore $X_o \in \mathfrak{q}^{H_o \cap K}$ and we are in the situation of Theorem 1.3.11. Since conjugation by an element of K does not move Z^0, we may assume that $Y_o = Y^0$ in the notation of that theorem. Decompose $X_o = \frac{1}{2}(X_+ + X_-)$ with $X_\pm \in (\mathfrak{q}^\pm)^{H_o\cap K}$ and $\theta(X_+) = -X_-$. As $Z^0 \in \mathfrak{q}_k$ and $\mathrm{ad}\,Y_o : \mathfrak{q} \to \mathfrak{q}$ is a linear involution, we obtain $Z^0 = \mathrm{ad}(Y^0)X_o = \frac{1}{2}(X_+ - X_-)$. Let $\{\gamma_1,\ldots,\gamma_r\}$ be a maximal system of strongly orthogonal roots with suitable root vectors $E_j = E_{\gamma_j}$ and co-roots $H_j = H_{\gamma_j}$ as in Appendix A.3. Then

$$X_\pm = X_o \pm Z^0 = \frac{1}{2}\sum_{j=1}^{r}(X_j \pm iH_j), \tag{2.21}$$

where $X_j = -i(E_j - E_{-j})$. By Theorem 2.1.21 there are minimal H_o-invariant cones $C_\pm \subset \mathfrak{q}^\pm$ such that $X_\pm \in C_\pm$ and $C_- = -\theta(C_+)$.

Lemma 2.6.1 *The cone C_+ is H-invariant if $X_o \in \mathfrak{q}^{H\cap K}$.*

Proof: Let $X = \operatorname{Ad}(h)X_+$ for some $h \in H_o$ and fix some $k \in H \cap K$. Then

$$\operatorname{Ad}(k)X = \operatorname{Ad}(kh)X_+ = \operatorname{Ad}(khk^{-1})\operatorname{Ad}(k)X_+ = \operatorname{Ad}(khk^{-1})X_+ \in C_+$$

because of $khk^{-1} \in H_o$ and the lemma follows by $C_+ = \overline{\operatorname{conv}\operatorname{Ad}(H_o)\mathbb{R}^+X_+}$ and $H = (H \cap K)H_o$ (cf. (1.8)). □

Remark 2.6.2 Consider H_o-stable cones

$$C_k := \{X + \theta(Y) \mid X, Y \in C_+\} = C_+ - C_- \tag{2.22}$$

and

$$C_p := \{X - \theta(Y) \mid X, Y \in C_+\} = C_+ + C_- . \tag{2.23}$$

They are pointed and generating in $\mathfrak{q}$.

The $(H_o \cap K)$-fixed points in the H_o-invariant cones C_k and C_p, guaranteed by Lemma 2.1.15, can be determined explicitly:

$$Z^0 = \frac{X_+ + \theta(X_+)}{2} \in \mathfrak{q}_k^{H_o\cap K} \cap C_k^o;$$

$$X_o = \frac{X_+ - \theta(X_+)}{2} \in \mathfrak{q}_p^{H_o\cap K} \cap C_p^o.$$

Moreover, we have

$$C_k \cap \mathfrak{p} = \{0\}, \quad C_p \cap \mathfrak{k} = \{0\}.$$ □

Proposition 2.6.3 *Suppose $X_o \in \mathfrak{q}^{H\cap K}$. Then the space $\mathcal{M} = G/H$ has a regular invariant causal structure.*

Proof: In view of Lemma 2.6.1, the assumption on X_o ensures that C_k and C_p are even H-invariant. □

Example 2.6.4 (cf. Example 1.3.15) Suppose that $X_o \in \mathfrak{q}^{H\cap K}$. Then the cone C_p defines an invariant causal structure on G/H but not on the symmetric space $\operatorname{Ad}(G)/\operatorname{Ad}(G)^\tau$, where $\operatorname{Ad}(G)^\tau = \{\varphi \in \operatorname{Ad}(G) \mid \varphi\tau = \tau\varphi\}$. In fact, note that by definition $C_k^o \cap \mathfrak{k} \neq \emptyset$ and $C_p^o \cap \mathfrak{p} \neq \emptyset$. Then $\theta \in \operatorname{Ad}(G)^\tau$, which implies the claim since C_p is not θ-invariant. □

We will now show that $\operatorname{Cone}_{H_o}(\mathfrak{q}) = \{\pm C_k, \pm C_p\}$.

Lemma 2.6.5 *$X_j + iH_j \in C_+$ for $j = 1, \ldots, r$.*

Proof: Recall from (2.21) $X_\pm = \frac{1}{2}\sum_{j=1}^r (X_j \pm iH_j)$. For $s_1, \ldots, s_r \in \mathbb{R}$, let $h = \exp s_1 Y_1 \cdots \exp s_r Y_r \in H$. Then

$$2\,\mathrm{Ad}(h)X_+ = e^{2s_1}(X_1 + iH_1) + \cdots + e^{2s_r}(X_r + iH_r)\,.$$

Let $s_j = 0$ and let the other s_k tend to $-\infty$. Then, as C_+ is closed, it follows that $X_j + iH_j \in C_+$. □

Lemma 2.6.6 *Let $C \in \mathrm{Cone}_{H_o}(\mathfrak{q}^+)$ such that $X_+ \in C^o$. Then C is self dual, $C = C_+$, and $C^o = \mathrm{Ad}(H_o)X_+$.*

Proof: By assumption, we have $C_+ \subset C \subset C_+^*$. Fix an $X \in C$. Then $X - \theta(X) \in \mathfrak{q} \cap \mathfrak{p}$. Applying $\tilde{\mathrm{C}}$, it follows from Lemma A.3.2 and Proposition A.3.3 that $\sum_{j=1}^r \mathbb{R}X_j$ is a maximal abelian subspace of $\mathfrak{p} \cap \mathfrak{q}$. Since $H_o \cap K$ is the group of θ-fixed points in the analytic subgroup of G with Lie algebra $\mathfrak{h}^a = \mathfrak{h} \cap \mathfrak{k} + \mathfrak{q} \cap \mathfrak{p}$, there exists a $k \in K \cap H_o$ such that

$$\mathrm{Ad}(k)(X - \theta(X)) = \sum_{j=1}^r t_j X_j \in \mathfrak{a}_q\,.$$

As $X_j = \frac{1}{2}(X_j + iH_j) + \frac{1}{2}(X_j - iH_j)$ and $X_j \pm iH_j \in \mathfrak{q}^\pm$, it follows that $\mathrm{Ad}(k)X = \frac{1}{2}\sum_{j=1}^r t_j(X_j + iH_j)$. Now $C \subset C_+^*$, and by Lemma 2.6.5,

$$0 \le (X_j + iH_j \mid \mathrm{Ad}(k)X)_\theta = \frac{1}{2} t_j |X_j + iH_j|^2\,.$$

Hence $t_j \ge 0$. Again by Lemma 2.6.5 it follows that $C \subset C_+$. Hence $C = C_+$. If we take $C = C_+^*$, this implies that $C_+^* = C_+$ and C_+ is self dual.

Now assume that $X \in C^o$. Then $t_j > 0$ for $j = 1, \ldots, r$. Define

$$h = \prod_{j=1}^r \exp\left(-\left(\frac{1}{2}\log t_j\right) Y_j\right) \in H_o\,.$$

Then $\mathrm{Ad}(hk)X = X_+$. Hence $C_+^o = \mathrm{Ad}(H_o)X_+$. □

Corollary 2.6.7 $\mathrm{Cone}_{H_o}(\mathfrak{q}^+) = \{C_+, -C_+\}$. *If, in addition, $X_o \in \mathfrak{q}^{H \cap K}$, then* $\mathrm{Cone}_H(\mathfrak{q}^+) = \{C_+, -C_+\}$. □

Theorem 2.6.8 *Let G/K be a tube domain, $G \subset G_\mathbb{C}$ with $G_\mathbb{C}$ simply connected, and τ the involution of $G_\mathbb{C}$ which on $\mathfrak{g}_\mathbb{C}$ is given by $\tau = e^{\mathrm{ad}\, i\pi Y_o}$. Further, let $H = G^\tau$ and $C_\pm$ the closed convex H_o-invariant cones in $\mathfrak{q}^\pm$ generated by $X_\pm = X_o \pm Z^0$. Then*

$$\mathrm{Cone}_{H_o}(\mathfrak{q}) = \{\pm(C_+ - C_-), \pm(C_+ + C_-)\}.$$

If, in addition, $X_o \in \mathfrak{q}^{H\cap K}$, *then*

$$\mathrm{Cone}_H(\mathfrak{q}) = \{\pm(C_+ - C_-), \pm(C_+ + C_-)\}.$$

Proof: We prove this in four steps. Fix $C \in \mathrm{Cone}_{H_o}(\mathfrak{q})$.

Step 1: Let $\mathrm{pr}_\pm : \mathfrak{q} \to \mathfrak{q}_\pm$ be the H-equivariant projection acording to the decomposition $\mathfrak{q} = \mathfrak{q}^+ \oplus \mathfrak{q}^-$. Fix an $X = \mathrm{pr}_+(X) + \mathrm{pr}_-(X) \in C$ and consider $h(t) = \exp tY_o$. Then

$$\mathrm{Ad}(h(t))X = e^t\, \mathrm{pr}_+(X) + e^{-t}\, \mathrm{pr}_-(X).$$

Hence

$$\lim_{t\to\infty} e^{-t}\, \mathrm{Ad}(h(t))X = \mathrm{pr}_+(X) \in C.$$

Similarly, we get $\mathrm{pr}_-(X) \in C$.

Step 2: By the first step it follows that

$$C(\pm) := \mathrm{pr}_\pm C \subset C.$$

Then $C \subset C(+) + C(-)$. We claim that $C(+)$ and $C(-)$ are H_o-invariant proper cones in $\mathfrak{q}^+$ and $\mathfrak{q}^-$, respectively. As C is generating, it follows that $C(+) \neq \{0\}$. If $X \in C(+) \cap -C(+)$, Step 1 implies that $X \in C \cap -C$. Hence $X = 0$ and $C(+)$ is pointed. Let $X \in \mathfrak{q}^+$. Then, as C is generating, there are $V, W \in C$ such that $X = V - W$. But then

$$X = \mathrm{pr}_+(V) - \mathrm{pr}_+(W) \in C(+) - C(+).$$

This shows that $C(+)$ is generating. As $C(+) \subset C$ and C, as well as $\mathfrak{q}^+$, are H_o-invariant, it follows that $\mathrm{Ad}(H_o)C(+) \subset C \cap \mathfrak{q}^+ \subset \mathrm{pr}_+(C) = C(+)$. Thus $C(+)$ is an H_o-invariant regular cone in $\mathfrak{q}^+$.

Step 3: By Corollary 2.6.7 we get $C(+) = \pm C_+$. Similarly, we find $C(-) = \pm\theta(C(+))$.

Step 4: As $C(\pm) \subset C$, we have $C(+) + C(-) \subset C$. The other inclusion is obvious, so we have

$$C = C(+) + C(-).$$

Step 3 now implies that either $C = \pm C_k = C_+ - C_-$ or $C = \pm C_p = C_+ + C_-$.

The last claim is an immediate consequence of Lemma 2.6.1. □

Remark 2.6.9 The cone $C_+ \subset \mathfrak{q}^+$ also has a geometric meaning, as G/K is biholomorphically equivalent to the tube domain $\mathfrak{q}^+ + i\Omega$, where $\Omega = C_+^o$. The idea of the proof is as follows. Let $Q_{\mathbb{C}}^+ = \exp(\mathfrak{q}^+)_{\mathbb{C}}$, $Q_{\mathbb{C}}^- = \exp(\mathfrak{q}^-)_{\mathbb{C}}$. Similarily, we let $K_{\mathbb{C}} \subset G_{\mathbb{C}}$ be the complexification of K. Let $c = \exp((i\pi/2)X_o)$. Then $c^{-1}K_{\mathbb{C}}c = H_{\mathbb{C}}$ and $G \subset Q_{\mathbb{C}}^+ cH_{\mathbb{C}}Q_{\mathbb{C}}^-$. Furthermore, $G \cap cH_{\mathbb{C}}Q_{\mathbb{C}}^-$. Thus G/K is diffeomorphic to the G-orbit through $cH_{\mathbb{C}}Q_{\mathbb{C}}^-$ in $G_{\mathbb{C}}/H_{\mathbb{C}}Q_{\mathbb{C}}^-$. This defines a complex structure on G/K. Now if $Z \in \mathfrak{q}^+ + i\Omega$, then $\exp Z \in GcH_{\mathbb{C}}(G_-)_{\mathbb{C}}$ and this defines a biholomorphic map onto G/K.

2.6.1 Boundary Orbits

We keep the asumptions from the last section. In particular, we assume that G is contained in the simply connected complex Lie group $G_{\mathbb{C}}$ with Lie algebra $\mathfrak{g}_{\mathbb{C}}$. By Theorem 1.1.11, $K_{\mathbb{C}} = G_{\mathbb{C}}^{\theta}$ and $H_{\mathbb{C}} = G_{\mathbb{C}}^{\tau}$. Recall the Cayley transform $\mathbf{C} = \mathrm{Ad}(c)$ with $c = \exp((\pi i/2)X_o)$, cf. p.255. Then $\mathbf{C} = \varphi_{iX_o}$ (cf. Lemma 1.2.1). By Lemma 1.2.5 we have $\tau\theta = \tau_{iX_o}$. Now Lemma 1.2.5 and Remark 2.6.9 yield

1) $\mathbf{C}^{-1} = \mathbf{C}^3 = \tau\theta \circ \mathbf{C} = \mathbf{C} \circ \tau\theta$.

2) $\tau \circ \mathbf{C} = \mathbf{C} \circ \theta$.

3) $c^{-1}H_{\mathbb{C}}c = K_{\mathbb{C}}$.

Lemma 2.6.10 $\mathfrak{q}^- = \mathbf{C}(\mathfrak{p}^+) \cap \mathfrak{g}$ *and* $\mathfrak{q}^+ = \mathbf{C}(\mathfrak{p}^-) \cap \mathfrak{g}$.

Proof: We will only prove the first statement, as the second follows in exactly the same way. By Lemma A.3.2, $\mathbf{C}(H_j) = X_j$. Therefore $-i\mathbf{C}(Z^0) = -Y_o$. It follows that for $Z \in \mathfrak{p}^+$:

$$[Y_o, \mathbf{C}(Z)] = \mathbf{C}([\mathbf{C}^{-1}(Y_o), Z]) = i\mathbf{C}([Z^0, Z]) = -\mathbf{C}(Z).$$

Thus $Z \in \mathfrak{q}_{\mathbb{C}}^-$. The same calculation shows that if $X \in \mathfrak{q}^-$, then $\mathbf{C}^{-1}(X) \in \mathfrak{p}^+$. From this the lemma follows. □

We recall that the realization of G/K as a bounded symmetric domain in $\mathfrak{p}^+$, cf. (A.23), p. 253; see also Lemma 5.1.4, Remark 5.1.9 and Example 5.1.10. The map

$$P^+ \times K_{\mathbb{C}} \times P^- \ni (p, k, q) \mapsto pkq \in G_{\mathbb{C}}$$

is a diffeomorphism onto an open dense submanifold of $G_{\mathbb{C}}$ and $G \subset P^+K_{\mathbb{C}}P^-$. If $g \in G$, then $g = p^+(g)k_{\mathbb{C}}(g)p^-(g)$ uniquely with $p^+(g) \in P^+$, $k_{\mathbb{C}}(g) \in K_{\mathbb{C}}$, and $p^-(g) \in P^-$. Let $\log := (\exp|_{\mathfrak{p}^+})^{-1} : P^+ \to \mathfrak{p}^+$. The bounded realization of G/K is given by

$$\Omega_+ = \{\log(p(g)) \mid g \in G\}.$$

For $g \in G_{\mathbb{C}}$ and $Z \in \mathfrak{p}^+$ such that $g \exp Z \in P^+K_{\mathbb{C}}P^-$, define $g \cdot Z \in \mathfrak{p}^+$ and $j(g, Z) \in K_{\mathbb{C}}$ by

$$g \exp Z \in \exp(g \cdot Z) j(g, Z) P^-. \tag{2.24}$$

Thus $\exp(g \cdot Z) = p^+(g \exp Z)$ and $j(g, Z) = k_{\mathbb{C}}(g \exp Z)$. We denote the map $P^+K_{\mathbb{C}}P^-/K_{\mathbb{C}}P^- \to \mathfrak{p}^+$, $pK_{\mathbb{C}}P^- \mapsto \log p$ by $p \mapsto \zeta(p)$. Define $E \in \mathfrak{p}^+$ by $E = \zeta(c)$. The *Shilov boundary* of Ω_+ is $S = G \cdot E$.

Lemma 2.6.11 *The stabilizer of E in G is HQ^+. In particular, $S = G/HQ^+ = K/K \cap H$ is compact.*

Proof: The stabilizer of $cK_{\mathbb{C}}P^- \in G_{\mathbb{C}}/K_{\mathbb{C}}P^-$ in $G_{\mathbb{C}}$ is $cK_{\mathbb{C}}P^-c^{-1} = H_{\mathbb{C}}Q^+_{\mathbb{C}}$. But by construction $\exp(E)K_{\mathbb{C}}P^- = cK_{\mathbb{C}}P^-$. □.

View G and $G_{\mathbb{C}}$ as subgroups of $G_1 := G \times G$, respectively $(G_1)_{\mathbb{C}} := G_{\mathbb{C}} \times G_{\mathbb{C}}$, by the diagonal embedding $g \mapsto (g, g)$. This induces a G-action on $\mathcal{S}_1 := \mathcal{S} \times \mathcal{S}$ and a $G_{\mathbb{C}}$-action on $G_{\mathbb{C}}/K_{\mathbb{C}}P^- \times G_{\mathbb{C}}/K_{\mathbb{C}}P^-$ as well as other homogeneous spaces of $G_1$and $(G_1)_{\mathbb{C}}$, respectively.

Lemma 2.6.12 *Suppose that $g\exp(-Z) \in P^+K_{\mathbb{C}}P^-$, $g \in G_{\mathbb{C}}$, and $Z \in \mathfrak{p}^+$. Then $\theta(g)\exp(Z) \in P^+K_{\mathbb{C}}P^-$ and*

$$\theta(g) \cdot Z = -[g \cdot (-Z)] .$$

Proof : Let $g\exp(-Z) = \exp(g \cdot (-Z))kp$, with $k \in K_{\mathbb{C}}$ and $p \in P^-$. Then

$$\begin{aligned} \theta(g)\exp Z &= \theta(g\exp -Z) \\ &= \exp(\theta(g \cdot (-Z)))kp^{-1} \\ &= \exp\left(-[g \cdot (-Z)]\right) kp^{-1} \end{aligned}$$

which proves the claim. □

Corollary 2.6.13 *$g \in P^+K_{\mathbb{C}}P^-$. Then $-\zeta(g) = \zeta(\theta(g))$. In particular, we have $\zeta(c^{-1}) = -E$.*

Proof: Take $Z = 0$ in Lemma 2.6.12 and note that $\theta(c) = c^{-1}$. □

Lemma 2.6.14 *Let the notation be as above. Then the following hold:*

1) The stabilizer of $-E$ is HQ^-.

2) $\mathcal{S}_1 := \mathcal{S} \times \mathcal{S} \simeq G/HQ^- \times G/HQ^+$.

3) Let $\xi_0 = (E, -E) \in \mathcal{S}_1$. Then $\mathcal{M} \simeq G \cdot \xi_0$.

Proof : Statement 1) follows from Lemma 2.6.12 and Corollary 2.6.13. The second claim is a consequence of 1) and Lemma 2.6.11, whereas the last claim follows immediately from 1) and 2). □

We list here the symmetric spaces, the corresponding Shilov boundary together with the real rank r, and the common dimension d of the restricted root spaces for short roots as provided by Moore's Theorem A.3.4. Here $k \geq 3$, $\mathbf{T}$ is the one-dimensional torus, Q_n is the real quadric in the real projective space $\mathbb{RP}^n$ defined by the quadratic form of signature $(1, n)$, and the subscript $+$ means positive determinant.

Symmetric Spaces Related to tube domains

$\mathcal{M} = G/H$	$\mathcal{S} = K/K\cap H$	r	d
$\mathrm{Sp}(n,\mathbb{R})/\mathrm{GL}(n,\mathbb{R})_+$	$U(n)/O(n)$	n	1
$\mathrm{SU}(n,n)/\mathrm{GL}(n,\mathbb{C})_+$	$U(n)$	n	2
$\mathrm{SO}^*(4n)/\mathrm{SU}^*(2n)\mathbb{R}^+$	$U(2n)/\mathrm{Sp}(2n)$	$2n$	4
$\mathrm{SO}(2,k)/\mathrm{SO}(1,k-1)\mathbb{R}^+$	Q_n	2	$k-2$
$E_{7(-25)}/E_{6(-26)}\mathbb{R}^+$	$E_6\mathbf{T}/F_4$	3	8

2.6.2 The Functions Ψ_m

Our aim now is to construct an analytic function Ψ on $\mathcal{S}_1$ such that $G/H = \{\xi \in \mathcal{S}_1 \mid \Psi(\xi) \neq 0\}$. This implies in particular that G/H is open and dense in $\mathcal{S}_1$. For this we need a few facts from representation theory. Let ρ_n be the half-sum of positive noncompact roots. Then

$$\rho_n = \frac{1}{2}\left[1 + \frac{d(r-1)}{2}\right](\gamma_1 + \cdots + \gamma_r). \tag{2.25}$$

Hence

$$\frac{2(\rho_n \mid \gamma_j)}{|\gamma_j|^2} = 1 + \frac{d(r-1)}{2} \quad \text{and} \quad \frac{2(\rho_n \mid \frac{1}{2}(\gamma_i + \gamma_j))}{|\frac{1}{2}(\gamma_i + \gamma_j)|^2} = 2 + d(r-1). \tag{2.26}$$

From the table we see that

$$\frac{2(\rho_n \mid \gamma_j)}{|\gamma_j|^2} \in \mathbb{Z}^+ \Leftrightarrow \mathfrak{g} \neq \mathfrak{sp}(2n,\mathbb{R}) \quad \text{or} \quad \mathfrak{g} \neq \mathfrak{so}(2,2k-1).$$

Theorem 2.6.15 *Fix the positive system $\Delta_n^+(\mathfrak{p}_\mathbb{C},\mathfrak{t}_\mathbb{C}) \cup -\Delta^+(\mathfrak{k}_\mathbb{C},\mathfrak{t}_\mathbb{C})$ on $\Delta = \Delta(\mathfrak{g}_\mathbb{C},\mathfrak{t}_\mathbb{C})$. Then the following hold:*

1) *There exists an irreducible finite-dimensional representation of $G_\mathbb{C}$ with lowest weight $-2\rho_n$.*

2) *Assume that $\mathfrak{g} \neq \mathfrak{sp}(2n,\mathbb{R})$, $\mathfrak{so}(2,2k+1)$, $n,k \geq 1$. Then there exists a finite-dimensional irreducible representation of $G_\mathbb{C}$ with lowest weight $-\rho_n$.*

Proof: We have to check the integrality of ρ_n and $2\rho_n$. Let $\{\gamma_1, \alpha_2, \ldots, \alpha_k\}$ be the set of simple roots for the positive system $\Delta_n^+(\mathfrak{p}_\mathbb{C},\mathfrak{t}_\mathbb{C}) \cup -\Delta^+(\mathfrak{k}_\mathbb{C},\mathfrak{t}_\mathbb{C})$. Then $\alpha_1 = \gamma_1$ is the only simple noncompact root. Furthermore, $(\rho_n|\alpha) = 0$

for $\alpha \in \Delta(\mathfrak{k}_{\mathbb{C}}, \mathfrak{t}_{\mathbb{C}})$. If γ is an arbitrary noncompact positive root, then $\gamma = \gamma_1 + \sum_{j>1} n_{\alpha_j}\alpha_j$. By (2.25),

$$\frac{2(k\rho_n|\gamma)}{|\gamma|^2} = \frac{2(k\rho_n|\gamma_1)}{|\gamma_1|^2}\frac{|\gamma_1|^2}{|\gamma|^2} = k\left[1 + \frac{d(r-1)}{2}\right]\frac{|\gamma_1|^2}{|\gamma|^2}.$$

From [46], p. 537, it follows that γ_1 is always a long noncompact root. Thus $|\gamma_1|^2/|\gamma|^2 \in \mathbb{Z}^+$. We have $(1 + [d(r-1)]/2) \in \mathbb{Z}$ if and only if d is even or r is odd. But in all cases $2(1 + [d(r-1)]/2) \in \mathbb{Z}$. The claim now follows from the table. □

Let $m \in \mathbb{Z}^+$ be such that there exists a finite-dimensional irreducible representation $(\pi_m, \mathbf{V}_m)$ of $G_{\mathbb{C}}$ with lowest weight $-m\rho_n$. Choose an inner product on $\mathbf{V}_m$ as in Lemma A.2.3, e.g.,

$$\pi_m(g)^* = \pi_m(\sigma\theta(g)^{-1})$$

where σ is the conjugation with respect to G. Let u_0 be a lowest-weight vector of norm 1. Define $\Phi_m : \mathfrak{p}^+ \to \mathbb{C}$ by

$$\Phi_m(Z) := (\pi_m(c^{-2}\exp Z)u_0 \mid u_0)$$

and set

$$\Psi_m(Z, W) := \Phi_m(Z - W).$$

Example 2.6.16 (The case SU(1,1)) Let us work out the special case $G = \mathrm{SU}(1,1)$ before we describe the general case. Now $G_{\mathbb{C}} = \mathrm{SL}(2,\mathbb{C})$. As an involution on SU(1,1) take $\tau(X) = \overline{X}$. Then τ is conjugation by $\mathrm{Int}\begin{pmatrix} 0 & 1 \\ 1 & 0 \end{pmatrix}$ and the holomorphic extension of τ is given by

$$\tau\left(\begin{pmatrix} a & b \\ c & -a \end{pmatrix}\right) = \begin{pmatrix} -a & c \\ b & a \end{pmatrix}.$$

Thus

$$H = \pm\left\{h(t) := \begin{pmatrix} \cosh t & \sinh t \\ \cosh t & \sinh t \end{pmatrix} \middle| t \in \mathbb{R}\right\}.$$

Let $Z^0 = \frac{i}{2}\begin{pmatrix} 1 & 0 \\ 0 & -1 \end{pmatrix}$. Then

$$\mathfrak{p}^+ = \left\{\begin{pmatrix} 0 & z \\ 0 & 0 \end{pmatrix} \middle| z \in \mathbb{C}\right\} \quad \text{and} \quad \mathfrak{p}^- = \left\{\begin{pmatrix} 0 & 0 \\ w & 0 \end{pmatrix} \middle| w \in \mathbb{C}\right\}.$$

The Cartan involution on SU(1,1) is given by conjugating by Z^0, and the holomorphic extension of θ to SL(2, $\mathbb{C}$) is given by

$$\begin{pmatrix} a & b \\ c & d \end{pmatrix} \mapsto \begin{pmatrix} a & -b \\ -c & d \end{pmatrix}.$$

Thus $\tau\theta(X) = {}^tX$. Identify $\mathfrak{p}^+$ with $\mathbb{C}$ by $z \mapsto z\begin{pmatrix} 0 & 1 \\ 0 & 0 \end{pmatrix}$. Similarily, $\mathfrak{a}_{\mathbb{C}} = \mathfrak{t}_{\mathbb{C}} = \mathfrak{k}_{\mathbb{C}} = \mathbb{C}\begin{pmatrix} 1 & 0 \\ 0 & -1 \end{pmatrix} \simeq \mathbb{C}$ and $K_{\mathbb{C}} \simeq \mathbb{C}^*$. A simple calculation now shows that

$$\begin{pmatrix} a & b \\ c & d \end{pmatrix} = \begin{pmatrix} 1 & b/d \\ 0 & 1 \end{pmatrix}\begin{pmatrix} 1/d & 0 \\ 0 & d \end{pmatrix}\begin{pmatrix} 1 & 0 \\ c/d & 1 \end{pmatrix}$$

if $d \neq 0$. Thus

$$P^+K_{\mathbb{C}}P^- = \left\{ \begin{pmatrix} a & b \\ c & d \end{pmatrix} \middle| d \neq 0 \right\} \quad \textit{and} \quad Z\left(\begin{pmatrix} a & b \\ c & d \end{pmatrix}\right) = \frac{b}{d}.$$

Thus $\Omega_+ = \{z \in \mathbb{C} \mid |z| < 1\}$ and $\mathcal{S} = \{z \in \mathbb{C} \mid |z| = 1\}$. Furthermore, $k_{\mathbb{C}}\left(\begin{pmatrix} a & b \\ c & d \end{pmatrix}\right) = 1/d$. Thus we recover the following well-known facts:

$$\begin{pmatrix} a & b \\ c & d \end{pmatrix} \cdot z = \frac{az+b}{cz+d}$$

and

$$j\left(\begin{pmatrix} a & b \\ c & d \end{pmatrix}, z\right) = (cz+d)^{-1}.$$

We have

$$\begin{aligned}
E_1 &= \begin{pmatrix} 0 & 1 \\ 0 & 0 \end{pmatrix}, \\
E_{-1} &= \begin{pmatrix} 0 & 0 \\ 1 & 0 \end{pmatrix}, \\
X_o &= \frac{i}{2}\begin{pmatrix} 0 & -1 \\ 1 & 0 \end{pmatrix}, \\
Y_o &= \frac{1}{2}\begin{pmatrix} 0 & 1 \\ 1 & 0 \end{pmatrix}, \\
Z^0 &= \frac{i}{2}\begin{pmatrix} 1 & 0 \\ 0 & -1 \end{pmatrix}, \\
\mathfrak{q}^+ &= \left\{ i\begin{pmatrix} r & -r \\ r & -r \end{pmatrix} \middle| r \in \mathbb{R} \right\},
\end{aligned}$$

and

$$\mathfrak{q}^- = \left\{ i \begin{pmatrix} r & r \\ -r & -r \end{pmatrix} \middle| r \in \mathbb{R} \right\}.$$

Thus $\exp itX_o = \begin{pmatrix} \cos\frac{t}{2} & \sin\frac{t}{2} \\ -\sin\frac{t}{2} & \cos\frac{t}{2} \end{pmatrix}$. Hence

$$c = \frac{1}{\sqrt{2}} \begin{pmatrix} 1 & 1 \\ -1 & 1 \end{pmatrix} \quad \text{and} \quad c^2 = \begin{pmatrix} 0 & 1 \\ -1 & 0 \end{pmatrix}.$$

We find that $c^{-2} \exp Z = \begin{pmatrix} 0 & -1 \\ 1 & z \end{pmatrix} \in P^+ K_{\mathbb{C}} P^-$ if and only if $z \neq 0$. Furthermore, we have

$$\zeta(c) = 1, \quad \zeta(c^{-1}) = -1 \quad \text{and} \quad j(c^{-2}, Z) = 1/z.$$

The functions Φ_m and Ψ_m are given by $\Phi_m(Z) = z^m$ and $\Psi_m(z,w) = (z-w)^m$. □

Define the homomorphisms $\varphi_j : \mathfrak{sl}(2,\mathbb{C}) \to \mathfrak{g}_{\mathbb{C}}$ by $E_{\pm 1} \mapsto E_{\pm j}$ as in Appendix A.3 and denote the corresponding homomorphisms $\mathrm{SL}(2,\mathbb{C}) \to G_{\mathbb{C}}$ by the same letters. Then $\varphi_j(\mathrm{SU}(1,1)) \subset G$ and since $\tau\sigma(\mathfrak{g}_{C\gamma_j}) = \mathfrak{g}_{C\gamma_j}$, we can choose the root vectors E_j in the construction such that $\sigma(E_j) = \tau(E_j) = E_{-j}$. Thus $\varphi_j \circ \bar{X} = \tau \circ \varphi_j(X)$.

Let ξ be a character of $K_{\mathbb{C}}$ and assume it is unitary on K. Let $\mathfrak{c} = \mathbb{R} Z^0$. Then $d\xi \in i\mathfrak{c}^*$. We write $k^{d\xi} := \xi(k)$, $k \in K_{\mathbb{C}}$.

Lemma 2.6.17 *Let π be a finite-dimensional irreduciblerepresentation of $G_{\mathbb{C}}$ with lowest weight μ. Let u_o be a nonzero vector of weight μ. Then the $K_{\mathbb{C}}$-module generated by u_o is irreducible. If $\mu \in i\mathfrak{c}^*$, then $K_{\mathbb{C}}$ acts on $\mathbb{C}u_o$ by the character $k \mapsto k^{\mu}$.*

Proof: Let $\mathbf{V}$ be the representation space of π and $\mathbf{W}$ be the $K_{\mathbb{C}}$-submodule generated by u_o. Then, as $K_{\mathbb{C}}$ normalizes $\mathfrak{p}^-$, we find $\mathbf{W} \subset \mathbf{V}^{\mathfrak{p}^-}$, where $\mathbf{V}^{\mathfrak{p}^-}$ is the space annihilated by $\mathfrak{p}^-$. For

$$(\mathfrak{n}_k)_{\mathbb{C}} = \bigoplus_{\alpha \in \Delta^+(\mathfrak{k}_{\mathbb{C}}, \mathfrak{t}_{\mathbb{C}})} (\mathfrak{k}_{\mathbb{C}})_{\alpha}$$

we obtain

$$\mathbf{W}^{(\mathfrak{n}_k)_{\mathbb{C}}} \subset \mathbf{V}^{(\mathfrak{n}_k)_{\mathbb{C}} \oplus \mathfrak{p}^-}.$$

But $\mathbf{V}^{(\mathfrak{n}_k)_{\mathbb{C}} \oplus \mathfrak{p}^-} = \mathbb{C}u_o$. Thus $\mathbf{W}^{(\mathfrak{n}_k)_{\mathbb{C}}}$ is one-dimensional, which shows that $\mathbf{W}$ is irreducible. Now the last claim is a consequence of the lowest-weight description of the irreducible holomorphic representations of $K_{\mathbb{C}}$. The claim follows. □

Lemma 2.6.18 *Let $\sigma : G_{\mathbb{C}} \to G_{\mathbb{C}}$ be the conjugation with respect to G. Let $\chi : K_{\mathbb{C}} \to \mathbb{C}^*$ be a character. If $\chi|_K$ is unitary, then $\overline{\chi(\sigma\tau(k))} = \chi(k)$.*

Proof : Both sides are holomorphic in k and agree on K as $\tau|_{\mathfrak{c}} = -1$. The lemma now follows as K is a real form of $K_{\mathbb{C}}$. □

Lemma 2.6.19 *Let $Z \in \mathfrak{p}^+$. Then the following hold:*

1) If $\exp Z \in c^2 P^+ K_{\mathbb{C}} P^-$ then

$$\Phi_m(Z) = (\pi_m(j(c^{-2}, Z))u_0 \mid u_0) = j(c^{-2}, Z)^{-m\rho_n}.$$

2) $\Phi_m(k \cdot Z) = k^{2m\rho_n}\Phi_m(Z)$ for every $k \in K_{\mathbb{C}}$.

3) Let $z_1, \ldots, z_r \in \mathbb{C}$. Then

$$\Phi_m(\mathrm{Ad}(k) \sum_{j=1}^{r} z_j E_j) = k^{2m\rho_n} \prod_{j=1}^{n} z_j^{m(1+d(r-1)/2)} .$$

Proof : 1) Write $c^{-2} \exp Z = pkq$ with $p \in P^+$, $k = j(c^{-2}, Z)$ and $q \in P^-$. As $\sigma(\mathfrak{p}^+) = \mathfrak{p}^-$, we get $\pi_m(p)^* u_0 = \pi_m(\sigma\theta(p)^{-1})u_0 = u_0$. Thus

$$\begin{aligned} \Phi_m(Z) &= (\pi_m(c^{-2} \exp Z)u_0 \mid u_0) \\ &= (\pi_m(k)\pi_m(q)u_0 \mid \pi_m(p)^* u_0) \\ &= (\pi_m(k)u_0 | u_0) \\ &= k^{-m\rho_n}, \end{aligned}$$

where the last equality follows from Lemma 2.6.17 .

2) We have $k \cdot Z = \mathrm{Ad}(k)Z$. By Lemma 2.6.17,

$$\begin{aligned} \Phi_m(k \cdot Z) &= (\pi_m(c^{-2} k \exp Z k^{-1})u_0 \mid u_0) \\ &= (\pi_m(\tau(k))\pi_m(c^{-2} \exp Z)\pi_m(k^{-1})u_0 \mid u_0) \\ &= k^{m\rho_n}(\pi_m(c^{-2} \exp Z)u_0 \mid \pi_m(\sigma\tau(k)^{-1})u_0) \\ &= k^{m\rho_n}\overline{(\sigma\tau(k))^{m\rho_n}}\Phi_m(Z) \\ &= k^{2m\rho_n}\Phi_m(Z) \end{aligned}$$

3) This follows from 2) and $\mathfrak{sl}_2$-reduction via φ_j. □

Theorem 2.6.20 *Let $k \in K_{\mathbb{C}}$, $g \in G_{\mathbb{C}}$ and $Z, W \in \mathfrak{p}^+$ be such that $g \cdot Z$ and $g \cdot W$ are defined. Then the following hold:*

1) $\Psi_m(g \cdot Z, g \cdot W) = j(g, Z)^{m\rho_n} \overline{j(g, W)^{m\rho_n}} \Psi_m(Z, W)$.

2) Let $z_j, w_j \in \mathbb{C}$, $j = 1, \ldots, r$. Then

$$\Psi_m(k \cdot \sum_{j=1}^{r} z_j E_j, k \cdot \sum_{j=1}^{r} w_j E_j) = k^{2m\rho_n} \prod_{j=1}^{r} (z_j - w_j)^{m(1+d(r-1)/2)} .$$

Proof: 1) We have, by definition,

$$\exp g \cdot Z = g \exp Z j(g, Z)^{-1} q, \quad q \in P^- ,$$

and similarily for W. Thus

$$\begin{aligned} \exp(g \cdot Z - g \cdot W) &= \exp(-g \cdot W) \exp(g \cdot Z) \\ &= (g \exp(W) j(g, W)^{-1} p)^{-1} g \exp(Z) j(g, Z)^{-1} q \\ &= p^{-1} j(g, W) \exp(-W) \exp(Z) j(g, Z)^{-1} \end{aligned}$$

for some $p \in P^-$. Now $\mathrm{Ad}(c^{-2}) = \mathrm{C}^2 = \theta\tau$. Thus $\mathrm{C}^2(P^-) = P^+$. As above, we get

$$\begin{aligned} \Psi_m(g \cdot Z, g \cdot W) &= (\pi_m(c^{-2} p^{-1} j(g, W) \exp(-W) \exp(Z) j(g, Z)^{-1}) u_0 \mid u_0) \\ &= j(g, Z)^{m\rho_n} j(g, W)^{m\rho_n} \Psi_m(Z, W) . \end{aligned}$$

2) We have $k \cdot \sum_j z_j E_j - k \cdot \sum_j w_j E_j = k \left(\sum_j (z_j - w_j) E_j \right)$, as $K_{\mathbb{C}}$ acts by linear transformations. The claim now follows from Lemma 2.6.19. □

2.6.3 The Causal Compactification of $\mathcal{M}$

A causal compactification of a causal manifold is an open dense embedding into a compact causal manifold preserving all structures. More precisely, we set the following definition.

Definition 2.6.21 Let $\mathcal{M}$ be a causal G-manifold. A *causal compactification* of $\mathcal{M}$ is a pair $(\mathcal{N}, \Phi)$ such that

1) $\mathcal{N}$ is a compact causal G-manifold.

2) The map $\Phi : \mathcal{M} \to \mathcal{N}$ is causal.

3) Φ is G-equivariant, i.e., $\Phi(g \cdot m) = g \cdot \Phi(m)$, for every $g \in G$ and every $m \in \mathcal{M}$.

4) $\Phi(\mathcal{M})$ is open and dense in $\mathcal{N}$. □

In this section we will show that the map $\mathcal{M} \ni gH \mapsto g\cdot(E,-E) \in \mathcal{S}\times\mathcal{S}$ is a causal compactification of $\mathcal{M}$ and that the image of this map is given by

$$\{\xi \in \mathcal{S}_1 \mid \Psi_m(\xi) \neq 0\}.$$

Identify the tangent space of $\mathcal{S}_1$ at ξ_0 with $(\mathfrak{g}\times\mathfrak{g})/((\mathfrak{h}+\mathfrak{q}^-)\times(\mathfrak{h}+\mathfrak{q}^+)) \simeq \mathfrak{q}^+\times\mathfrak{q}^-$. Let D be the image of $C_+\times -C_-$ in $T_{\xi_0}\mathcal{S}_1$ under this identification. Then D is an $(H\times H)$-invariant regular cone in $T_{\xi_0}(\mathcal{S}_1)$.

Lemma 2.6.22 *D is an $(HQ^-\times HQ^+)$-invariant cone in $T_{\xi_0}(\mathcal{S}_1)$.*

Proof: Let $q = \exp X \in Q^+$, $p = \exp Y \in Q^-$, and $(R,T) \in \mathfrak{q}^+\times\mathfrak{q}^-$. Then

$$\begin{aligned}\mathrm{Ad}(p,q)(R,T) &= (e^{\mathrm{ad}\,Y}R, e^{\mathrm{ad}\,X}T)\\ &= (R+[Y,R]+[Y,[Y,R]], T+[X,T]+[X,[X,T]]).\end{aligned}$$

But $[Y,R] \in \mathfrak{h}$ and $[Y,[Y,R]] \in \mathfrak{q}^-$. Thus $R+[Y,R]+[Y,[Y,R]] = R \bmod(\mathfrak{h}+\mathfrak{q}^-)$. Similarily, $T+[X,T]+[X,[X,T]] = T \bmod(\mathfrak{h}+\mathfrak{q}^+)$. It follows that $\mathrm{Ad}(p,q)|_{T_{\xi_0}(\mathcal{S}_1)} = \mathrm{id}$. This implies the claim. □

It follows that D defines an invariant causal structure on $\mathcal{S}_1$. Recall the maximal abelian subalgebra $\sum_{j=1}^r \mathbb{R}X_j$ of $\mathfrak{p}$ from Proposition A.3.3. Write $X_j = X_j^+ + X_j^-$ with $X_j^\pm \in \mathfrak{q}^\pm$.

Proposition 2.6.23 *1)* $\mathfrak{q}^- = \mathrm{Ad}(H\cap K)\sum_{j=1}^r \mathbb{R}X_j^-$.

2) $Q^-\cdot E = \mathrm{Ad}(K\cap H)\{\sum_{j=1}^r z_jE_j \in \mathcal{S} \mid |z_j| = 1,\ z_j \neq -1\}$.

3) $\mathcal{S} = (K\cap H)\left\{\sum_{j=1}^r z_jE_j \in \mathcal{S} \,\middle|\, |z_j| = 1\right\}$.

Proof: 1) Let $X \in \mathfrak{q}^-$, then $X-\theta(X) \in \mathfrak{q}\cap\mathfrak{p}$. Therefore we can find $k \in K\cap H$ and $x_j \in \mathbb{R}$ such that

$$\mathrm{Ad}(k)(X-\theta(X)) = \sum_{j=1}^r x_jX_j = \sum_{j=1}^r x_jX_j^- + \sum_{j=1}^r x_jX_j^+ \in \mathfrak{a}_q.$$

As $\theta(X), X_j^+ \in \mathfrak{q}^+$ it follows that $\mathrm{Ad}(k)X = \sum_{j=1}^r x_jX_j^-$

2) Assume first that $G = \mathrm{SU}(1,1)$. Then $X_1^- = i\begin{pmatrix}1 & 1\\ -1 & -1\end{pmatrix}$. Thus

$$(\exp tX_1)\cdot 1 = \begin{pmatrix}1+it & it\\ -it & 1-it\end{pmatrix} = \frac{1+2it}{1-2it}.$$

So if $|z| = 1$, $z \neq -1$, we choose $r = \frac{1}{2i}\frac{z-1}{z+1}$. The general case now follows from 1) and $\mathfrak{sl}_2$-reduction.

3) $Q^+ \cdot E$ is dense in $\mathcal{S}$ as Q^+HQ^- is dense in G. The claim follows from that, as $(K \cap H)\{\sum_{j=1}^r z_j E_j \in \mathcal{S} \mid |z_j| = 1\}$ is closed and contained in $\mathcal{S}$. □

Theorem 2.6.24 *Define $\Phi : \mathcal{M} \to \mathcal{S}_1$ by $\Phi(gH) := g \cdot \xi_0$. Then $(\mathcal{S}_1, \Phi)$ is a causal compactification of $\mathcal{M}$ with $\Phi(\mathcal{M}) = \{\xi \in \mathcal{S}_1 \mid \Psi_m(\xi) \neq 0\}$.*

Proof: The G-equivariance of the function Φ is clear. Let us show that Φ is causal. As both the causal structures on $\mathcal{M}$ and that on $\mathcal{S}_1$ are G-invariant, and because Φ is G-equivariant, we only have to show that $(d\Phi)_o(C) \subset D_{\xi_0}$. But this is obvious from the definition of D.

To show that the image of Φ is dense, it suffices to show that it is given as stated. It follows from Theorem 2.6.20 that the left-hand side is contained in the right-hand side. Assume now that $\Psi_m(Z, W) \neq 0$, $\xi = (Z, W) \in \mathcal{S}_1$. Let $g \in G$ be such that $g \cdot W = -E$ and then choose $k \in K \cap H$ such that $k \cdot (g \cdot Z) = \sum z_j E_j$. Then

$$(kg) \cdot \xi = (\sum_{j=1}^{r} z_j E_j, -E).$$

By Theorem 2.6.20 we have $\Psi_m((kg) \cdot \xi) \neq 0$. By the second part of that theorem we have $z_j \neq -1$ for $j = 1, \ldots, r$. By using Proposition 2.6.23 we now find $q \in Q^+$ such that $q \cdot (kg \cdot \xi) = \xi_0$. Hence $\xi = (qkg)^{-1} \cdot \xi_0$. □

Remark 2.6.25 The compactification in Theorem 2.6.24 is also causal with respect to the causal structure on G/H coming from the cone field $C_+ + C_-$.

2.6.4 SU(n, n)

Let $n = p + q$. Then

$$\begin{aligned} \mathrm{SU}(p,q) &= \{a \in \mathrm{SL}(n, \mathbb{C}) \mid a^* I_{p,q} a = I_{p,q}\} \\ &= \left\{ \begin{pmatrix} A & B \\ C & D \end{pmatrix} \middle| \begin{array}{l} A^*A - C^*C = I_p \\ D^*D - B^*B = I_q \\ B^*A - D^*C = 0 \end{array} \right\}. \end{aligned}$$

The conjugation in SL$(n, \mathbb{C})$ with respect to SU(p, q) is given by

$$\sigma(a) = I_{p,q}\theta(a)I_{p,q},$$

where θ is the Cartan involution $a \mapsto (a^*)^{-1}$. If $a \in \mathrm{SU}(p, q)$, then $a^{-1} = I_{p,q} a^* I_{p,q}$. Hence

$$\begin{pmatrix} A & B \\ C & D \end{pmatrix}^{-1} = \begin{pmatrix} A^* & -C^* \\ -B^* & D^* \end{pmatrix}. \tag{2.27}$$

The Lie algebra of $\mathrm{SU}(p,q)$ is given by

$$\mathfrak{su}(p,q) = \left\{ \begin{pmatrix} X & Y \\ Y^* & Z \end{pmatrix} \middle| \begin{array}{l} Y \in M(p \times q, \mathbb{C}),\ X \in \mathfrak{u}(p), \\ Z \in \mathfrak{u}(q),\ \operatorname{Tr} X + \operatorname{Tr} Z = 0 \end{array} \right\}. \tag{2.28}$$

The maximal compact subgroup K is given by $S(\mathrm{U}(p) \times \mathrm{U}(p))$. Furthermore,

$$\mathfrak{k} = \left\{ \begin{pmatrix} A & 0 \\ 0 & B \end{pmatrix} \middle| A \in \mathfrak{u}(p),\ B \in \mathfrak{u}(q),\ \operatorname{Tr}(A) + \operatorname{Tr}(B) = 0 \right\}$$

and

$$\mathfrak{p} = \left\{ \begin{pmatrix} 0 & Y \\ Y^* & 0 \end{pmatrix} \middle| Y \in M(p \times q, \mathbb{C}) \right\}.$$

We choose $Z^0 = \begin{pmatrix} \frac{ip}{n} I_p & 0 \\ 0 & -\frac{iq}{n} I_q \end{pmatrix} \in \mathfrak{k}$. Then

$$\operatorname{ad}(Z^0) \begin{pmatrix} A & Y \\ Y^* & B \end{pmatrix} = \begin{pmatrix} 0 & iY \\ -iY^* & 0 \end{pmatrix},$$

which implies that

$$\mathfrak{p}_{\mathbb{C}}^+ = \left\{ \begin{pmatrix} 0 & Y \\ 0 & 0 \end{pmatrix} \middle| Y \in M(p \times q, \mathbb{C}) \right\} \simeq M(p \times q, \mathbb{C}).$$

Suppose that $\begin{pmatrix} A & B \\ C & D \end{pmatrix} \in \mathrm{SU}(p,q)$. Then D is invertible and we have a decomposition

$$\begin{pmatrix} A & B \\ C & D \end{pmatrix} = \begin{pmatrix} I_p & BD^{-1} \\ 0 & I_q \end{pmatrix} \begin{pmatrix} A - BD^{-1}C & 0 \\ 0 & D \end{pmatrix} \begin{pmatrix} I_p & 0 \\ D^{-1}C & I_q \end{pmatrix}.$$

Thus the Harish-Chandra embedding $G/K \hookrightarrow \mathfrak{p}^+ \simeq M(p \times q, \mathbb{C})$ is given by

$$z\left(\begin{pmatrix} A & B \\ C & D \end{pmatrix} K_{\mathbb{C}} P^- \right) = \begin{pmatrix} 0 & BD^{-1} \\ 0 & 0 \end{pmatrix} \mapsto BD^{-1},$$

inducing a biholomorphic isomorphism

$$\mathrm{SU}(p,q)/\mathrm{S}(\mathrm{U}(p) \times \mathrm{U}(q)) \simeq D_{p,q},$$

where $D_{p,q} := \{Z \in M(p \times q, \mathbb{C}) \mid I_q - Z^*Z > 0\}$. Here the action of $\mathrm{SU}(p,q)$ on $D_{p,q}$ is given by

$$\begin{pmatrix} A & B \\ C & D \end{pmatrix} \cdot Z = (AZ + B)(CZ + D)^{-1}.$$

G/K is of tube type if and only if $p = q$. In that case we have

$$\begin{aligned} Y_o &= \frac{1}{2}\begin{pmatrix} 0 & I_n \\ I_n & 0 \end{pmatrix}, \\ X_o &= \frac{i}{2}\begin{pmatrix} 0 & -I_n \\ I_n & 0 \end{pmatrix}, \\ c &= \frac{1}{\sqrt{2}}\begin{pmatrix} I_n & I_n \\ -I_n & I_n \end{pmatrix}, \\ c^{-2} &= \begin{pmatrix} 0 & I_n \\ -I_n & 0 \end{pmatrix}, \\ E &= I_n, \\ \mathcal{S} &= \mathrm{U}(n). \end{aligned}$$

The involution $\tau = \tau_{Y_o}$ is conjugation by $2Y_o$. Thus

$$\tau\left(\begin{pmatrix} A & B \\ C & D \end{pmatrix}\right) = \begin{pmatrix} D & C \\ B & A \end{pmatrix}.$$

Therefore

$$\begin{aligned} H &= \left\{ h(A,B) := \begin{pmatrix} A & B \\ B & A \end{pmatrix} \middle| \begin{array}{c} A, B \in M_n(\mathbb{C}) \\ A^*A - B^*B = I_n,\ B^*A = A^*B \end{array} \right\}, \\ \mathfrak{h} &= \{ h(X,Y) \in \mathfrak{su}(n,n) \mid X^* = -X,\ Y^* = Y,\ \mathrm{Tr}\, X = 0 \}, \end{aligned}$$

and

$$\mathfrak{q} = i\left\{ q(X,Y) := \begin{pmatrix} X & Y \\ -Y & -X \end{pmatrix} \middle| {}^tX = X,\ {}^tY = Y \right\}.$$

Define $\varphi_\pm : H \cup \mathfrak{h} \to M(n,\mathbb{C})$ by

$$\varphi_\pm(h(A,-B)) = A \pm B.$$

We leave the simple proof of the following assertions to the reader:

1) $\varphi_\pm : H \to \mathrm{GL}(n,\mathbb{C})_+$ is an isomorphism of groups.

2) $\varphi_\pm : \mathfrak{h} \to \mathfrak{sl}(n,\mathbb{C}) + \mathbb{R}\, I_n$ is an isomorphism of Lie algebras.

Note that the Cartan involution on $\mathrm{GL}(n,\mathbb{C})$ is $\theta(a) = (a^*)^{-1}$. By (2.27), $\varphi_- = \theta \circ \varphi_+$. We choose $\frac{1}{2}\begin{pmatrix} 0 & I_n \\ I_n & 0 \end{pmatrix}$ as Y_o. Then the $\mathfrak{h}$-module structure of $\mathfrak{q}$ is described by

$$\mathfrak{q}^+ = i\{q(X,-X) \mid {}^tX = X\} \ni iq(X,-X) \overset{\frac{1}{2i}\varphi_+}{\longmapsto} X \in \mathrm{H}(n,\mathbb{C})$$

and

$$\mathfrak{q}^- = i\{q(X,X) \mid {}^tX = X\} \ni iq(X,X) \overset{\frac{1}{2i}\varphi_-}{\mapsto} X \in \mathrm{H}(n,\mathbb{C}).$$

Obviously, both φ_+ and φ_-, are isomorphisms. By (2.27) we get for $X \in \mathfrak{q}^{\pm}$ and $a \in H$,

$$\varphi_{\pm}(\mathrm{Ad}(a)X) = \varphi_{\pm}(a)\left[\frac{i}{2i}\varphi_{\pm}(X)\right]\varphi_{\pm}(a)^*.$$

In this case $C_{\pm} = \overline{\frac{1}{2i}\varphi_{\pm}^{-1}(\mathrm{H}^+(n,\mathbb{C}))}$, cf. Example 2.1.14.

In the bounded realization we have $G/K = D_{n,n}$. The space $\sum_{j=1}^r \mathbb{C}E_j$ corresponds to the diagonal matrices and $E = I_n$. In particular, $\mathcal{S} = \mathrm{U}(n)$. By Lemma 2.6.19 and the table on p. 58, $\Phi_m(Z) = \det(Z)^{mn}$. Thus

$$\mathrm{SU}(n,n)/\,\mathrm{GL}(n,\mathbb{C})_+ \simeq \{(Z,W) \in \mathrm{U}(n) \times \mathrm{U}(n) \mid \det(Z-W) \neq 0\}.$$

2.6.5 $\mathrm{Sp}(n,\mathbb{R})$

We realize $G = \mathrm{Sp}(n,\mathbb{R})$ inside $\mathrm{SU}(n,n)$ as

$$\mathrm{Sp}(n,\mathbb{R}) = \left\{\begin{pmatrix} A & B \\ \overline{B} & \overline{A}\end{pmatrix} \,\middle|\, A^*A - {}^tB\overline{B} = I_n,\ {}^t(B^*A) = B^*A\right\}. \qquad (2.29)$$

Its Lie algebra is given by

$$\mathfrak{sp}(n,\mathbb{R}) = \left\{\begin{pmatrix} X & Y \\ \overline{Y} & \overline{X}\end{pmatrix} \,\middle|\, X,Y \in M(n,\mathbb{C}),\ X^* = -X,\ {}^tY = Y\right\}. \qquad (2.30)$$

The involution τ leaves G and $\mathfrak{g}$ stable and is also given by complex conjugation. Therefore H, $\mathfrak{h}$ are just the real points of the corresponding object for $\mathrm{SU}(n,n)$. We also note that the above X_o and Y_o are in $\mathfrak{sp}(n,\mathbb{R})$. Thus

$$\begin{aligned}
H &= \{h(A,B) \in \mathrm{Sp}(n,\mathbb{R}) \mid A,B \in M(n,\mathbb{R})\} \\
&\overset{\varphi_+}{\simeq} \mathrm{GL}(n,\mathbb{R})_+, \\
\mathfrak{h} &= \{h(X,Y) \in \mathfrak{sp}(n,\mathbb{R}) \mid X,Y \in M(n,\mathbb{R}),\ {}^tX = -X,\ {}^tY = Y\} \\
&\overset{\varphi_+}{\simeq} \mathfrak{gl}(n,\mathbb{R}), \\
\mathfrak{q} &= i\{h(X,Y) | X,Y \in M(n,\mathbb{R}), \quad {}^tX = X,\ {}^tY = Y\}, \\
\mathfrak{q}^+ &= i\{q(X,-X) \mid X \in M_n(\mathbb{R}),\ {}^tX = X\} \\
&\ni iq(X,-X) \overset{\frac{1}{2i}\varphi_+}{\mapsto} X \in \mathrm{H}(n,\mathbb{R}), \\
\mathfrak{q}^- &= i\{q(X,X) \mid X \in M(n,\mathbb{R}),\ {}^tX = X\} \\
&\ni iq(X,X) \overset{\frac{1}{2i}\varphi_-}{\mapsto} X \in \mathrm{H}(n,\mathbb{R}).
\end{aligned}$$

The Cartan involution on $\mathrm{GL}(n,\mathbb{R})$ is $\theta(a) = {}^t a^{-1}$. From (2.27) we obtain $\varphi_- = \theta \circ \varphi_+$ and, with $X \in \mathfrak{q}^{\pm}$ and $a \in H$,

$$\varphi_{\pm}(\mathrm{Ad}(a)X) = \varphi_{\pm}(a)\left[\frac{1}{2i}\varphi_{\pm}(X)\right]{}^t\varphi_{\pm}(a).$$

In this case $C_{\pm} = 2i\overline{\varphi_{\pm}^{-1}(\mathrm{H}^+(n,\mathbb{R}))}$, cf. Example 2.1.14.

The bounded realization of G/K is

$$\{Z \in M_n(\mathbb{C}) \mid I_n - Z^*Z > 0,\, {}^tZ = Z\}.$$

The space $\sum_{j=1}^{r} \mathbb{C}E_j$ corresponds to the diagonal matrices and $E = I_n$. In particular,

$$\mathcal{S} = \mathrm{U}(n)/\mathrm{O}(n) \simeq \{A \in \mathrm{U}(n) \mid {}^tZ = Z\}.$$

Lemma 2.6.19 and the table on p. 58 imply

$$\Phi_m(Z) = \det(Z)^{\frac{m(n+1)}{2}}.$$

Thus

$$\mathrm{Sp}(n,\mathbb{R})/\,\mathrm{GL}(n,\mathbb{R})_+ \simeq \left\{(Z,W) \in \mathrm{O}(n) \times \mathrm{O}(n) \,\middle|\, \begin{matrix} Z, W \quad \text{symmetric} \\ \det(Z-W) \neq 0 \end{matrix}\right\}.$$

Notes for Chapter 2

Cones have been used in different parts of mathematics for a long time and are related to concepts such as the Laplace transform [24, 28], Hardy spaces over tube domains, and Hermitian symmetric spaces [83, 84]. The concept of causal orderings associated to cone fields has also been used for a long time implicitly in the context of Lorentzian geometry and relativity (e.g., in [4, 37, 42]). Group invariant cone fields appear in Segal's book [157]. Vinberg, Paneitz, and Ol'shanskii considered the special case of bi-invariant cone fields on Lie groups in [166, 147, 148]. The first article on invariant cone fields on semisimple symmetric spaces was [138]. A systematic study of invariant cone fields on homogeneous spaces was started in [47] and [50]. Later it was taken up in the work of Lawson [93], Ólafsson [129, 130], and Neeb [112]-[115], [52].

The algebraic side of the theory, i.e., a closer inspection of the cones that appear in the study of causal orderings, was also initiated by Vinberg in [166] and then taken up by many authors [48, 50, 129, 137, 148].

The order compactification was introduced in [55], motivated by the study of Wiener-Hopf operators on ordered homogeneous spaces.

The results in the last section are taken partly from [136], where further information about this class of spaces can be found. This causal compactification has also been considered in [86, 87]. The compactly causal hyperboloids were studied in [107]. In [7] and [6], causal compactifications for a more general class of causal symmetric spaces are given. Compactifications without the causal structure have also been obtained in [98] and [76].

Chapter 3

Irreducible Causal Symmetric Spaces

In this chapter we determine the irreducible semisimple causal symmetric spaces. The crucial observation is that the existence of causal structures on $\mathcal{M} = G/H$ is closely connected to the existence of $(H \cap K)$-fixed vectors in the tangent spaces of $\mathcal{M}$. With that tie established, one can use the results of Chapter 1 to single out which irreducible non-Riemannian semisimple symmetric spaces admit causal structures.

As we have seen already in Chapters 1 and 2, the existence of causal structures may depend on the fundamental group of the space. So we include a discussion on how causal structures behave with respect to coverings. In Section 3.2 we give a list of all the irreducible semisimple symmetric pairs $(\mathfrak{g}, \tau)$ for which the universal symmetric space admits a causal structure.

3.1 Existence of Causal Structures

In this section we assume that $\mathcal{M} = G/H$ is a non-Riemannian semisimple symmetric space such that the corresponding symmetric pair $(\mathfrak{g}, \tau)$ is irreducible and effective. We fix a Cartan involution θ commuting with τ and use the notation introduced in Remark 1.1.15.

Lemma 3.1.1 *Let $0 \neq X \in \mathfrak{q}^{H \cap K}$ and C the smallest H_o-invariant convex closed cone in $\mathfrak{q}$ containing X. Then C is H-invariant.*

Proof: We mimick the proof of Lemma 2.6.1: If $h \in H$, then h is of the form $h = h_o k$ with $k \in H \cap K$ and $h_o \in H_o$ (cf. (1.8)). Further, let

$Y = \sum \lambda_j \operatorname{Ad}(h_j)X \in C$, $\lambda_j \geq 0, h_j \in H_o$. Thus

$$\begin{aligned} \operatorname{Ad}(h)Y &= \sum \lambda_j \operatorname{Ad}(hh_j)X \\ &= \sum \lambda_j \operatorname{Ad}(hh_jh^{-1}) \operatorname{Ad}(h_o) \operatorname{Ad}(k)X \\ &= \sum \lambda_j \operatorname{Ad}(hh_jh^{-1}) \operatorname{Ad}(h_o)X \in C \end{aligned}$$

since $hh_jh^{-1} \in H_o$. As C is closed and the set of elements of the form $\sum \lambda_j \operatorname{Ad}(h_j)X$ is dense, it follows that C is H-invariant. □

Lemma 3.1.2 *If $\mathcal{M}$ admits a G-invariant causal structure, then we have:*

1) There exists an H-invariant proper closed convex cone in $\mathfrak{q}$.

2) $\mathfrak{q}$ is a completely reducible H-module with either one or two irreducible components.

3) $\dim(\mathfrak{q}^{H\cap K})$ is equal to the number of irreducible H-submodules of $\mathfrak{q}$.

4) If $\dim(\mathfrak{q}^{H\cap K}) < \dim(\mathfrak{q}^{H_o\cap K})$, then there exists an element $h \in H\cap K$ such that $\operatorname{Ad}(h)Y = -Y$ for all $Y \in \mathfrak{z}(\mathfrak{h})$.

Proof: 1) Let C be an H-invariant nontrivial closed convex cone in $\mathfrak{q}$. Then space $\mathfrak{q}^C = C \cap (-C)$ is H-invariant and not equal to $\mathfrak{q}$. According to Lemma 1.3.4, $\mathfrak{q}^C$ is either trivial or otherwise equal to $\mathfrak{q}^+$ or $\mathfrak{q}^-$, where $\mathfrak{q} = \mathfrak{q}^+ + \mathfrak{q}^-$ is the decompostion of $\mathfrak{q}$ into irreducible $\mathfrak{h}$-modules. In each case we find a proper H_o-invariant cone in C.

2) Since H_o is normal in H, the action of H on $\mathfrak{q}$ maps H_o-submodules to H_o-submodules. Thus $\mathfrak{q}$ is a reducible H-module if and only if it is a reducible H_o-module and $\mathfrak{q}^\pm$ are both H-invariant. In this case $\mathfrak{q}^\pm$ are both irreducible H-modules, which implies the claim.

3) If $\mathfrak{q}$ is a reducible H-module, then we consider the projection $\operatorname{pr}\colon \mathfrak{q} \to \mathfrak{q}^+$, which is H-equivariant. Therefore $\operatorname{pr}(C) \subseteq \mathfrak{q}^+$ and $\theta \circ \operatorname{pr}(C) \subseteq \mathfrak{q}^-$ are H-invariant proper cones. Thus Theorem 1.3.11.4) implies that $\dim(\mathfrak{q}^{H\cap K}) = \dim(\mathfrak{q}^{H_o\cap K}) = 2$.

If $\mathfrak{q}$ is irreducible as an H-module, then $\mathfrak{q}^C = \{0\}$, i.e., C is proper. Then Lemma 2.1.15 shows that $\mathfrak{q}^{H\cap K} \neq \{0\}$. It remains to be shown that $\dim(\mathfrak{q}^{H\cap K}) = 1$. We have two cases to consider.

Case 1: $\dim(\mathfrak{q}^{H_o\cap K}) = 1$. In this case, obviously, $\dim(\mathfrak{q}^{H\cap K}) = 1$ as well.

Case 2: $\dim(\mathfrak{q}^{H_o\cap K}) = 2$. Then we are in the situation of Theorem 1.3.11 and, as far as the symmetric pair $(\mathfrak{g}, \tau)$ is concerned, Section 2.6. Recall the θ-stable subgroup H_1 of index 2 in H_o from Lemma 1.3.14. According to this lemma, we can find an $h \in H \cap K$, not contained in

H_1, such that $\mathrm{Ad}(h)\mathfrak{q}^+ = \mathfrak{q}^-$. Since $\mathfrak{z}(\mathfrak{h})$ is one-dimensional, we have $\mathrm{Ad}(h)Y_o = rY_o$ with $r \in \mathbb{R}$. Moreover, $h^2 \in H_1$, whence $r^2 = 1$. If $r = 1$, then $X_o, Z^0 \in \mathfrak{q}^{H\cap K}$, which shows that the H_o-invariant cones generated by $X_\pm = X_o \pm Z^0 \in \mathfrak{q}^\pm$ are H-invariant (Lemma 3.1.1). But this contradicts the H-irreducibility of $\mathfrak{q}$. Thus we have $r = -1$. This means that $\mathrm{Ad}(h)X_o = -X_o$ which, together with $h\,\mathrm{Ad}(h)Z^0 = Z^0$, shows that the only H-invariant vectors in $\mathfrak{q}^{H_o\cap K} = \mathbb{R}\,X_o + \mathbb{R}\,Z^0$ are the multiples of Z^0. This implies $\dim(\mathfrak{q}^{H\cap K}) = 1$. Finally, we note that $\mathrm{Ad}(h)Y_o = -Y_o$ by the above, so that assertion 4) follows, too. □

Theorem 3.1.3 *Let $\mathcal{M} = G/H$ be a non-Riemannian semisimple symmetric space. If $\mathcal{M}$ is irreducible, then the following statements are equivalent:*

1) *$\mathcal{M}$ admits a G-invariant causal structure.*

2) $\dim(\mathfrak{q}^{H\cap K}) > 0$.

If these conditions hold, then $\mathrm{Cone}_H(\mathfrak{q}) \neq \emptyset$, *i.e., $\mathcal{M}$ even admits a regular G-invariant causal structure.*

Proof: Lemma 3.1.2 shows that the existence of G-invariant causal structure on $\mathcal{M}$ implies that $\mathfrak{q}^{H\cap K} \neq \{0\}$. Assume conversely that $\mathfrak{q}^{H\cap K} \neq \{0\}$. We have to consider two cases.

Case 1: Suppose that $\mathfrak{q}$ is irreducible as an $\mathfrak{h}$-module. Then by Theorem 1.3.11, H_o is semisimple, and by Lemma A.2.5, $\dim \mathfrak{q}^{H_o\cap K} = 1$. Thus $\mathfrak{q}^{H\cap K} = \mathfrak{q}^{H_o\cap K}$. Let $0 \neq X \in \mathfrak{q}^{H\cap K}$ and $C_{\min}$ be the H_o-invariant cone generated by X (cf. Theorem 2.1.21). Then Lemma 3.1.1 shows that $C_{\min}$ is H-invariant.

Case 2: Suppose that $\mathfrak{q}$ is not irreducible as an $\mathfrak{h}$-module. Then we are in the situation of Theorem 1.3.11 and Section 2.6. Let C be the H_o-invariant cone in $\mathfrak{q}$ generated by Z^0. Then the group case described in Section 2.5.1 shows that C is proper. It is also H-invariant by Lemma 3.1.1.

It remains to show that the existence of proper H-invariant cones imply the existence of regular H-invariant cones. If $\mathfrak{q}$ is H-irreducible this is obvious, since the span of an H-invariant cone is H-invariant. If $\mathfrak{q}$ is not H-irreducible, then the cones $C_\pm$ constructed in Section 2.6 are H-invariant, proper, and generating in $\mathfrak{q}^\pm$. Thus $C_+ + C_-$ is an H-invariant regular cone. □

Lemma 3.1.4 *Let $\mathcal{M} = G/H$ be an irreducible non-Riemannian semisimple symmetric space and C an H-invariant cone in $\mathfrak{q}$. If either*

$$C^o \cap \mathfrak{k} \neq \emptyset \quad \textit{and} \quad C \cap \mathfrak{p} \neq \{0\}$$

or

$$C^o \cap \mathfrak{p} \neq \emptyset \quad \textit{and} \quad C \cap \mathfrak{k} \neq \{0\},$$

then C *contains a line.*

Proof: We will only show that the first assumption implies that C contains a line. The second part then follows from c-duality.

Assume that C is proper and $C^o \cap \mathfrak{k} \neq \emptyset$ and $C \cap \mathfrak{p} \neq \{0\}$. Let $Z_1 \in C^o \cap \mathfrak{k}$ and $X_1 \in C \cap \mathfrak{p}$, $Z_1, X_1 \neq 0$. Define

$$Z := \int_{\mathrm{Ad}_G(K\cap H)} k \cdot Z_1 \, dk \quad \text{and} \quad X := \int_{\mathrm{Ad}_G(K\cap H)} k \cdot X_1 \, dk .$$

Then (cf. Lemma 2.1.15) Z, X are nonzero and $(K\cap H)$-fixed. Furthermore, as θ commutes with $\mathrm{Ad}(K\cap H)$, $Z \in C^o \cap \mathfrak{q}_k^0$ and $X \in C \cap \mathfrak{q}_p^0$. Thus we are in the situation of Theorem 1.3.11 and Section 2.6. In particular, Remark 2.6.2 shows that $Z \in \mathbb{R}\, Z^0$ and $X \in \mathbb{R}\, X_o$. Normalize Z and X such that $\mathrm{ad}\, Z$ has the eigenvalues $\pm i$ and 0 and $\mathrm{ad}\, X$ has the eigenvalues ± 1 and 0. In particular,

$$\mathrm{ad}(Z)^2|_{\mathfrak{p}} = -\,\mathrm{id} \qquad \text{and} \qquad \mathrm{ad}(X)^2|_{\mathfrak{q}^a} = \mathrm{id}\,.$$

Let $Y = [Z, X] \in \mathbb{R}\, Y_o \subset \mathfrak{z}(\mathfrak{h}) \cap \mathfrak{p}$. Then

$$[Y, Z] = [[Z, X], Z] = -\,\mathrm{ad}(Z)^2 X = X$$

and similarly, $[Y, X] = Z$. Thus we have reduced the problem to one on $\mathfrak{sl}(2, \mathbb{R})$. A short direct argument goes as follows:

$$\begin{aligned} e^{t\,\mathrm{ad}\,Y}(Z + X) &= e^{t}(Z + X), \\ e^{t\,\mathrm{ad}\,Y}(Z - X) &= e^{-t}(Z - X). \end{aligned}$$

In particular, we get

$$\begin{aligned} e^{t\,\mathrm{ad}\,Y} Z &= \cosh t\,[Z + (\tanh t) X]\,, \\ e^{t\,\mathrm{ad}\,Y} X &= \cosh t\,[(\tanh t) Z + X]\,. \end{aligned}$$

Dividing by $\cosh t$ and letting $t \to \pm\infty$ shows that $\pm(Z \pm X) \in C$. Thus C contains a line. □

Theorem 3.1.5 *Let* $\mathcal{M} = G/H$ *be a non-Riemannian semisimple symmetric space. Suppose that* $\mathcal{M}$ *is irreducible and admits a* G*-invariant causal structure. Then the following cases may occur:*

1) $\dim(\mathfrak{q}^{H\cap K}) = \dim(\mathfrak{q}^{H_o \cap K}) = 1$. *In this case* $\mathfrak{q}$ *is irreducible as* H*- and* $\mathfrak{h}$*-module. There are two possibilities which are c-dual to each other:*

1.k) $\mathfrak{q}^{H\cap K} \subset \mathfrak{q}_k$. *In this case, for every cone* $C \in \mathrm{Cone}_H(\mathfrak{q})$ *we have*

$$C^o \cap \mathfrak{q}_k^{H\cap K} \neq \emptyset, \quad C \cap \mathfrak{p} = \{0\}.$$

1.p) $\mathfrak{q}^{H\cap K} \subset \mathfrak{q}_p$. *In this case, for every cone* $C \in \mathrm{Cone}_H(\mathfrak{q})$ *we have*

$$C^o \cap \mathfrak{q}_p^{H\cap K} \neq \emptyset, \quad C \cap \mathfrak{k} = \{0\}.$$

2) $\dim(\mathfrak{q}^{H\cap K}) = \dim(\mathfrak{q}^{H_o\cap K}) = 2$. *In this case* $\mathfrak{q}$ *is neither* H*- nor* $\mathfrak{h}$*-irreducible and we have*

$$\mathrm{Cone}_H(\mathfrak{q}) = \{\pm C_k, \pm C_p\}$$

(cf. Remark 2.6.2 *).*

3) $\dim(\mathfrak{q}^{H\cap K}) = 1, \dim(\mathfrak{q}^{H_o\cap K}) = 2$. *In this case,* $\mathfrak{q}$ *is* H*-irreducible but not* $\mathfrak{h}$*-irreducible and we have*

$$\mathrm{Cone}_H(\mathfrak{q}) = \{\pm C_k\}.$$

Proof: Note first that Theorem 1.3.11 and Theorem 3.1.3 show that no more than these three cases are possible. Moreover, Lemma 3.1.2 shows the claims about H- and $\mathfrak{h}$-irreducibility.

1) Recall that $\mathfrak{q}^{H\cap K}$ is θ-invariant. This proves the dichotomy of (1.k) and (1.p). Now the claim follows from Lemma 3.1.4 in view of Lemma 2.1.15.

2) In view of Theorem 1.3.11, this is a consequence of Theorem 2.6.8.

3) Theorems 1.3.11 and 2.6.8 show that $\mathrm{Cone}_{H_o}(\mathfrak{q}) = \{\pm C_k, \pm C_p\}$. This proves that C_k is the H_o-invariant cone generated by Z^0. Since Z^0 is an $(H \cap K)$-fixed point, Lemma 3.1.1 implies that C_k is H-invariant. On the other hand, Lemma 3.1.2 shows that we can find an $h \in H \cap K$ with $\mathrm{Ad}(h)Y_o = -Y_o$ so that also $\mathrm{Ad}(h)X_o = -X_o$ and hence does not leave C_p invariant. □

The following corollary is an immediate consequence of Theorem 3.1.5 and Remark 2.6.2.

Corollary 3.1.6 *If* $\mathcal{M}$ *is an irreducible non-Riemannian semisimple symmetric space and* $C \in \mathrm{Cone}_H(\mathfrak{q})$, *then we have either*

$$C^o \cap \mathfrak{q}_k^{H\cap K} \neq \emptyset, \quad C \cap \mathfrak{p} = \{0\}$$

or

$$C^o \cap \mathfrak{q}_p^{H\cap K} \neq \emptyset, \quad C \cap \mathfrak{k} = \{0\}.$$ □

Remark 3.1.7 1) Theorem 1.3.8 shows that in Case (1.k) of Theorem 3.1.5 the Riemannian symmetric space G/K is a bounded Hermitian domain and τ induces an antiholomorphic involution on G/K. Moreover, Theorem 1.3.11 implies that G/K is not of tube type in this case.

2) Under *c*-duality $\mathfrak{k}$ corresponds to $\mathfrak{h}^a$ and Hermitian to *para-Hermitian* structures. Therefore in Case (1.p) of Theorem 3.1.5 the space G/H^a carries a para-Hermitian structure.

3) Theorem 1.3.11 shows that in the cases 2) and 3) of Theorem 3.1.5 the Riemannian symmetric space G/K is a bounded Hermitian domain of tube type.

4) Theorem 2.6.8 implies that Case 3) in Theorem 3.1.5 cannot occur if H is connected. □

Definition 3.1.8 Let $\mathcal{M}$ be an irreducible non-Riemannian semisimple symmetric space. Then we call $\mathcal{M}$

CC) a *compactly causal symmetric space* if there exists a $C \in \mathrm{Cone}_H(\mathfrak{q})$ such that $C^o \cap \mathfrak{k} \neq \emptyset$,

NCC) a *noncompactly causal symmetric space* if there is a $C \in \mathrm{Cone}_H(\mathfrak{q})$ such that $C^o \cap \mathfrak{p} \neq \emptyset$, and

CT) a *symmetric space of Cayley type* if both (CC) and (NCC) hold.

CAU) a *causal symmetric space* if either (CC) or (NCC) holds.

The pair $(\mathfrak{g}, \tau)$ is called *compactly causal* (*noncompactly causal*, of *Cayley type*) if the corresponding universal symmetric space $\tilde{\mathcal{M}}$ has that property. Finally, $(\mathfrak{g}, \tau)$ is called *causal* if it is either noncompactly causal or compactly causal. □

Remark 3.1.9 1) It follows directly from the definitions that $(\mathfrak{g}, \tau)$ is noncompactly causal if and only if $(\mathfrak{g}^c, \tau)$ is compactly causal.

2) In view of (1), Lemma 1.2.1 implies that a noncompactly causal symmetric pair $(\mathfrak{g}, \tau)$ is of Cayley type if $(\mathfrak{g}, \tau) \cong (\mathfrak{g}^c, \tau)$.

3) If $(\mathfrak{g}, \tau)$ is compactly causal, then Theorem 1.3.8 shows that either $\mathfrak{g}$ is simple Hermitian or of the form $\mathfrak{g}_1 \times \mathfrak{g}_1$ with $\mathfrak{g}_1$ simple Hermitian and $\tau(X, Y) = (Y, X)$.

4) Example 1.2.2 now shows that if $(\mathfrak{g}, \tau)$ is noncompactly causal, then either $\mathfrak{g}^c$ is simple Hermitian or of the form $\mathfrak{h}_{\mathbb{C}}$ with $\tau = \sigma$. Note that in both cases $\mathfrak{g}$ is a simple Lie algebra.

5) If $(\mathfrak{g}, \tau)$ is of Cayley type, then 3) and 4) imply that $\mathfrak{g}$ is simple Hermitian.

6) Let $\mathfrak{g}$ be a simple noncompact Lie algebra. Then $(\mathfrak{g}, \mathrm{id})$ is an irreducible symmetric pair. It is *not* causal since it does not belong to a non-Riemannian semisimple symmetric space (cf. Remark 1.1.15). □

Definition 3.1.10 Let $(\mathfrak{g}, \tau)$ be noncompactly causal symmetric pair. An element $X^0 \in \mathfrak{q}_p$ is called *cone-generating* if $\mathrm{spec}(\mathrm{ad}\, X^0) = \{-1, 0, 1\}$ and the centralizer of X^0 in $\mathfrak{g}$ is $\mathfrak{h}^a$. □

Proposition 3.1.11 *Suppose that $(\mathfrak{g}, \tau)$ is a noncompactly causal symmetric pair.*

1) Cone-generating elements exist and are unique up to sign.

2) Let $\mathfrak{b}$ be an abelian subspace of $\mathfrak{p}$ containing a cone-generating element X^0. Then $\mathfrak{b} \subset \mathfrak{q}_p$.

3) Let $\mathfrak{a}$ be a maximal abelian subspace of $\mathfrak{q}_p$. Then $\mathfrak{a}$ is maximal abelian in $\mathfrak{p}$ and maximal abelian in $\mathfrak{q}$. Moreover, $\mathfrak{a}$ contains X^0.

4) Let $\mathfrak{a}$ be maximal abelian in $\mathfrak{q}$ and assume that $X^0 \in \mathfrak{a}$. Then $\mathfrak{a} \subset \mathfrak{q}_p$.

5) Let $X^0 \in \mathfrak{q}_p$ be a cone-generating element. Then $\mathbb{R}X^0 = \mathfrak{z}(\mathfrak{h}^a)$.

Proof: 1) According to Theorem 3.1.5, the centralizer $\mathfrak{z}_{\mathfrak{q}_p}(\mathfrak{h})$ of $\mathfrak{h}_k$ in $\mathfrak{q}$ is nontrivial. Then, in view of Theorem 1.3.11, it is one-dimensional, say of the form $\mathbb{R}X$. Lemma 1.3.5 says that $\mathfrak{z}_{\mathfrak{q}_p}(\mathfrak{h}) = \mathfrak{z}(\mathfrak{h}^a) \cap \mathfrak{q}$. But then $\mathfrak{z}_{\mathfrak{g}}(X)$ is θ-invariant and contains $\mathfrak{h}^a$, so Lemma 1.3.2 implies that $\mathfrak{h}^a = \mathfrak{z}_{\mathfrak{g}}(X)$. Note that the spectrum of $\mathrm{ad}(X)$ is real and pick the largest eigenvalue r of X. Then $-r$ is also an eigenvalue and $\mathfrak{g}(-r, X) + \mathfrak{h}^a + \mathfrak{g}(r, X)$ is a θ-invariant subalgebra strictly containing $\mathfrak{h}^a$, hence equal to $\mathfrak{g}$, again by Lemma 1.3.2. Thus there are no more eigenvalues of $\mathrm{ad}(X)$ than $-r, 0, r$ and this implies the claim.

2) If $\mathfrak{b}$ is abelian and contains X^0, then $\mathfrak{b} \subset \mathfrak{z}_{\mathfrak{g}}(X^0) = \mathfrak{h}^a$. Thus $\mathfrak{b} \subset \mathfrak{q}_p$.

3) If $\mathfrak{a}$ is maximal abelian in $\mathfrak{q}_p$, then $X^0 \in \mathfrak{a}$ is abelian since $X^0 \in \mathfrak{z}(\mathfrak{h}^a)$. Let $\mathfrak{b}$ be maximal abelian in $\mathfrak{p}$ containing $\mathfrak{a}$. Then 2) implies $\mathfrak{b} \subset \mathfrak{q}_p$, so $\mathfrak{b} = \mathfrak{a}$. That $\mathfrak{a}$ is maximal abelian in $\mathfrak{q}$ follows from $\mathfrak{z}_{\mathfrak{q}}(X^0) = \mathfrak{q}_p$.

4) This again follows from $\mathfrak{z}_{\mathfrak{q}}(X^0) = \mathfrak{q}_p$.

5) We have seen already that $\mathfrak{h}^a$ is the centralizer of X^0. Therefore it only remains to show that $\dim(\mathfrak{z}(\mathfrak{h}^a)) \leq 1$. But that follows from Lemma 1.3.10 applied to $(\mathfrak{g}, \mathfrak{h}^a)$. □

The analog of Proposition 3.1.11 for a compactly causal symmetric pair follows via c-duality. We only record the first part, which will be used later.

Proposition 3.1.12 *Suppose that $(\mathfrak{g}, \tau)$ is compactly causal. Then there exists an, up to sign unique, element $Z^0 \in \mathfrak{q}_k$ such that* $\operatorname{spec}(\operatorname{ad} Z^0) = \{-i, 0, i\}$ *and the centralizer of Z^0 in $\mathfrak{g}$ is $\mathfrak{k}$.* □

Example 3.1.13 Recall the $\mathrm{SL}(2, \mathbb{R})$ Example 1.3.12 for which one has $G^\tau \cap G^\theta = \{\pm 1\}$ so that $\mathfrak{q}^{G^\tau \cap G^\theta} = \mathfrak{q}$ is two-dimensional. This shows that the one-sheeted hyperboloid is of Cayley type. Note that $\mathfrak{a} := \mathfrak{q}_p$ is abelian itself. The corresponding cone C_p is $\mathbb{R}^+(X^0 + Z^0) + \mathbb{R}^+(X^0 - Z^0)$. □

The following proposition gives some useful isomorphisms between dual spaces of causal symmetric pairs.

Proposition 3.1.14 *1) Let $(\mathfrak{g}, \tau)$ be a compactly causal symmetric pair. Fix $Z^0 \in \mathfrak{z}(\mathfrak{k})_q$ such that $\operatorname{ad}_\mathfrak{p} Z^0$ is a complex structure on $\mathfrak{p}$. Let $\psi_k = \varphi_{Z^0}$ (cf.* Lemma 1.2.1*). Then the following hold:*

a) $\psi_k^2 = \theta$.

b) $\psi_k^{-1} = \psi_k \circ \theta = \theta \circ \psi_k$.

c) $\tau \circ \psi_k = \psi_k \circ \tau^a$.

d) $\psi_k : (\mathfrak{g}, \tau, \theta) \to (\mathfrak{g}, \tau^a, \theta)$ *is an isomorphism.*

2) Let $(\mathfrak{g}, \tau)$ be a noncompactly causal symmetric pair. Fix $X^0 \in \mathfrak{z}_{\mathfrak{q}_p}(\mathfrak{h}_k)$ such that $\mathfrak{g} = \mathfrak{g}(0, X^0) \oplus \mathfrak{g}(+1, X^0) \oplus \mathfrak{g}(-1, X^0)$. Let $\psi_p = \varphi_{iX^0}$. Then the following hold:

a) $\psi_p^2 = \tau^a$.

b) $\tau \circ \psi_p = \psi_p \circ \theta$.

c) ψ_p *defines an isomorphism* $\psi_p : (\mathfrak{g}, \theta, \tau) \to (\mathfrak{g}, \tau, \theta)^r$.

3) Let $(\mathfrak{g}, \tau)$ be a symmetric pair of Cayley type. Fix an element $Y^0 \in \mathfrak{z}(\mathfrak{h})$ such that $\mathfrak{g} = \mathfrak{g}(0, Y^0) \oplus \mathfrak{g}(+1, Y^0) \oplus \mathfrak{g}(-1, Y^0)$. Define $\psi_c = \varphi_{iY^0}$. Then the following hold:

a) $\psi_c^2 = \tau$.

b) $\psi_c \circ \tau = \tau \circ \psi_c$,

c) $\psi_c \circ \theta = \tau^a \circ \psi_c$.

d) ψ_c *defines an isomorphism* $\psi_c : (\mathfrak{g}, \tau, \theta) \to (\mathfrak{g}^c, \tau, \tau^a)$.

Proof: 1.a) follows from Lemma 1.2.1.

1.b) and 1.c): By 1.a), we have $\psi^4 = \mathrm{id}$. Thus $\varphi_{-Z^0} = \psi^{-1} = \psi^3 = \psi \circ \theta = \theta \circ \psi$. As $\tau Z^0 = -Z^0$, it follows that $\tau \circ \psi = \psi^{-1} \circ \tau$. This implies 1.b) and 1.c).

1.d): This is an immediate consequence of 1.c).

Parts 2) and 3) can be proved in the same way as Part 1). □

Given a causal symmetric pair, it is not clear which, if any, symmetric space $\mathcal{M}$ associated to $(\mathfrak{g}, \tau)$ is causal (cf. Section 2.5). With the structure theory just established, we are in a position to clarify the situation. The following proposition shows that compactly causal symmetric pairs do not pose any problems in this respect.

Proposition 3.1.15 *If* $(\mathfrak{g}, \tau)$ *is compactly causal, then every symmetric space associated to* $(\mathfrak{g}, \tau)$ *is causal.*

Proof: Let $\mathcal{M} = G/H$ be a symmetric space associated to $(\mathfrak{g}, \tau)$. Choose $Z^0 \in \mathfrak{z}(\mathfrak{k}) \cap \mathfrak{q}$ (cf. Proposition 3.1.12). Then $\mathrm{Ad}(k)Z^0 = Z^0$ for every $k \in K$. In particular, $Z^0 = \mathfrak{q}^{H \cap K}$, which proves the claim in view of Theorem 3.1.3. □

Note here that Proposition 3.1.15 does not say that *any* H_o-invariant cone in $\mathfrak{q}$ is H-invariant, i.e., gives a causal structure on $\mathcal{M}$.

As we have seen before (cf. e.g. Theorem 3.1.5), in the noncompactly causal case the existence of causal structures is related to the nature of the component group H/H_o of H. The right concept to study in our context is that of *essential connectedness.*

Definition 3.1.16 Let $\mathcal{M} = G/H$ be a non-Riemannian semisimple symmetric space and $(\mathfrak{g}, \tau)$ corresponding symmetric pair. Further, let $\mathfrak{a}$ be a maximal abelian subalgebra in $\mathfrak{q}_p$. Then H is called *essentially connected* in G if

$$H = Z_{K \cap H}(\mathfrak{a}) H_o. \qquad \square$$

We note that this definition is independent of the choice of $\mathfrak{a}$, since the maximal abelian subspaces in $\mathfrak{q}_p$ are conjugate under $H_o \cap K$.

Remark 3.1.17 Let $(\mathfrak{g}, \tau)$ be a noncompactly causal symmetric pair. We fix a cone generating element $X^0 \in \mathfrak{q}_p$ and a maximal abelian subspace $\mathfrak{a}$ of $\mathfrak{q}_p$ containing X^0. Then $\mathfrak{a}$ is a maximal abelian subspace of $\mathfrak{p}$ and we denote the set $\Delta(\mathfrak{g}, \mathfrak{a})$ of restricted roots of $\mathfrak{g}$ w.r.t. $\mathfrak{a}$ by Δ. Further, we set

$$\Delta_0 = \{\alpha \in \Delta \mid \alpha(X^0) = 0\}. \qquad (3.1)$$

Since $\mathfrak{h}^a$ is the centralizer of X^0, we get

$$\Delta_0 = \Delta(\mathfrak{h}^a, \mathfrak{a}). \tag{3.2}$$

Let

$$\Delta_+ := \{\alpha \in \Delta \mid \alpha(X^0) = 1\} \quad \text{and} \quad \Delta_- := \{\alpha \in \Delta \mid \alpha(X^0) = -1\}. \tag{3.3}$$

Choose a positive system Δ_0^+ in Δ_0. Then a positive system Δ^+ for Δ can be defined via

$$\Delta^+ := \Delta_+ \cup \Delta_0^+ . \tag{3.4}$$

Set

$$\mathfrak{n}_\pm = \sum_{\alpha(X^0)=\pm 1} \mathfrak{g}_\alpha, \quad \mathfrak{n}_0 = \sum_{\alpha\in\Delta_0^+} \mathfrak{g}_\alpha .$$

Then $\mathfrak{n} = \mathfrak{n}_+ + \mathfrak{n}_0$, $[\mathfrak{n}_+, \mathfrak{n}_+] = \{0\}$,$[\mathfrak{n}_-, \mathfrak{n}_-] = \{0\}$ and$[\mathfrak{h}^a, \mathfrak{n}_\pm] \subset \mathfrak{n}_\pm$. □

Theorem 3.1.18 *Let $\mathcal{M} = G/H$ be a symmetric space such that the corresponding symmetric pair $(\mathfrak{g}, \tau)$ is noncompactly causal. Then $\mathcal{M}$ is noncompactly causal if and only if H is essentially connected in G.*

Proof: Choose a cone-generating element $X^0 \in \mathfrak{q}_p$. Then X^0 centralizes $\mathfrak{h}_k$ and hence is contained in $\mathfrak{q}_p^{H_o\cap K}$. Next we choose a maximal abelian subspace $\mathfrak{a}$ of $\mathfrak{p}$ containing X^0. If H is essentially connected in G, then obviously $X^0 \in \mathfrak{q}^{H\cap K}$ and G/H is noncompactly causal by Theorem 3.1.3.

Assume conversely that G/H is noncompactly causal. Then, in view of Theorem 1.3.11, Theorem 3.1.5 implies that $\mathfrak{q}_p^{H\cap K} = \mathfrak{q}_{\mathfrak{p}}^{H_o\cap K} = \mathbb{R} X^0$. Let $\mathfrak{a}$, Δ, and Δ^+ be as in Remark 3.1.17 and fix some $k \in K \cap H$. Then $\mathrm{Ad}(k)\mathfrak{a}$ is a maximal abelian subalgebra in $\mathfrak{q}_p$. Since all such algebras are $H_o \cap K$-conjugate, we can find an $h \in H_o \cap K$ such that $\mathrm{Ad}(hk)$ normalizes $\mathfrak{a}$. But k and hk are contained in the same connected component of H, so we may as well assume that $\mathrm{Ad}(k)$ normalizes $\mathfrak{a}$. Since $H \cap K$ fixes X^0, it leaves Δ_0 invariant. Therefore $k \cdot \Delta_0^+$ is again a positive system in Δ_0 and we can find a $k_o \in H_o \cap K$ such that $k_o(k\Delta_0^+) = \Delta_0^+$. But then Δ^+ is invariant under $k_o k$, so that $k_o k \in M \cap H = Z_{H\cap K}(\mathfrak{a})$. Thus $k \in Z_{H\cap K}(\mathfrak{a})H_o$ and H is essentially connected. □

We use Theorem 3.1.18 to show that the symmetric space $\check{\mathcal{M}} = G/H$ is noncompactly causal if the corresponding symmetric pair $(\mathfrak{g}, \tau)$ is noncompactly causal. To do this we need one more lemma.

Lemma 3.1.19 *Let $(\mathfrak{g}, \tau)$ be a noncompactly causal symmetric pair and $G_{\mathbb{C}}$ be a simply connected complex Lie group with Lie algebra $\mathfrak{g}_{\mathbb{C}}$. Choose a cone-generating element $X^0 \in \mathfrak{q}_p$. Then*

1) $Z_{G_{\mathbb{C}}}(X^0)$ *is connected and equal to* $G_{\mathbb{C}}^{\tau\theta}$.

2) $Z_G(X^0) = G^{\tau^a}$.

3) $Z_K(X^0) = K^\tau$, *where* $K = G^\theta$.

Proof: 1) Let $\varphi = \tau_{iX^0}$ (cf. Lemma 1.2.1). Then φ defines an involution on $G_{\mathbb{C}}$ as $G_{\mathbb{C}}$ is simply connected. Obviously $Z_{G_{\mathbb{C}}}(X^0) \subset G_{\mathbb{C}}^\varphi$ and both groups have the same Lie algebra $\mathfrak{h}_{\mathbb{C}}^a = (\mathfrak{h}_k)_{\mathbb{C}} \oplus (\mathfrak{q}_p)_{\mathbb{C}}$. This shows that $\varphi = \tau^a = \tau\theta$. By Theorem 1.1.11, $G_{\mathbb{C}}^\varphi$ is connected. Hence we have $Z_{G_{\mathbb{C}}}(X^0) = G_{\mathbb{C}}^\varphi = G_{\mathbb{C}}^{\tau\theta}$.
2) $Z_G(X^0) = G_{\mathbb{C}}^{\tau\theta} \cap G = G^{\tau^a}$ because of 1).
3) $Z_K(X^0) = G^{\tau^a} \cap K = K^\tau$ because of 2) and $\theta|_K = \mathrm{id}$. □

Theorem 3.1.20 *Let* $(\mathfrak{g}, \tau)$ *be a noncompactly causal symmetric pair and* $G_{\mathbb{C}}$ *be a simply connected complex Lie group with Lie algebra* $\mathfrak{g}_{\mathbb{C}}$. *Further, let* G *be the analytic subgroup of* $G_{\mathbb{C}}$ *with Lie algebra* $\mathfrak{g}$ *and* H *a subgroup of* G^τ *containing* G_o^τ. *Then* H *is essentially connected and* $\mathcal{M} = G/H$ *is a noncompactly causal symmetric space. In particular* $\check{\mathcal{M}} = G/G^\tau$ *is noncompactly causal.*

Proof: Fix a cone-generating element $X^0 \in \mathfrak{q}_p$. Then Lemma 3.1.19 shows that $X^0 \in \mathfrak{q}_p^{G^\tau \cap K}$. Now Theorem 3.1.3 and Theorem 3.1.5 imply that $\mathcal{M} = G/G^\tau$ is noncompactly causal. Therefore Theorem 3.1.18 shows that G^τ is essentially connected. But then all open subgroups of G^τ are essentially connected as well, so that the converse direction of Theorem 3.1.18 proves the claim. □

Remark 3.1.21 Theorem 3.1.20 shows that in the situation of Section 2.6, i.e., for spaces related to tube domains, the assumptions made to ensure the H-invariance of the various H_o-invariant cones are automatically satisfied. In fact, one can choose $X^0 = X_o$ in that context, so one sees that $X_o \in \mathfrak{q}_p^{H \cap K}$. □

The results of Lemma 3.1.19 can be substantially extended.

Lemma 3.1.22 *Let* $(\mathfrak{g}, \tau)$ *be a noncompactly causal symmetric pair and* $G_{\mathbb{C}}$ *be a simply connected complex Lie group with Lie algebra* $\mathfrak{g}_{\mathbb{C}}$. *Further, let* G *be the analytic subgroup of* $G_{\mathbb{C}}$ *with Lie algebra* $\mathfrak{g}$ *and* $K = G^\theta$. *If* $\mathfrak{a} \subset \mathfrak{q}_p$ *is a maximal abelian subspace, then*

1) $G^{\tau^a} = M(G^{\tau^a})_o$, *where* $M = Z_K(\mathfrak{a})$.

2) $M = Z_{G^\tau}(\mathfrak{a}) = Z_{K^\tau}(\mathfrak{a})$.

3) Let $F = K \cap \exp i\mathfrak{a}$. *Then* $F = \{g \in \exp i\mathfrak{a} \mid g^2 = 1\} \subset M$, $K^\tau = F(K^\tau)_o$, *and* $G^\tau = F(G^\tau)_o$.

4) $H^a = F(H^a)_o = F(G^{\tau^a})_o = G^{\tau^a}$, *where* $H^a = (H \cap K)\exp \mathfrak{q}_p$ *(cf. (1.11))*

Proof: 1) Let $X^0 \in \mathfrak{a} \subset \mathfrak{q}_p$ be as in Proposition 3.1.11. Then Lemma 3.1.19 implies $M \subset G^{\tau^a}$, since $X^0 \in \mathfrak{a}$. Conversely, recall from Theorem 3.1.20 that G^τ is essentially connected, so that

$$G^{\tau^a} \cap K = G^{\tau\theta} \cap G^\theta \subset [Z_{K\cap G^\tau}(\mathfrak{a})(G^\tau)_o] \cap G^\theta \subset M[G^{\tau^a}]_o.$$

2) According to Lemma 3.1.19, we have

$$Z_{G^\tau}(\mathfrak{a}) \subset Z_{G^\tau}(X^0) \subset G^\tau \cap G_{\mathbb{C}}^{\tau\theta} \subset G^\tau \cap G^\theta \subset K$$

which implies $Z_{G^\tau}(\mathfrak{a}) \subset M$. Conversely, $M \subset G^{\tau^a} \cap G^\theta = G^\tau \cap K$ by 1), so $M \subset Z_{G^\tau}(\mathfrak{a})$.

3) We recall that the involutions $\tau, \sigma, \tau\sigma$, and θ_u on $\mathfrak{g}_{\mathbb{C}}$ with fixed point algebras $\mathfrak{h}_{\mathbb{C}}, \mathfrak{g}, \mathfrak{g}^c$ and $\mathfrak{k} + i\mathfrak{p}$, respectively, all integrate to involutions on $G_{\mathbb{C}}$ and have connected sets of fixed points $H_{\mathbb{C}}, G, G^c$ and U in $G_{\mathbb{C}}$ (cf. Theorem 1.1.11). The involution θ_u induces Cartan involutions on G, G^c, and $G_{\mathbb{C}}$. Let K^c be the corresponding maximal compact subgroup of G^c. Then

$$(K^c)^\tau = (U^{\sigma\tau})^\tau = U^\sigma \cap U^\tau = K^\tau.$$

Now assume that $k \in K \cap \exp i\mathfrak{a}$. Then $k = \sigma(k) = k^{-1}$, so that $k^2 = 1$. Conversely, if $k \in \exp i\mathfrak{a}$ and $k^2 = 1$, we have $\sigma(k) = k^{-1} = k$, i.e., $k \in G$. But we also have $\theta_u(k) = k$, whence $k \in G \cap U = K$.

Note that $\tau(k) = k^{-1} = k$ implies that $F \subset K^\tau$. It is clear that $F \subset Z_K(\mathfrak{a})$, so it only remains to show that $G^\tau \subset F(G^\tau)_0$. To this end we fix $h \in G^\tau$ and write it as

$$h = k \exp X \in K \exp \mathfrak{p}.$$

Now the τ-invariance of the Cartan decomposition shows that $k \in K^\tau = (K^c)^\tau$ and $X \in \mathfrak{h}_p$. Note that $(\mathfrak{k}^c, \tau)$ is a compact Riemannian symmetric Lie algebra and $i\mathfrak{a}$ is a maximal abelian subspace in $\mathfrak{k}^c \cap \mathfrak{q}_{\mathbb{C}} = i(\mathfrak{q}_p)$. According to [44], Chapter 7, Theorem 8.6, we have $K^c = (K^\tau)_o(\exp i\mathfrak{a})(K^\tau)_o$, so we can write $k = lal'$ with $a \in \exp i\mathfrak{a}$ and $l, l' \in (K^\tau)_o$. Applying τ, this yields $lal' = la^{-1}l'$ and thus $a = a^{-1} \in F$. Now the claim follows from

$$k \in (K^\tau)_o F (K^\tau)_o (K^\tau)_o \subset F(K^\tau)_o.$$

4) Lemma 3.1.19 implies $H^a = K^\tau Z_G(X^0)_o$, so 3) proves the first two equalities. For the last equality, note that 3) implies $M = FM_o$ and hence $M \subset H^a$. □

3.2 The Classification of Causal Symmetric Pairs

In this section we give a classification of the causal symmetric pairs $(\mathfrak{g}, \tau)$. Remark 3.1.9 shows that in order to do that it suffices to classify the noncompactly causal symmetric pairs and then apply c-duality to find the compactly causal symmetric pairs. The same remark also shows that we may assume $\mathfrak{g}$ to be a simple Lie algebra.

Let $\mathfrak{g}$ be a noncompact simple Lie algebra with Cartan involution θ and corresponding Cartan decomposition $\mathfrak{g} = \mathfrak{k} + \mathfrak{p}$. As was noted in Remark 1.1.15, to each involution τ on $\mathfrak{g}$ there exists a Cartan involution θ_1 on $\mathfrak{g}$ commuting with τ. Let $\mathfrak{g} = \mathfrak{k}_1 + \mathfrak{p}_1$ be the Cartan decomposition belonging to θ_1. According to [44], p. 183, there exists a $\varphi \in \mathrm{Aut}(\mathfrak{g})_o$ such that $\varphi(\mathfrak{k}) = \mathfrak{k}_1$ and $\varphi(\mathfrak{p}) = \mathfrak{p}_1$. But then

$$\theta = \varphi^{-1} \circ \theta_1 \circ \varphi,$$

and $\varphi^{-1} \circ \tau \circ \varphi$ commutes with θ. Thus, in order to classify the causal symmetric pairs $(\mathfrak{g}, \tau)$ up to isomorphism, it suffices to classify those causal involutions on $\mathfrak{g}$ that commute with the fixed Cartan involution θ.

Proposition 3.2.1 *Let $\mathfrak{g}$ be a simple Lie algebra with Cartan involution θ and $\tau: \mathfrak{g} \to \mathfrak{g}$ be an involution commuting with θ. If $(\mathfrak{g}, \tau)$ is irreducible, then the following statements are equivalent:*

1) $(\mathfrak{g}, \tau)$ is noncompactly causal.

2) There exists an $X \in \mathfrak{q}_p$ such that

$$\mathfrak{g} = \mathfrak{g}(0, X) \oplus \mathfrak{g}(+1, X) \oplus \mathfrak{g}(-1, X)$$

and $\tau = \theta\tau_{iX}$ (cf. Lemma 1.2.1).

Proof: 1) $\Rightarrow$ 2) is an immediate consequence of Proposition 3.1.11 and Theorem 3.1.14. For the converse, note that comparing the eigenspaces of $\mathrm{ad}\, X$ and τ_{iX}, condition (2) implies that $X \in \mathfrak{z}_{\mathfrak{q}_p}(\mathfrak{h}_k)$ and hence $X \in \mathfrak{q}_p^{H_o \cap K}$ for any symmetric $\mathcal{M} = G/H$ associated to $(\mathfrak{g}, \tau)$. But then Theorem 3.1.3 and Theorem 3.1.5 imply that $\tilde{\mathcal{M}}$ is noncompactly causal, and this proves the claim. □

Proposition 3.2.2 *Let $(\mathfrak{g}, \tau)$ be a noncompactly causal symmetric pair with $\tau = \theta\tau_{iX}$, where $X \in \mathfrak{q}_p$ such that*

$$\mathfrak{g} = \mathfrak{g}(0, X) \oplus \mathfrak{g}(+1, X) \oplus \mathfrak{g}(-1, X).$$

Further, let $\mathfrak{a}$ be a maximal abelian subspace of $\mathfrak{p}$ containing X and $\Delta = \Delta(\mathfrak{g}, \mathfrak{a})$ the corresponding set of restricted roots. Pick a system Δ^+ of positive roots in Δ such that $\alpha(X) \geq 0$ for all $\alpha \in \Delta^+$ and denote the corresponding set of simple roots by Σ. Then there exists a unique $\alpha_X \in \Sigma$ with $\alpha_X(X) = 1$. In particular, α_X determines X completely.

Proof: Let $\Delta_0 = \{\alpha \in \Delta \mid \alpha(X) = 0\}$ and $\Delta_0^+ := \Delta_0 \cap \Delta^+$. Then we have $\Delta_0 = \Delta(\mathfrak{g}(0, X), \mathfrak{a}) = \Delta(\mathfrak{h}^a, \mathfrak{a})$ and

$$\Delta^+ = \{\alpha \in \Delta \mid \alpha(X) = 1\} \cup \Delta_0^+.$$

Consider the set $\Sigma_0 \subset \Delta_0$ of simple roots for Δ_0^+. We claim that

$$\Sigma_0 \subset \Sigma \quad \text{and} \quad \#(\Sigma \setminus \Sigma_0) = 1. \tag{3.5}$$

In fact, let $\alpha \in \Delta_0^+$. Assume that $\alpha = \beta + \gamma$ with $\beta, \gamma \in \Delta^+$. Then $\alpha(X) = \beta(X) + \gamma(X) = 0$. As $\beta(X) \geq 0$ and $\gamma(X) \geq 0$, this implies $\beta(X) = \gamma(X) = 0$ and hence $\beta, \gamma \in \Delta_0^+$. Thus we have $\Sigma_0 \subset \Sigma$. Proposition 3.1.11 shows that $\mathbb{R} X = \mathfrak{z}(\mathfrak{h}^a)$ and

$$\dim(\mathfrak{a} \cap [\mathfrak{h}^a, \mathfrak{h}^a]) = \dim(\mathfrak{a}) - 1.$$

But $\mathfrak{h}^a = \mathfrak{g}(0, X)$, so $\dim(\mathfrak{a}) = \#(\Sigma)$ and $\dim(\mathfrak{a} \cap [\mathfrak{h}^a, \mathfrak{h}^a]) = \#(\Sigma_0)$. This proves (3.5) and the claim follows if we let α_X be the only root in Σ which is not contained in Σ_0. □

Remark 3.2.3 Let $\mathfrak{g}$ be a noncompact simple Lie algebra with Cartan involution θ. Consider a maximal abelian subspace $\mathfrak{a}$ of $\mathfrak{p}$ and $\Delta = \Delta(\mathfrak{g}, \mathfrak{a})$, the corresponding set of restricted roots. Pick a system Δ^+ of positive roots in Δ and denote the corresponding set of simple roots by Σ. Let δ be the highest root of Δ^+ (cf. [44], p. 475) and denote by $d(\alpha)$ the multiplicity of $\alpha \in \Delta^+$ in δ. This means that $\delta = \sum_{\alpha \in \Sigma} d(\alpha)\alpha$. Given $\alpha \in \Sigma$ with $d(\alpha) = 1$, define $X(\alpha) \in \mathfrak{a}$ via

$$\beta(X(\alpha)) = \begin{cases} 1 & \text{for } \beta = \alpha \\ 0 & \text{otherwise.} \end{cases}$$

Suppose that $\alpha \in \Sigma$ and $d(\alpha) = 1$. We claim that

$$\mathfrak{g} = \mathfrak{g}(0, X(\alpha)) \oplus \mathfrak{g}(-1, X(\alpha)) \oplus \mathfrak{g}(+1, X(\alpha)).$$

In fact, if $\gamma \in \Delta$, then $\gamma = \sum_{\beta \in \Sigma} m_\beta(\gamma)\beta$ with $m_\beta(\gamma) \in \mathbb{Z}$, and $\gamma(X(\alpha)) = m_\alpha(\gamma)$. But $d(\alpha) = m_\alpha(\delta) = 1$ and $|m_\alpha(\gamma)| \leq d(\alpha)$, since δ is the highest root. There we have $m_\alpha(\gamma) = 0, 1$, or -1, and this implies the claim. Now we can apply Proposition 3.2.2 to $X(\alpha)$ and obtain

$$\alpha = \alpha_{X(\alpha)}.$$

We denote the involution on $\mathfrak{g}$ obtained from $X(\alpha)$ via Proposition 3.2.1 by $\tau(\alpha)$ and its algebra of fixed points by $\mathfrak{h}(\alpha)$. □

Theorem 3.2.4 *Let $\mathfrak{g}$ be a noncompact simple Lie algebra with Cartan involution θ. Let $\mathfrak{a}$ be a maximal abelian subspace of $\mathfrak{p}$ and $\Delta = \Delta(\mathfrak{g}, \mathfrak{a})$ the corresponding set of restricted roots. Pick a system Δ^+ of positive roots in Δ and denote the corresponding set of simple roots by Σ. Then the following statements are equivalent:*

1) *There exists an involution $\tau: \mathfrak{g} \to \mathfrak{g}$ commuting with θ such that $(\mathfrak{g}, \tau)$ is noncompactly causal.*

2) *There exists an element $X \in \mathfrak{p}$, $X \neq 0$ such that*

$$\mathfrak{g} = \mathfrak{g}(-1, X) \oplus \mathfrak{g}(0, X) \oplus \mathfrak{g}(+1, X).$$

3) *Δ is a reduced root system and there exists an $\alpha \in \Sigma$ such that the multiplicity $d(\alpha)$ of α in the highest root $\delta \in \Delta$ is 1.*

Proof: 1) $\Rightarrow$ 2): This follows from Proposition 3.1.11.

2) $\Rightarrow$ 1): Given $X \in \mathfrak{p}$ as in 2), we apply Lemma 1.2.1 to $iX \in i\mathfrak{p}$ and find that τ_{iX} is an involution on $\mathfrak{g}$ commuting with θ. Then $\tau := \theta\tau_{iX}$ also commutes with θ and $\tau(X) = -X$. Moreover, $(\mathfrak{g}, \tau)$ is irreducible since $\mathfrak{g}$ is simple. Thus we can apply Proposition 3.2.1 and conclude that $(\mathfrak{g}, \tau)$ is non-compactly causal.

2) $\Rightarrow$ 3): Conjugating by an element of K, we may assume that $X \in \mathfrak{a}$. If Δ is nonreduced, then [44], Theorem 3.25, p. 475, says that Δ is of type $(\mathfrak{bc})_n$, i.e., of the form

$$\Delta(\mathfrak{g}, \mathfrak{a}) = \pm\{\tfrac{1}{2}\alpha_j, \alpha_j, \tfrac{1}{2}(\alpha_i \pm \alpha_k) \mid 1 \leq i, j, k \leq r; i < k\}$$

(cf. also Moore's Theorem A.3.4). But this contradicts the fact that

$$\operatorname{spec}(\operatorname{ad} X) = \{-1, 0, 1\}.$$

Hence Δ is reduced.

Now let $\alpha_X \in \Sigma$ be the element determined by Proposition 3.2.2. Then $\alpha_X(X) = 1$ and $\delta(X) = d(\alpha_X)$. As $\delta(X) \in \{-1, 0, 1\}$, it follows that $d(\alpha_X) = 1$. Thus 3) follows.

3) $\Rightarrow$ 2): This follows from Remark 3.2.3. □

Remark 3.2.5 The maximal abelian subspaces of $\mathfrak{p}$ are conjugate under K and the positive systems for $\Delta = \Delta(\mathfrak{g}, \mathfrak{a})$ are conjugate under the normalizer $N_K(\mathfrak{a})$ of $\mathfrak{a}$ in K. Therefore Theorem 3.2.4, Propositions 3.2.1 and 3.2.2, and Remark 3.2.3 show that the following procedure gives a complete list of representatives $(\mathfrak{g}, \tau)$ for the isomorphy classes of noncompactly causal symmetric spaces. Some of the symmetric pairs obtained will be isomorphic due to outer automorphisms of the relevant diagrams.

Step 1: List the simple noncompact real Lie algebras $\mathfrak{g}$ together with the Dynkin diagrams of the restricted root systems and the multiplicies of the highest root δ.

Step 2: Given a simple root α whose multiplicity in δ is 1, construct the symmetric pair $(\mathfrak{g}, \tau(a))$.

Note that after removing α from the set Σ of simple roots one obtains the Dynkin diagram for the restricted roots of the commutator algebra of $\mathfrak{h}(\alpha)^a$ and the corresponding set of simple roots is $\Sigma_0 = \Sigma \setminus \{\alpha\}$. If one wants to read of $\mathfrak{h}(\alpha)^a$ and $\mathfrak{h}(\alpha)$ from diagrams directly, one has to use the full Satake diagram instead of the Dynkin diagram of the restricted root system. □

Example 3.2.6 We show how the procedure of Remark 3.2.5 works in the case that $\mathfrak{g}$ has a complex structure.

The type $\mathbf{A}_n$: $(\mathfrak{sl}(n+1, \mathbb{C}))$

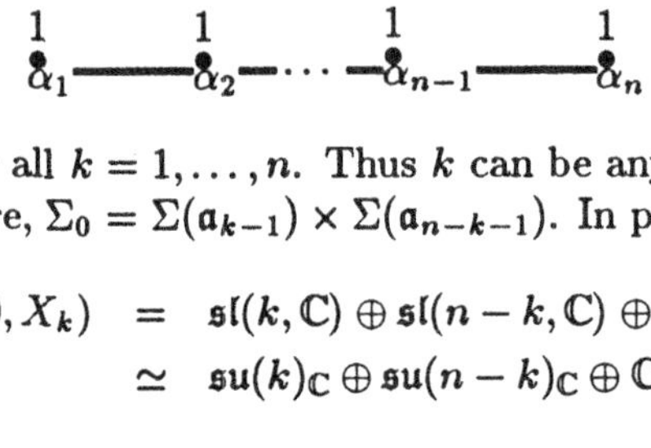

Here $d(\alpha_k) = 1$ for all $k = 1, \ldots, n$. Thus k can be any number between 1 and n. Furthermore, $\Sigma_0 = \Sigma(\mathfrak{a}_{k-1}) \times \Sigma(\mathfrak{a}_{n-k-1})$. In particular,

$$\begin{aligned} \mathfrak{g}_{\mathbb{C}}(0, X_k) &= \mathfrak{sl}(k, \mathbb{C}) \oplus \mathfrak{sl}(n-k, \mathbb{C}) \oplus \mathbb{C}X_k \\ &\simeq \mathfrak{su}(k)_{\mathbb{C}} \oplus \mathfrak{su}(n-k)_{\mathbb{C}} \oplus \mathbb{C}X_k\,. \end{aligned}$$

Hence $\mathfrak{h} = \mathfrak{su}(k, n-k)$.

The type $\mathbf{B}_n$: $(\mathfrak{so}(2n+1, \mathbb{C}))$

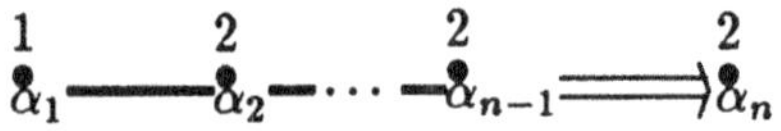

In this case $k = 1$. Then $\Sigma_0 = \Sigma(\mathfrak{b}_{n-1})$ and

$$\begin{aligned} \mathfrak{g}(0, X_1) &= \mathfrak{so}(2n-1, \mathbb{C}) \oplus \mathbb{C}X_1 \\ &\simeq \mathfrak{so}(2n-1)_{\mathbb{C}} \oplus \mathbb{C}X_1\,. \end{aligned}$$

Thus $\mathfrak{h} = \mathfrak{so}(2, 2n-1)$.

The type C_n: $(\mathfrak{sp}(n,\mathbb{C}))$

Hence $k = n$, $\Sigma_0 = \Sigma(\mathfrak{a}_{n-1})$ and

$$\begin{aligned}\mathfrak{g}(0, X_n) &= \mathfrak{sl}(n,\mathbb{C}) \oplus \mathbb{R}X_n \\ &\simeq \mathfrak{su}(n)_{\mathbb{C}} \oplus \mathbb{R}X_n\,.\end{aligned}$$

Thus $\mathfrak{h} = \mathfrak{sp}(n,\mathbb{R})$.

The type D_n: $(\mathfrak{so}(2n,\mathbb{C}))$

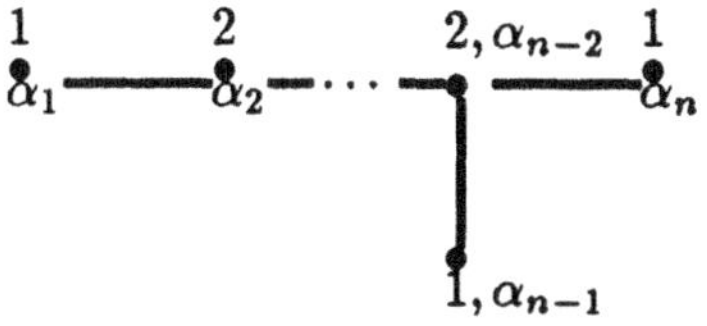

Here we can take $k = 1, n-1, n$. But $k = n-1$ and $k = n$ give isomorphic $\mathfrak{h}$'s. Thus we only have to look at $k = 1$ and $k = n$. For $k = 1$ we get Σ_0 of type $\mathfrak{d}_{n-1}$ and

$$\begin{aligned}\mathfrak{g}(0, X_1) &= \mathfrak{so}(2n-2,\mathbb{C}) \oplus \mathbb{C}X_1 \\ &\simeq \mathfrak{so}(2n-2)_{\mathbb{C}} \oplus \mathbb{C}X_1\,.\end{aligned}$$

Thus $\mathfrak{h} = \mathfrak{so}(2, 2n-2)$.

For $k = n$ we get $\Sigma_0 = \Sigma(\mathfrak{a}_{n-1})$ and

$$\begin{aligned}\mathfrak{g}(0, X_n) &= \mathfrak{sl}(n,\mathbb{C}) \oplus \mathbb{C}X_n \\ &\simeq \mathfrak{u}_{\mathbb{C}}\,.\end{aligned}$$

Thus $\mathfrak{h} = \mathfrak{so}^*(2n)$.

The type E_6:

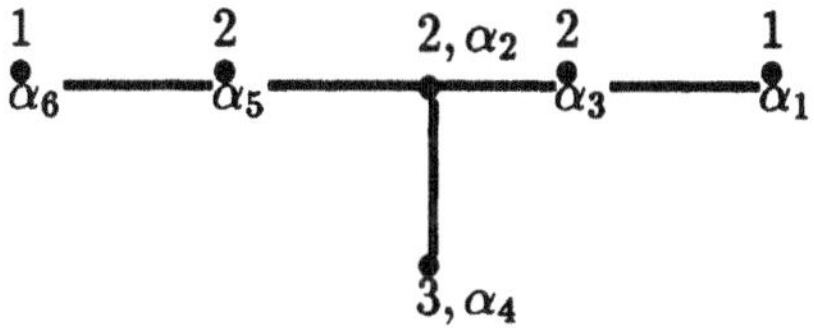

Here $k = 1, 6$. As both give isomorphic $\mathfrak{h}$, we may assume that $k = 1$. Then $\Sigma_0 = \Sigma(\mathfrak{d}_5)$. Thus

$$\begin{aligned}\mathfrak{g}(0, X_1) &= \mathfrak{so}(10,\mathbb{C}) \oplus \mathbb{C}X_1 \\ &\simeq \mathfrak{so}(10)_{\mathbb{C}} \oplus \mathbb{C}X_1\,.\end{aligned}$$

Hence $\mathfrak{h} = \mathfrak{e}_{6(-14)}$.

The type $\mathbf{E}_7$:

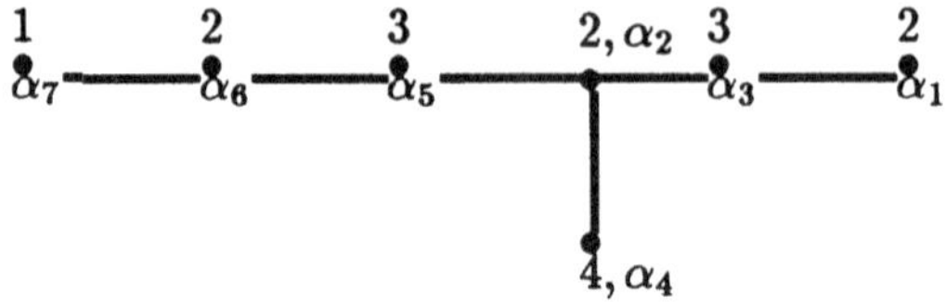

Thus $k = 1$, Σ_0 is of type $\mathfrak{e}_6$ and $\mathfrak{g}(0, X_7) = \mathfrak{e}_6 \oplus \mathbb{C}X_7$. Hence $\mathfrak{h} = \mathfrak{e}_{7(-25)}$. □

Example 3.2.7 To conclude this chapter we work out the real groups of type AI, AII, and $AIII$, i.e., $\mathrm{SL}(n, \mathbb{R})$, $\mathrm{SU}^*(2n)$, and $\mathrm{SU}(p, q)$. We fix a maximal abelian subalgebra $\mathfrak{a}$ of $\mathfrak{p}$ and denote a set of positive roots in $\Delta = \Delta(\mathfrak{g}, \mathfrak{a})$ by Δ^+. Let $\Sigma = \{\alpha_1, \ldots, \alpha_r\}$ be the set of simple restricted roots.

$\mathrm{SL}(n, \mathbb{R})$:. In this case Σ is of type A_{n-1}, so we can take out any α_j. It follows that $\mathfrak{h}^a$ is of type $A_{p-1} \times A_{q-1}$, with $p + q = n$. In particular,

$$\mathfrak{g}(0, X_k) = \mathfrak{sl}(p, \mathbb{R}) \times \mathfrak{sl}(q, \mathbb{R}) \times \mathbb{R}X_k .$$

Thus $\mathfrak{h}_k = \mathfrak{so}(p) \times \mathfrak{so}(q)$. There are thus two possibilities for $\mathfrak{h}$. Either $\mathfrak{h} = \mathfrak{g}(0, X_k)$ or $\mathfrak{h} = \mathfrak{so}(p, q)$. We can exclude the first case, as $\mathfrak{h} \simeq \mathfrak{h}^a$ is possible only for $n = 2$, in which case both $\mathfrak{a}$ and $\mathfrak{a}^a$ are one-dimensional and abelian.

$\mathrm{SU}^*(2n)$: In this case $\Sigma = \Sigma(\mathfrak{a}_{n-1})$ and $m_\lambda = \dim \mathfrak{g}_\lambda = 4$ for every $\lambda \in \Sigma$. Once again we can take out any one of the simple roots. It follows that $\Sigma_o = \Sigma(\mathfrak{a}_{p-1}) \times \Sigma(\mathfrak{a}_{q-1})$ and multiplicities equal 4. It follows that $\mathfrak{g}(0, X_k) = \mathfrak{su}^*(2p) \times \mathfrak{su}^*(2q) \times \mathbb{R}X_k$. Thus $\mathfrak{k}_k = \mathfrak{sp}(p) \times \mathfrak{sp}(q)$. We can exclude that $\mathfrak{h} \simeq \mathfrak{h}^a$, therefore $\mathfrak{h} = \mathfrak{sp}(p, q)$.

$\mathrm{SU}(p, q)$: The root system Δ is nonreduced if $p \neq q$. So the only possibility is $p = q$. In that case $\Sigma = \Sigma(\mathfrak{c}_n)$ and the multiplicities are $m_{\lambda_j} = 2$, $j < n$ and $m_{\lambda_n} = 1$. The only possibility is to take out γ_n, and we are left with $\Sigma(\mathfrak{a}_{n-1})$ and all multipicities equal 2. But then $\mathfrak{g}(0, X_n) = \mathfrak{sl}(n, \mathbb{C}) \times \mathbb{R}X_n$. This leaves us with $\mathfrak{h} = \mathfrak{sl}(n, \mathbb{C}) \times \mathbb{R}$ except in the case $n = 8$. In that case $\mathfrak{h} = \mathfrak{e}_{7(7)}$ would be another possibility. But $\mathrm{ad}\, X_k$ is an isomorphism $\mathfrak{h}_p \to \mathfrak{q}_k$, which shows that $\dim H/(H \cap K) = \dim K/(K \cap H)$. The dimension of $\mathfrak{e}_{7(7)}/\mathfrak{su}(8)$ is 70, which is bigger than $\dim \mathfrak{u}_n = \mathfrak{s}(\mathfrak{u}_n \times \mathfrak{u}_n)/\mathfrak{s}_n$. □

The procedure described in Remark 3.2.5 yields the following theorem.

Theorem 3.2.8 (The Causal Symmetric Pairs) *The irreducible semisimple causal symmetric spaces are given up to covering by the following symmetric pairs.*

$\mathfrak{g}$ with Complex Structure

$\mathfrak{g} = \mathfrak{h}_{\mathbb{C}}$ noncompactly causal	$\mathfrak{g}^c = \mathfrak{h} \times \mathfrak{h}$ compactly causal	$\mathfrak{h}$
$\mathfrak{sl}(p+q, \mathbb{C})$	$\mathfrak{su}(p,q) \times \mathfrak{su}(p,q)$	$\mathfrak{su}(p,q)$
$\mathfrak{so}(2n, \mathbb{C})$	$\mathfrak{so}^*(2n) \times \mathfrak{so}^*(2n)$	$\mathfrak{so}^*(2n)$
$\mathfrak{so}(n+2, \mathbb{C})$	$\mathfrak{so}(2,n) \times \mathfrak{so}(2,n)$	$\mathfrak{so}(2,n)$
$\mathfrak{sp}(n, \mathbb{C})$	$\mathfrak{sp}(n, \mathbb{R}) \times \mathfrak{sp}(n, \mathbb{R})$	$\mathfrak{sp}(n, \mathbb{R})$
$\mathfrak{e}_6$	$\mathfrak{e}_{6(-14)} \times \mathfrak{e}_{6(-14)}$	$\mathfrak{e}_{6(-14)}$
$\mathfrak{e}_7$	$\mathfrak{e}_{7(-25)} \times \mathfrak{e}_{7(-25)}$	$\mathfrak{e}_{7(-25)}$

$\mathfrak{g}$ without Complex Structure

$\mathfrak{g}$ noncompactly causal	$\mathfrak{g}^c$ compactly causal	$\mathfrak{h}$
$\mathfrak{sl}(p+q, \mathbb{R})$	$\mathfrak{su}(p,q)$	$\mathfrak{so}(p,q)$
$\mathfrak{su}(n,n)$	$\mathfrak{su}(n,n)$	$\mathfrak{sl}(n, \mathbb{C}) \times \mathbb{R}$
$\mathfrak{su}^*(2(p+q))$	$\mathfrak{su}(2p, 2q)$	$\mathfrak{sp}(p,q)$
$\mathfrak{so}(n,n)$	$\mathfrak{so}^*(2n)$	$\mathfrak{so}(n, \mathbb{C})$
$\mathfrak{so}^*(4n)$	$\mathfrak{so}^*(4n)$	$\mathfrak{su}^*(2n) \times \mathbb{R}$
$\mathfrak{so}(p+1, q+1)$	$\mathfrak{so}(2, p+q)$	$\mathfrak{so}(p,1) \times \mathfrak{so}(1,q)$
$\mathfrak{sp}(n, \mathbb{R})$	$\mathfrak{sp}(n, \mathbb{R})$	$\mathfrak{sl}(n, \mathbb{R}) \times \mathbb{R}$
$\mathfrak{sp}(n,n)$	$\mathfrak{sp}(2n, \mathbb{R})$	$\mathfrak{sp}(n, \mathbb{C})$
$\mathfrak{e}_{6(6)}$	$\mathfrak{e}_{6(-14)}$	$\mathfrak{sp}(2,2)$
$\mathfrak{e}_{6(-26)}$	$\mathfrak{e}_{6(-14)}$	$\mathfrak{f}_{4(-20)}$
$\mathfrak{e}_{7(-25)}$	$\mathfrak{e}_{7(-25)}$	$\mathfrak{e}_{6(-26)} \times \mathbb{R}$
$\mathfrak{e}_{7(7)}$	$\mathfrak{e}_{7(-25)}$	$\mathfrak{su}^*(8)$

Notes for Chapter 3

Most of the material in Section 3.1 appeared for the first time in [129, 130]. Definition 3.1.16 is due to E. van den Ban [1]. Theorem 3.1.18 can be found in [131, 136]. The idea of using a simple root with $d(\alpha) = 1$ for classifying the causal symmetric spaces was pointed out to us by S. Sahi, cf. [85]. There are other ways of classification. In [129] this was done by reducing it, as in Theorem 3.2.4, to the classification of para-Hermitian symmetric pairs. The para-Hermitian and para-Kähler spaces were introduced by Libermann in 1951-1952 [95], [96]. The para-Hermitian symmetric spaces were classified by S. Kaneyuki [75] by reducing it to the classification of graded Lie algebra of the first kind by S. Kobayashi and T. Nagano in [80] (see [77] for the general classification). A list may be found in [78], [111], and also [138]. Another method is to use Lemma 1.3.8 to reduce it to the classification of real forms of bounded symmetric domains. This was done by H. Jaffee using homological methods in the years 1975 [69] and 1978 [70]. Compactly causal symmetric spaces were also introduced by Matsumoto [103] via the root structure. These spaces were classified the same year by Doi in [21]. Later, B. Ørsted and one of the authors introduced the symmetric spaces of Hermitian type in [133] with applications to representation theory in mind. The connection with causal spaces was pointed out in [129, 130].

Chapter 4

Classification of Invariant Cones

Let $\mathcal{M} = G/H$ be causal symmetric space with H essentially connected. In this chapter we classify the H-invariant regular cones in $\mathfrak{q}$, i.e., all possible causal structures on $\mathcal{M}$. Because of c-duality, we can restrict ourselves to noncompactly causal symmetric spaces.

The crucial observation is that regular H-invariant cones in $\mathfrak{q}$ are completely determined by their intersections with a suitable Cartan subspace $\mathfrak{a}$. We give a complete description of the cones in $\mathfrak{a}$ which occur in this way. Further, we show how H-invariant cones can be reconstructed from the intersection with an $\mathfrak{a}$. An important fact in this context is that the intersection of a cone with $\mathfrak{a}$ is the same as the orthogonal projection onto $\mathfrak{a}$. In order to prove this, one needs a convexity theorem saying that for X in an appropriate maximal cone $c_{\max}$ in $\mathfrak{a}$, $\mathrm{pr}(\mathrm{Ad}(h)X) \in \mathrm{conv}(W_0 \cdot X) + c_{\max}^*$.

We also prove an extension theorem for H-invariant cones saying that these cones are all traces of G^c-invariant cones in $i\mathfrak{g}^c$.

An important basic tool in this chapter is $\mathfrak{sl}(2, \mathbb{R})$ reduction, which is compatible with the involution τ.

4.1 Symmetric SL(2, R) Reduction

Let $\mathcal{M} = G/H$ be a noncompactly causal symmetric space with involution τ. In this section we describe a version of the usual $\mathrm{SL}(2, \mathbb{R})$ reduction that commutes with the involutions on G and $\mathrm{SL}(2, \mathbb{R})$.

Recall the decomposition $\Delta = \Delta_- \cup \Delta_0 \cup \Delta_+$ associated to the choice of a cone-generating element $X^0 \in \mathfrak{q}_p$ and the corresponding nilpotent subalgebras $\mathfrak{n}_\pm$ from Remark 3.1.17.

Lemma 4.1.1 *Let $(\mathfrak{g}, \tau)$ be a noncompactly causal symmetric pair. Then we have $\tau = -\theta$ on $\mathfrak{n}_+ \oplus \mathfrak{n}_-$.*

Proof: From $\mathfrak{g}(0, X^0) = \mathfrak{h}_k \oplus \mathfrak{q}_p$ and $\mathrm{ad}(X^0)(\mathfrak{q}_k \oplus \mathfrak{h}_p) \subset \mathfrak{q}_k \oplus \mathfrak{h}_p$ we obtain

$$\mathfrak{n}_+ \oplus \mathfrak{n}_- = \mathfrak{q}_k \oplus \mathfrak{h}_p, \tag{4.1}$$

and this implies the claim. □

Recall that $\mathfrak{a} \subset \mathfrak{q}_p$ is maximal abelian in $\mathfrak{p}$. Therefore the Killing form and the inner product $(\cdot \mid \cdot) := (\cdot \mid \cdot)_\theta$ agree on $\mathfrak{a}$. We use this inner product to identify $\mathfrak{a}$ and $\mathfrak{a}^*$. This means that

$$B(X, \lambda) = (X \mid \lambda) = \lambda(X)$$

for all $X \in \mathfrak{a}$ and $\lambda \in \mathfrak{a}^*$. For $\lambda \neq 0$ we set

$$X^\lambda := \frac{\lambda}{|\lambda|^2} \in \mathfrak{a}, \tag{4.2}$$

where $|\cdot|$ denotes the norm corresponding to $(\cdot \mid \cdot)$. We obviously have $\lambda(X^\lambda) = 1$.

Lemma 4.1.2 *Let $\alpha \in \Delta$ and $X \in \mathfrak{g}_\alpha$. Then $[X, \theta(X)] = -|X|^2 \alpha$.*

Proof: Note first that $\theta(X) \in \mathfrak{g}_{-\alpha}$. Hence $[X, \theta(X)] \in \mathfrak{z}_\mathfrak{g}(\mathfrak{a}) \cap \mathfrak{p} = \mathfrak{a}$. Let $Y \in \mathfrak{a}$. Then

$$\begin{aligned} B(Y, [X, \theta(X)]) &= B([Y, X], \theta(X)) \\ &= \alpha(Y) B(X, \theta(X)) \\ &= -|X|^2 \alpha(Y) \\ &= B(Y, -|X|^2 \alpha), \end{aligned}$$

and the claim follows. □

Let $\alpha \in \Delta_+$ and choose $Y_\alpha \in \mathfrak{g}_\alpha$ such that

$$|Y_\alpha|^2 = \frac{2}{|\alpha|^2}. \tag{4.3}$$

We set $Y_{-\alpha} := \tau(Y_\alpha)$. By Lemma 4.1.1 we have $Y_{-\alpha} = \tau(Y_\alpha) = -\theta(Y_\alpha)$. Thus Lemma 4.1.2 implies

$$[Y_\alpha, Y_{-\alpha}] = 2X^\alpha. \tag{4.4}$$

Finally, we introduce

$$Y^\alpha := \frac{1}{2}(Y_\alpha + Y_{-\alpha}) \in \mathfrak{h}_p, \tag{4.5}$$

$$Z^\alpha := \frac{1}{2}(Y_{-\alpha} - Y_\alpha) \in \mathfrak{q}_k, \tag{4.6}$$

and

$$X_{\pm\alpha} := X^\alpha \pm Z^\alpha \in \mathfrak{q}. \tag{4.7}$$

Example 4.1.3 Let $\mathcal{M} = \mathrm{SL}(2,\mathbb{R})/\mathrm{SO}_o(1,1)$. We use the notation from Example 1.3.12. Then $\mathfrak{a} := \mathfrak{q}_p = \mathbb{R}\,X^0$ is abelian and the corresponding roots are $\Delta = \{\alpha, -\alpha\}$, where $\alpha(X^0) = 1$. We choose α to be the positive root. As root spaces we obtain

$$\mathfrak{g}_\alpha = \mathfrak{g}(+1, X^0) = \mathbb{R}\,Y_+ \quad \text{and} \quad \mathfrak{g}_{-\alpha} = \mathfrak{g}(-1, X^0) = \mathbb{R}\,Y_- .$$

The Killing form on $\mathfrak{sl}(2,\mathbb{R})$ is given by $B(X,Y) = 4\mathrm{tr}(XY)$. In particular, we find

$$|xX^0 + zZ^0|^2 = 2(x^2 + z^2)$$

and

$$|Y_+|^2 = 4, \quad |X^0|^2 = 2.$$

This shows that $\alpha = \frac{1}{2}X^0$ and $|\alpha|^2 = \frac{1}{2}$. Now we obtain

$$Y_{\pm\alpha} = Y_\pm, \quad X_{\pm\alpha} = X_\pm,$$

and

$$X^\alpha = X^0, \quad Y^\alpha = Y^0, \quad Z^\alpha = Z^0.$$

□

Remark 4.1.4 Note that we can rescale the inner product without changing X^α. Also, the norm condition on Y_α is invariant under rescaling. On the other hand, the construction of Y^α, Z^α and $X_{\pm\alpha}$ depends on the choice of Y_α (recall that in general $\dim(\mathfrak{g}_\alpha) > 1$). □

Define a linear map $\varphi_\alpha : \mathfrak{sl}(2,\mathbb{R}) \to \mathfrak{g}$ by

$$\frac{1}{2}\begin{pmatrix} 1 & 0 \\ 0 & -1 \end{pmatrix} \mapsto X^\alpha, \quad \begin{pmatrix} 0 & 1 \\ 0 & 0 \end{pmatrix} \mapsto Y_\alpha \quad \text{and} \quad \begin{pmatrix} 0 & 0 \\ 1 & 0 \end{pmatrix} \mapsto Y_{-\alpha}. \tag{4.8}$$

Then φ_α is a Lie algebra monomorphism such that θ induces the usual Cartan involution $X \mapsto -{}^tX$ on $\mathfrak{sl}(2,\mathbb{R})$, whereas τ induces the involution on

$\mathfrak{sl}(2,\mathbb{R})$ described in Example 1.3.12 (denoted also by τ there). In particular, $\mathfrak{s}_\alpha := \operatorname{Im}\varphi_\alpha$ is τ and θ-stable. As a consequence, we have $\varphi_\alpha(Y^0) = Y^\alpha$ and $\varphi_\alpha(Z^0) = Z^\alpha$.

Of course, φ_α exponentiates to a homomorphism of $\tilde{\mathrm{SL}}(2,\mathbb{R})$, the universal covering group of $\mathrm{SL}(2,\mathbb{R})$, into G. Similarly, φ_α defines a homomorphism $\varphi_\alpha : \mathrm{SL}(2,\mathbb{C}) \to G_\mathbb{C}$, since $\mathrm{SL}(2,\mathbb{C})$ is simply connected.

The notation in Example 1.3.12 was set up in such a way that $X^0 \in \mathfrak{sl}(2,\mathbb{R})$ plays the role in the noncompactly causal space $(\mathfrak{sl}(2,\mathbb{R}),\tau)$ that it should play in Proposition 3.1.11. Unfortunately, this property is not carried over by the maps φ_α just constructed. In other words, X^α is in general not a cone-generating element. All we have is the following remark:

Remark 4.1.5 Let $X^0 \in \mathfrak{z}(\mathfrak{h}^a)$ with $\alpha(X^0) = 1$. Then

$$X^\alpha - X^0 \in \alpha^\perp = \{X \in \mathfrak{a} \mid \alpha(X) = 0\}. \qquad \square$$

Theorem 4.1.6 *Let $\mathcal{M} = G/H$ be a noncompactly causal symmetric space and $C \in \mathrm{Cone}_H(\mathfrak{q})$ the cone defining the causal structure.*

1) *There exists a unique cone-generating element $X^0 \in C \cap \mathfrak{q}_\mathfrak{p}^{H\cap K} \subset \mathfrak{z}(\mathfrak{h}^a)$.*

2) *Let $\mathfrak{a}$ be a maximal abelian subspace of $\mathfrak{q}_\mathfrak{p}$ and $\Delta = \Delta(\mathfrak{g},\mathfrak{a})$ the corresponding set of restricted roots. Then for all $\alpha \in \Delta_+ = \{\beta \in \Delta \mid \beta(X^0) = 1\}$ and for any choice of $Y_\alpha \in \mathfrak{g}_\alpha$ satisfying* (4.3), *we have $X^\alpha, X_\alpha, X_{-\alpha} \in C$.*

Proof: 1) According to Corollary 3.1.6, we can find an element of $X \in \mathfrak{q}_\mathfrak{p}^{H\cap K}$ in the interior C^o of C. But then Proposition 3.1.11 shows that a multiple of X satisfies the conditions of 1), since $\mathfrak{q}_\mathfrak{p}^{H\cap K}$ is one-dimensional and contains $\mathfrak{z}(\mathfrak{h}^a)$.

2) Note first that $[X^0 - X^\alpha, Y^\alpha] = 0$ for all $\alpha \in \Delta_+$ since $Y^\alpha \in \mathfrak{g}_\alpha + \mathfrak{g}_{-\alpha}$. Hence by $\mathfrak{sl}(2,\mathbb{R})$-reduction we have

$$\begin{aligned}\mathrm{Ad}(\exp tY^\alpha)X^0 &= \mathrm{Ad}(\exp tY^\alpha)[\tfrac{1}{2}(X_\alpha + X_{-\alpha}) + (X^0 - X^\alpha)] \\ &= \frac{1}{2}(e^t X_\alpha + e^{-t}X_{-\alpha}) + (X^0 - X^\alpha). \qquad (4.9)\end{aligned}$$

Thus

$$2\lim_{t\to\infty} e^{-t}\,\mathrm{Ad}(\exp tY^\alpha)X^0 = X_\alpha \in C$$

and

$$2\lim_{t\to-\infty} e^{t}\,\mathrm{Ad}(\exp tY^\alpha)X^0 = X_{-\alpha} \in C.$$

As $2X^\alpha = X_\alpha + X_{-\alpha}$, the lemma follows. □

Let $(\mathfrak{g}, \tau)$ be a noncompactly causal symmetric pair. We choose a cone generating element $X^0 \in \mathfrak{a} \subset \mathfrak{q}_p$. According to Remark 3.1.9, $(\mathfrak{g}^c, \tau)$ is compactly causal and either simple Hermitian or the product of a simple Hermitian algebra with itself. In either case we can choose a Cartan subalgebra $\mathfrak{t}^c$ of $\mathfrak{g}^c$ containing $i\mathfrak{a}$ and contained in $\mathfrak{k}^c = \mathfrak{h}_k + i\mathfrak{q}_p$. Note that $Z^c := iX^0 \in \mathfrak{z}(\mathfrak{k}^c)$ and the centralizer of Z^c in $\mathfrak{g}_\mathbb{C}$ is $\mathfrak{k}^c_\mathbb{C}$ (cf. Proposition 3.1.12). Let $(\mathfrak{p}^c)^\pm$ be the $\pm i$-eigenspaces of $\operatorname{ad} Z^c$ in $\mathfrak{p}^c_\mathbb{C} = (\mathfrak{h}_p + i\mathfrak{q}_k)_\mathbb{C}$. Then

$$(\mathfrak{p}^c)^\pm \cap \mathfrak{g} = \mathfrak{n}_\pm. \tag{4.10}$$

In addition to the notation from Remark 3.1.17, we use the following abbreviations

$$\tilde{\Delta} := \Delta(\mathfrak{g}_\mathbb{C}, \mathfrak{t}^c_\mathbb{C}), \quad \tilde{\Delta}_\pm := \Delta((\mathfrak{p}^c)^\pm, \mathfrak{t}^c_\mathbb{C}), \quad \text{and} \quad \tilde{\Delta}_0 := \Delta(\mathfrak{k}^c_\mathbb{C}, \mathfrak{t}^c_\mathbb{C}). \tag{4.11}$$

Then we obtain

$$\Delta = \{\tilde{\alpha}|_\mathfrak{a} \mid \tilde{\alpha} \in \tilde{\Delta}, \tilde{\alpha}|_\mathfrak{a} \neq 0\},$$
$$\Delta_\pm = \{\tilde{\alpha}|_\mathfrak{a} \mid \tilde{\alpha} \in \tilde{\Delta}_\pm\},$$

and

$$\Delta_0 := \{\tilde{\alpha}|_\mathfrak{a} \mid \tilde{\alpha} \in \tilde{\Delta}_0, \tilde{\alpha}|_\mathfrak{a} \neq 0\}.$$

Moreover, we can choose a positive system $\tilde{\Delta}^+$ for $\tilde{\Delta}$ such that $\tilde{\Delta}_0^+ := \Delta^+ \cap \tilde{\Delta}_0$ is a positive system in Δ_0 and

$$\Delta_0^+ = \{\tilde{\alpha}|_\mathfrak{a} \mid \tilde{\alpha} \in \tilde{\Delta}_0^+, \tilde{\alpha}|_\mathfrak{a} \neq 0\} \quad \text{and} \quad \Delta^+ = \{\tilde{\alpha}|_\mathfrak{a} \mid \tilde{\alpha} \in \tilde{\Delta}^+, \tilde{\alpha}|_\mathfrak{a} \neq 0\}. \tag{4.12}$$

Lemma 4.1.7 *Let $\tilde{\alpha} \in \tilde{\Delta}_+$ be such that $-\tau\tilde{\alpha} \neq \tilde{\alpha}$. Then $\tilde{\alpha}$ and $-\tau\tilde{\alpha}$ are strongly orthogonal.*

Proof: Let $\tilde{\alpha} \in \tilde{\Delta}_+$ be such that $-\tau\tilde{\alpha} \neq \tilde{\alpha}$. Then $-\tau\tilde{\alpha} \in \tilde{\Delta}_+$ and $\tilde{\alpha} - \tau\tilde{\alpha}$ is not a root. Assume that $\gamma := \tilde{\alpha} + \tau\tilde{\alpha}$ is a root. Since $\gamma|_\mathfrak{a} = 0$, we have $(\mathfrak{g}_\mathbb{C})_\gamma \subset \mathfrak{z}_{\mathfrak{g}_\mathbb{C}}(\mathfrak{a}) \subset \mathfrak{a}_\mathbb{C} \oplus (\mathfrak{h}_k)_\mathbb{C}$. It follows that $(\mathfrak{g}_\mathbb{C})_\gamma \subset \mathfrak{k}_\mathbb{C} \cap \mathfrak{h}_\mathbb{C}$ because $(\mathfrak{g}_\mathbb{C})_\gamma$ is τ-invariant. Let $X \in (\mathfrak{g}_\mathbb{C})_{\tilde{\alpha}}$, $X \neq 0$. As $\mathfrak{t}^c_\mathbb{C}$ is a Cartan subalgebra of $\mathfrak{g}_\mathbb{C}$, it follows that $\dim_\mathbb{C}(\mathfrak{g}_\mathbb{C})_{\tilde{\alpha}} = 1$ and $0 \neq [X, \tau(X)] \in (\mathfrak{g}_\mathbb{C})_\gamma$. But then $[X, \tau(X)] \in \mathfrak{q}_\mathbb{C}$ gives a contradiction. □

Remark 4.1.8 The τ-invariance of $\mathfrak{t}^c$ shows that also $\tilde{\Delta}$ is invariant. Let $\sigma^c = \sigma\theta$ be the complex conjugation of $\mathfrak{g}_\mathbb{C}$ w.r.t. $\mathfrak{g}^c$. Then $\mathfrak{t}^c_\mathbb{C}$ is σ^c invariant. Since the elements of $\tilde{\Delta}$ take real values on $\mathfrak{t}^c$, it follows that $\sigma^c\tilde{\alpha} = -\tilde{\alpha}$ for all $\tilde{\alpha} \in \tilde{\Delta}$. Therefore we can choose $\tilde{E}_{\tilde{\alpha}} \in (\mathfrak{g}_\mathbb{C})_{\tilde{\alpha}}$ such that

$$\tilde{E}_{-\tilde{\alpha}} = \tilde{E}_{\sigma^c\tilde{\alpha}} = \sigma^c\tilde{E}_{\tilde{\alpha}} \quad \text{and} \quad \tilde{E}_{\tau\tilde{\alpha}} = \tau\tilde{E}_{\tilde{\alpha}}$$

for all $\tilde{\alpha} \in \tilde{\Delta}$. Moreover, the normalization can be chosen such that the element

$$\tilde{H}_{\tilde{\alpha}} = [\tilde{E}_{\tilde{\alpha}}, \tilde{E}_{-\tilde{\alpha}}] = [\tilde{E}_{-\tau\tilde{\alpha}}, \tilde{E}_{\tau\tilde{\alpha}}] = \tilde{H}_{-\tau\tilde{\alpha}}$$

satisfies $\tilde{\alpha}(\tilde{H}_{\tilde{\alpha}}) = 2$ (cf. Appendix A.3).

Fix a maximal set

$$\tilde{\Gamma} = \{\tilde{\gamma}_1, \ldots, \tilde{\gamma}_{r^c}\} \subset \tilde{\Delta}_+ \tag{4.13}$$

of strongly orthogonal roots (cf. Appendix A.3). In fact, Lemma 4.1.7 shows that we may assume $\tilde{\Gamma}$ to be invariant by $-\tau$ simply by adding $-\tau(\tilde{\gamma}_k)$ after each inductive step.

Recall from (A.27) and Lemma A.3.3 that the space

$$\mathfrak{a}^c = \mathfrak{p}^c \cap \sum_{\tilde{\gamma} \in \tilde{\Gamma}} \mathbb{R}\tilde{Y}_{\tilde{\gamma}},$$

where $\tilde{Y}_{\tilde{\gamma}} = -i(\tilde{E}_{\tilde{\gamma}} - \tilde{E}_{-\tilde{\gamma}})$, is maximal abelian in $\mathfrak{p}^c$. Note that $\tilde{\Delta}_+$ is invariant under $-\tau$ and renormalize $\tilde{Y}_{\tilde{\gamma}}$:

$$Y_{\tilde{\gamma}} := r_{\tilde{\gamma}} \tilde{Y}_{\tilde{\gamma}} \in \mathfrak{g}_{\gamma}, \tag{4.14}$$

where $\gamma = \tilde{\gamma}|_{\mathfrak{a}}$. Here we choose $r_{\tilde{\gamma}}$ in such a way that $Y_{\tilde{\gamma}}$ satisfies the condition (4.3), i.e.,

$$r_{\tilde{\gamma}} = \frac{2}{|\gamma|^2 |\tilde{Y}_{\tilde{\gamma}}|^2}.$$

Now we see that

$$Y^{\tilde{\gamma}} := \frac{1}{2}(Y_{\tilde{\gamma}} + \tau Y_{\tilde{\gamma}}) \in \mathfrak{h}_p \cap \mathfrak{a}_h^c$$

and

$$\mathfrak{a}_h^c := \mathfrak{a}^c \cap \mathfrak{g} = \sum_{\tilde{\gamma} \in \tilde{\Gamma}} \mathbb{R} Y^{\tilde{\gamma}}. \qquad \square$$

Lemma 4.1.9 *The space $\mathfrak{a}_h^c$ is maximal abelian in $\mathfrak{h}_p$.*

Proof: Let $G_{\mathbb{C}}$ be a simply connected Lie group with Lie algebra $\mathfrak{g}_{\mathbb{C}}$ and $G, G^c, H, K, K^c, A^c, A_h^c$, etc., the analytic subgroups of $G_{\mathbb{C}}$ corresponding to $\mathfrak{g}, \mathfrak{g}^c, \mathfrak{h}, \mathfrak{k}, \mathfrak{k}^c, \mathfrak{a}^c, \mathfrak{a}_h^c$, etc. Recall the Cartan decomposition

$$K^c A^c K^c = G^c$$

from [44], p. 402. The A^c-component is unique up to a conjugation by a Weyl group element. Let $\sigma: G_{\mathbb{C}} \to G_{\mathbb{C}}$ be the complex conjugation with fixed point set G. Then G^c, K^c and A^c are σ-invariant. Thus we have

$$(G \cap K^c)(G \cap A^c)(G \cap K^c) = G \cap G^c.$$

The Lie algebras of $G \cap G^c$, $G \cap K^c$, and $G \cap A^c$ are $\mathfrak{h}$, $\mathfrak{h}_k$, and $\mathfrak{a}_h^c$. Since $\mathfrak{a}_h^c$ is an abelian subspace of $\mathfrak{h}_p$ and $\mathfrak{h}_k$ is maximal compact in $\mathfrak{h}$, this shows that

$$(K \cap H)A_h^c(K \cap H) = H \tag{4.15}$$

is a Cartan decomposition and hence $\mathfrak{a}_h^c$ is maximal abelian in $\mathfrak{h}_p$. □

Remark 4.1.10 We set

$$\Gamma := \{\tilde{\gamma}|_{\mathfrak{a}} \mid \tilde{\gamma} \in \tilde{\Gamma}\} \subset \Delta_+. \tag{4.16}$$

View $\tilde{\gamma}$ as an element of $(i\mathfrak{t}^c)^*$ and write $\tilde{\gamma} = \gamma + \gamma'$ with $\gamma = \tilde{\gamma}|_{\mathfrak{a}}$ and $\gamma' = \tilde{\gamma}|_{(i\mathfrak{t}^c)\cap\mathfrak{h}}$. Then, under the identification of dual spaces via $(\cdot \mid \cdot)$, the restriction means orthogonal projection to the respective space. Note that $-\tau\tilde{\gamma} = \gamma - \gamma' \in \tilde{\Gamma}$, so the orthogonality of the elements of $\tilde{\Gamma}$ implies also that their restrictions to $\mathfrak{a}$ are orthogonal.

Γ actually consists of strongly orthogonal roots. To see this, suppose that $\gamma_i = \tilde{\gamma}_i|_{\mathfrak{a}}$ and $\gamma_j = \tilde{\gamma}_j|_{\mathfrak{a}}$ with $\gamma_i - \gamma_j \in \Delta$. Since γ_i and γ_j are orthogonal we have $s_{\gamma_j}(\gamma_i - \gamma_j) = \gamma_i + \gamma_j$, where s_{γ_j} is the reflection in $\mathfrak{a}^*$ at the hyperplane orthogonal to γ_j. But s_{γ_j} is an element of the Weyl group of Δ and hence leaves Δ invariant. Therefore we have $\gamma_i + \gamma_j \in \Delta$, a contradiction.

Suppose that $\gamma_o \in \Delta_+$ is strongly orthogonal to all $\gamma \in \Gamma$. Then there exists a $\tilde{\gamma}_o \in \tilde{\Delta}_+$ such that $\gamma_o = \tilde{\gamma}_o|_{\mathfrak{a}}$ and $0 \neq \gamma_j - \gamma_o \in \Delta$ whenever $\tilde{\gamma}_j - \tilde{\gamma}_o \in \tilde{\Delta}$. Therefore $\tilde{\gamma}_o$ is strongly orthogonal to all $\tilde{\gamma}_j$, in contradiction to the maximality of $\tilde{\Gamma}$. Thus Γ is a maximal set of strongly orthogonal roots in Δ_+.

Note that the orthogonality of the elements of $\tilde{\Gamma}$ together with $\gamma = \frac{1}{2}(\tilde{\gamma} - \tau\tilde{\gamma})$ shows that the only elements of $\tilde{\Gamma}$ restricting to a given $\gamma = \tilde{\gamma}|_{\mathfrak{a}} \in \Gamma$ on $\mathfrak{a}$ are $\tilde{\gamma}$ and $-\tau\tilde{\gamma}$.

The definition of $\mathfrak{a}_h^c$ shows that each element $Y \in \mathfrak{a}_h^c$ can be written in the form

$$Y = \sum_{\gamma \in \Gamma} r_\gamma Y^\gamma \tag{4.17}$$

constructed via (4.5) from pairwise commuting elements $Y_\alpha \in \mathfrak{g}_\alpha$ satisfying the normalization condition (4.3). Moreover, the images of the corresponding $\mathfrak{sl}(2, \mathbb{R})$-embeddings φ_α commute, since the images of the $\mathfrak{sl}(2, \mathbb{C})$ embeddings corresponding to the different elements of $\tilde{G}$ (cf. Appendix A.3) commute. □

Lemma 4.1.11 *1)* $L := X^0 - \sum_{\gamma \in \Gamma} X^\gamma \in \bigcap_{\gamma \in \Gamma} \ker \gamma$.

2) If $Y = \sum_{\gamma \in \Gamma} t_\gamma Y^\gamma \in \mathfrak{a}_h^c$ *is chosen as in* Remark 4.1.10, *then we have*

$$e^{\operatorname{ad} Y} X^0 = L + \sum_{\gamma \in \Gamma} \cosh(t_\gamma) X^\gamma + \sum_{\gamma \in \Gamma} \sinh(t_\gamma) Z^\gamma.$$

Proof: 1) Fix $\gamma \in \Gamma$ and note that $\gamma(X^0 - X^\gamma) = 0$ because of the normalization of X^γ. On the other hand, $\gamma(X^\beta) = 0$ for all $\gamma \neq \beta \in \Gamma$ by Remark 4.1.10. This proves 1).

2) Using $\mathfrak{sl}(2, \mathbb{R})$ reduction we calculate

$$\begin{aligned} e^{\operatorname{ad} Y} X^0 &= e^{\operatorname{ad} Y}(L + \sum_{\gamma\in\Gamma} X^\gamma) \\ &= L + \left(\sum_{\gamma\in\Gamma} e^{\operatorname{ad} t_\gamma Y^\gamma} X^\gamma \right) \\ &= L + \sum_{\gamma\in\Gamma} \left[e^{t_\gamma}(X^\gamma + Z^\gamma) + e^{-t_\alpha}(X^\gamma - Z^\gamma) \right] \end{aligned}$$

as $X_{\pm\gamma} = X^\gamma \pm Z^\gamma$. From this the claim now follows. □

4.2 The Minimal and Maximal Cones

In this section we study certain convex cones which will turn out to be minimal and maximal H-invariant cones in $\mathfrak{q}$, respectively their intersections with $\mathfrak{a}$.

Definition 4.2.1 Let $\mathcal{M} = G/H$ be a noncompactly causal symmetric space and $(\mathfrak{g}, \tau)$ the corresponding symmetric pair. Further, let $X^0 \in \mathfrak{q}_p$ be a cone-generating element. Then the closed convex cones

$$C_{\min}(X^0) := C_{\min} := \overline{\operatorname{conv}\left[\operatorname{Ad}(H_o)(\mathbb{R}^+ X^0)\right]}. \tag{4.18}$$

and

$$C_{\max}(X^0) := C_{\max} := \{X \in \mathfrak{q} \mid \forall Y \in C_{\min} : B(X,Y) \geq 0\} \tag{4.19}$$

in $\mathfrak{q}$ are called the *minimal* and the *maximal* cone in $\mathfrak{q}$ determined by the choice of X^0. A reference to the space $\mathcal{M} = G/H$ is not necessary, since the definitions depend only on the group generated by $e^{\operatorname{ad}\mathfrak{h}}$ in $\mathrm{GL}(\mathfrak{q})$. □

Definition 4.2.2 Let $(\mathfrak{g}, \tau)$ be a noncompactly causal symmetric pair and $\mathfrak{a} \subset \mathfrak{q}_p$ a maximal abelian subspace. Choose a cone-generating element $X^0 \in \mathfrak{a}$ and recall the set Δ_+ of restricted roots taking the value 1 on X^0. Then the closed convex cones

$$c_{\min}(X^0) := c_{\min} := \sum_{\alpha\in\Delta_+} \mathbb{R}_0^+ X^\alpha = \sum_{\alpha\in\Delta_+} \mathbb{R}_0^+ \alpha \tag{4.20}$$

and

$$c_{\max}(X^0) := c_{\max} := \{X \in \mathfrak{a} \mid \forall \alpha \in \Delta_+ \, : \, \alpha(X) \geq 0\} = c_{\min}^* \qquad (4.21)$$

in $\mathfrak{a}$ are called the *minimal* and the *maximal* cone in $\mathfrak{a}$ determined by the choice of X^0. □

It follows from Proposition 3.1.11 that there are only two minimal and maximal cones.

Remark 4.2.3 It follows from the definition of $C_{\min}$ that $\theta(C_{\min}) = -C_{\min}$. This shows that we can replace the definition of $C_{\max}$ by

$$C_{\max} = \{X \in \mathfrak{q} \mid \forall Y \in C_{\min} \, : \, (X \mid Y) \geq 0\}. \qquad □$$

Lemma 4.2.4 *Let $\mathcal{M} = G/H$ is a noncompactly causal symmetric space. Then $C_{\min}$ is minimal in $\mathrm{Cone}_H(\mathfrak{q})$ and $C_{\max}$ is maximal in $\mathrm{Cone}_H(\mathfrak{q})$.*

Proof: Note first that, by duality via the Killing form, we only have to show the assertion concerning $C_{\min}$.

Theorem 4.1.6 implies that X^0 is $(H \cap K)$-invariant, so that $C_{\min}$ is indeed H-invariant by Lemma 3.1.1. Since any element of $\mathrm{Cone}_H(\mathfrak{q})$ contains either X^0 or $-X^0$, again by Theorem 4.1.6, it only remains to show that $C_{\min}$ is regular. If $\mathfrak{q}$ is $\mathfrak{h}$-irreducible this follows from Theorem 2.1.21. If $\mathfrak{q}$ is not $\mathfrak{h}$ irreducible, then the only elements in $\mathrm{Cone}_H(\mathfrak{q})$ intersecting $\mathfrak{p}$ nontrivially are $\pm C_p$ (cf. Section 2.6). But the contruction of C_p shows that it contains X^0. This implies $C_p = C_{\min}$ and hence the claim. □

The following proposition is an immediate consequence of Theorem 4.1.6.

Proposition 4.2.5 *Let $\mathcal{M} = G/H$ be a noncompactly causal symmetric space and $(\mathfrak{g}, \tau)$ the corresponding symmetric pair. Suppose that the causal structure is given by $C \in \mathrm{Cone}_H(\mathfrak{q})$ and let X^0 be the unique cone generating in C. Then*

$$c_{\min}(X^0) \subset C. \qquad □$$

Lemma 4.2.6 *Assume that $\mathcal{M}$ is a noncompactly causal symmetric space. Let $\mathfrak{a}$ be maximal abelian in $\mathfrak{q}_p$. Let $\Delta_0 = \Delta(\mathfrak{h}^a, \mathfrak{a})$ and W_0 be the Weyl group of the root system Δ_0. Then*

$$W_0 = N_{H\cap K}(\mathfrak{a})/Z_{H\cap K}(\mathfrak{a}).$$

Proof: Recall from [44], p. 289, that $W_0 = N_{H_o \cap K}(\mathfrak{a})/Z_{H_o \cap K}(\mathfrak{a})$. But Theorem 3.1.18 implies that $H \cap K = Z_{H \cap K}(\mathfrak{a})(H_o \cap K)$ and that proves the lemma. □

Proposition 4.2.7 *$c_{\min}$ and $c_{\max}$ are regular W_0-invariant cones in $\mathfrak{a}$. Furthermore, $c_{\min} \subset c_{\max}$ and*

$$c_{\max}^o = \{X \in \mathfrak{a} \mid \forall \alpha \in \Delta_+ \, : \, \alpha(X) > 0\}.$$

Proof: As the Weyl group W_0 fixes X^0, it follows that W_0 permutes Δ_+. We have $w(X^\alpha) = X^{w\alpha}$ and hence $c_{\min}$ is W_0-invariant. By duality also $c_{\max}^*$ is W_0-invariant.

If $\alpha, \beta \in \Delta_+$, then $\alpha(X^\beta) = (\alpha \mid \beta)/(\beta \mid \beta) \geq 0$ for otherwise $\alpha + \beta$ would be a root. Hence $c_{\min} \subset c_{\min}^*$.

The equality $c_{\max}^o = \{X \in \mathfrak{a} \mid \forall \alpha \in \Delta_+ \, : \, \alpha(X) > 0\}$ is an immediate consequence of the definitions. It shows that $c_{\max}$ is generating and hence that $c_{\min}$ is proper (cf. Lemma 2.1.3 and Lemma 2.1.4). It only remains to show that $c_{\min}$ is generating. If $< c_{\min} > \neq \mathfrak{a}$, then we can find a non-zero element $X \in \mathfrak{a}$ with $\alpha(X) = 0$ for all $\alpha \in \Delta_+$. Thus $iX \in i\mathfrak{a} \subset \mathfrak{g}^c$ centralizes $(\mathfrak{n}_+ \oplus \mathfrak{n}_-)_{\mathbb{C}} = (\mathfrak{q}_k \oplus \mathfrak{h}_p)_{\mathbb{C}} = \mathfrak{p}_{\mathbb{C}}^c$, which is absurd. □

Lemma 4.2.8 *$X^0 \in c_{\min}(X^0)^o$.*

Proof: Let $X \in c_{\min}(X^0)^o$ and define $\tilde{X} := [1/\#W_0] \sum_{w \in W_0} w \cdot X$. Then $\tilde{X} \neq 0$ is W_0-invariant and contained in $c_{\min}(X^0)^o$. Let $\alpha \in \Delta_0$ and $s_\alpha \in W_0$ the reflection at the hyperplane orthogonal to α. Then

$$\alpha(\tilde{X}) = \alpha(s_\alpha(\tilde{X})) = < s_\alpha(\alpha), \tilde{X} > = -\alpha(\tilde{X}),$$

i.e., $\alpha(\tilde{X}) = 0$ for all $\alpha \in \Delta_0$. Therefore $\tilde{X} \in \mathfrak{z}(\mathfrak{h}^a)$ and hence $\tilde{X}$ is a multiple of X^0. On the other hand, $X^0 \in c_{\max}(X^0)^o$ so that $c_{\min}(X^0)^o \subset c_{\max}(X^0)^o$ implies the claim. □

For later use we record an application a convexity theorem due to Kostant (cf. [45], p. 473) to the Lie algebra $\mathfrak{h}^a$.

Proposition 4.2.9 *Let $\mathcal{M} = G/H$ be a noncompactly causal symmetric space and* pr: $\mathfrak{q}_p \to \mathfrak{a}$ *the orthogonal projection. Then for $X \in \mathfrak{a}$ we have*

$$\mathrm{pr}(\mathrm{Ad}(K \cap H_o)X) = \mathrm{conv}(W_0 \cdot X).$$

Proposition 4.2.10 *Let $\mathcal{M} = G/H$ be a noncompactly causal symmetric space. Choose a maximal abelian subspace $\mathfrak{a}$ of $\mathfrak{q}_p$ and a cone-generating element $X^0 \in \mathfrak{a}$. Denote the orthogonal projection $\mathfrak{q} \to \mathfrak{a}$ by* pr. *Then*

1) $\mathrm{pr}(C_{\min}) \subset c_{\min}$.

2) $\mathrm{pr}(C_{\max}) \subset c_{\max}$.

Proof: 1) Let $X = \mathrm{Ad}(h)X^0$ for some $h \in H_o$. The Cartan decomposition (4.15)

$$H_o = (H_o \cap K)A_h^c(H_o \cap K)$$

shows that we can write $h = k(\exp Y)k_1$ with $k, k_1 \in K \cap H_o$ and $Y \in \mathfrak{a}_h^c$. We can write $Y = \sum_{\gamma\in\Gamma} t_\gamma Y^\gamma$ as in Remark 4.1.10 and then Lemma 4.1.11 shows that

$$\begin{aligned}
\mathrm{pr}(\mathrm{Ad}(h)X^0) &= \mathrm{pr}\left(\mathrm{Ad}(k)e^{\mathrm{ad}\,Y}X^0\right) \\
&= \mathrm{pr}\left(\mathrm{Ad}(k)\left[L + \sum_{\gamma\in\Gamma}\cosh(t_\gamma)X^\gamma\right]\right) \\
&= \mathrm{pr}\left(\mathrm{Ad}(k)X^0 + \mathrm{Ad}(k)\sum_{\gamma\in\Gamma}[\cosh(t_\gamma) - 1]X^\gamma\right) \\
&= X^0 + \mathrm{pr}\left(\mathrm{Ad}(k)\sum_{\gamma\in\Gamma}[\cosh(t_\gamma) - 1]X^\gamma\right)
\end{aligned}$$

It follows that $\mathrm{pr}(\mathrm{Ad}(h)X^0) \in c_{\min}^o$ as $\mathrm{pr}(\mathrm{Ad}(k)X^\gamma) \in \mathrm{conv}\, W_0 \cdot X^\gamma \subset c_{\min}$ by Proposition 4.2.9. Since $C_{\min}$ is the closed convex cone generated by $\mathrm{Ad}(H_o)$, the claim follows.

2) Let $Y \in c_{\min}$. By Lemma 4.2.5 we have $Y \in C_{\min}$. Thus $(Y|\mathrm{pr}(X)) = (Y|X) \geq 0$, which implies that $\mathrm{pr}(X) \in c_{\min}^* = c_{\max}$. □

Recall the intersection and projection operations from Section 2.1. We set $I := I_{\mathfrak{a}}^{\mathfrak{q}}$ and $P := P_{\mathfrak{a}}^{\mathfrak{q}}$.

Proposition 4.2.11 *Let $\mathcal{M} = G/H$ be a noncompactly causal symmetric space with cone-generating element $X^0 \in \mathfrak{q}_p^{H\cap K}$. Then*

1) $c_{\min} = I(C_{\min}) = P(C_{\min})$.

2) $c_{\max} = I(C_{\max}) = P(C_{\max})$.

Proof: 1) According to Proposition 4.2.5 we have

$$c_{\min} \subset I(C_{\min}) \subset P(C_{\min}).$$

Proposition 4.2.10 now shows that $P(C_{\min}) = c_{\min}$.

2) Lemma 2.1.8 and part 1) show

$$I(C_{\max}) = P(C_{\max}^*)^* = P(C_{\min})^* = c_{\min}^* = c_{\max}$$

and hence the claim again follows from Proposition 4.2.10. □

Recall that for any noncompactly causal symmetric pair $(\mathfrak{g}, \tau)$, the symmetric pair $(\mathfrak{g}_{\mathbb{C}}, \sigma^c)$, where $\sigma^c: \mathfrak{g}_{\mathbb{C}} \to \mathfrak{g}_{\mathbb{C}}$ is the complex conjugation w.r.t. $\mathfrak{g}^c$, is either noncompactly causal (if $\mathfrak{g}$ carries no complex structure) or the direct sum of two isomorphic noncompactly causal pairs. In particular, any symmetric space $G_{\mathbb{C}}/G^c$ associated to $(\mathfrak{g}_{\mathbb{C}}, \sigma^c)$ admits a causal structure. We assume for the moment that $\mathfrak{g}$ carries no complex structure. Then any cone-generating element for $(\mathfrak{g}, \tau)$ is automatically a cone-generating element for $(\mathfrak{g}_{\mathbb{C}}, \sigma^c)$. Fix a Cartan subalgebra $\mathfrak{t}^c$ of $\mathfrak{k}^c$ containing $\mathfrak{a}$, which then is also a Cartan subalgebra of $\mathfrak{g}^c$. Further, we choose a cone-generating element $X^0 \in \mathfrak{q}_p \subset i\mathfrak{h}_k + \mathfrak{q}_p$ and a positive system $\tilde{\Delta}^+$ for $\tilde{\Delta} = \Delta(\mathfrak{g}_{\mathbb{C}}, \mathfrak{t}^c)$ as in (4.12). The corresponding minimal and maximal cones in the maximal abelian subspace $i\mathfrak{t}^c$ of $i\mathfrak{h}_k + \mathfrak{q}_p$ are then given by

$$\tilde{c}_{\min} = \sum_{\tilde{\alpha} \in \tilde{\Delta}_+} \mathbb{R}_0^+ \tilde{\alpha} \tag{4.22}$$

and

$$\tilde{c}_{\max} = \{X \in i\mathfrak{t}^c \mid \forall \tilde{\alpha} \in \tilde{\Delta}_+ : \tilde{\alpha}(X) \geq 0\} = \tilde{c}_{\min}^*. \tag{4.23}$$

Proposition 4.2.12 *The cones $c_{\min}(\tilde{\Delta}_+)$ and $c_{\max}(\tilde{\Delta}_+)$ are $-\tau$-invariant. Moreover,*

1) $I_{\mathfrak{a}}^{i\mathfrak{t}^c}(\tilde{c}_{\min}) = P_{\mathfrak{a}}^{i\mathfrak{t}^c}(\tilde{c}_{\min}) = c_{\min}$;

2) $I_{\mathfrak{a}}^{i\mathfrak{t}^c}(\tilde{c}_{\max}) = P_{\mathfrak{a}}^{i\mathfrak{t}^c}(\tilde{c}_{\max}) = c_{\max}$.

Proof. The $-\tau$-invariance follows from the $-\tau$-invariance of $\tilde{\Delta}_+$. In view of duality and Lemma 2.1.9, it only remains to show that $P_{\mathfrak{a}}^{i\mathfrak{t}^c}(\tilde{c}_{\min}) = c_{\min}$. But that is clear, since $P_{\mathfrak{a}}^{i\mathfrak{t}^c}(\tilde{\alpha}) = \tilde{\alpha}|_{\mathfrak{a}}$. □

Lemma 4.2.13 *Let $X \in c_{\max}^o$. Then*

1) $\mathfrak{z}_{\mathfrak{g}}(X) \subset \mathfrak{q}_p$.

2) Let $h \in H$ be such that $\mathrm{Ad}(h)X \in \mathfrak{a}$. Then $h \in K \cap H$.

Proof. 1) Let $Y = \sum_{\alpha \in \Delta_+} [L_\alpha - \tau(L_\alpha)]$ with $L_\alpha \in \mathfrak{g}_\alpha$. As $X \in c_{\max}^o$ we have

$$[X, Y] = \sum_{\alpha \in \Delta_+} \alpha(X)(L_\alpha + \tau(L_\alpha)) \neq 0$$

which implies the claim.

2) Let $Y \in \mathfrak{a}$. Then

$$[\mathrm{Ad}(h^{-1})Y, X] = \mathrm{Ad}(h^{-1})[Y, \mathrm{Ad}(h)X] = 0.$$

Thus $\mathrm{Ad}(h^{-1})Y \in \mathfrak{h}^a \cap \mathfrak{q} = \mathfrak{q}_p$. In particular, $\mathrm{Ad}(h^{-1})\mathfrak{a}$ is a maximal abelian subalgebra of $\mathfrak{q}_p$. Thus there is a $k \in K \cap H$ such that $\mathrm{Ad}(kh^{-1})\mathfrak{a} = \mathfrak{a}$. This implies that

$$kh^{-1} \in N_H(\mathfrak{a}) \subset K \cap H.$$

Thus $h \in K \cap H$ as claimed. □

Let $G_{\mathbb{C}}/G^c$ be any symmetric space with corresponding symmetric pair $(\mathfrak{g}_{\mathbb{C}}, \sigma^c)$. Then the maximal and the minimal cones in $i\mathfrak{g}^c$ are given by

$$\tilde{C}_{\min}(X^0) = \tilde{C}_{\min} = \overline{\mathrm{conv}\,[\mathrm{Ad}(G_o^c)\,(\mathbb{R}^+X^0)]} \tag{4.24}$$

and

$$\tilde{C}_{\max}(X^0) = \tilde{C}_{\max} = \{X \in \mathfrak{q} \mid \forall Y \in \tilde{C}_{\min} : B(X, Y) \geq 0\}. \tag{4.25}$$

Remark 4.2.14 In order to be able to treat the cases of complex and non-complex $\mathfrak{g}$ simultaneously, we make the following definitions. Suppose that $\mathfrak{g} = \mathfrak{l}_{\mathbb{C}}$ and τ the corresponding complex conjugation, so that $\mathfrak{g}^c = \mathfrak{l} \times \mathfrak{l}$ and τ^c is the switch of factors. More precisely, $\mathfrak{g}_{\mathbb{C}} = \mathfrak{l}_{\mathbb{C}} \times \mathfrak{l}_{\mathbb{C}}$ with the opposite complex structure on the second factor and the embedding

$$\mathfrak{g} = \mathfrak{l}_{\mathbb{C}} \ni X \mapsto \big(X, \tau(X)\big) \in \mathfrak{l}_{\mathbb{C}} \times \mathfrak{l}_{\mathbb{C}} = \mathfrak{g}_{\mathbb{C}}.$$

The involution $\sigma^c\colon \mathfrak{g}_{\mathbb{C}} \to \mathfrak{g}_{\mathbb{C}}$ is given by

$$\sigma^c(X, Y) = (\tau(X), \tau(Y)).$$

The algebra $\mathfrak{l}$ is Hermitian and we can choose the maximal abelian subspace of $\mathfrak{q}_p$ to be $\mathfrak{a} = i\mathfrak{t}$, where $\mathfrak{t}$ is a Cartan subalgebra of $\mathfrak{l} \cap \mathfrak{k}$. Then $\mathfrak{t}^c := \mathfrak{t} \times \mathfrak{t}$ is a Cartan subalgebra of $\mathfrak{g}^c$. Choose a cone-generating element $X^0 \cong (X^0, \tau(X^0)) = (X^0, -X^0) \in \mathfrak{a}$. We set $\tilde{\Delta} := \Delta(\mathfrak{l}_{\mathbb{C}} \times \mathfrak{l}_{\mathbb{C}}, \mathfrak{t}_{\mathbb{C}} \times \mathfrak{t}_{\mathbb{C}})$ and

$$\tilde{\Delta}_+ := \{(\alpha, \beta) \in \mathfrak{t}_{\mathbb{C}}^* \times \mathfrak{t}_{\mathbb{C}}^* \mid \alpha(X^0) = 1 = \beta(-X^0)\} = \Delta_+ \times \Delta_-,$$

where $\Delta = \Delta(\mathfrak{l}_{\mathbb{C}}, i\mathfrak{t})$ and $\Delta_+ = \{\alpha \in \Delta \mid \alpha(X^0) = 1\}$. Now the formulas (4.22) and (4.23) make sense and yield

$$\tilde{c}_{\min}(X^0, -X^0) = c_{\min}(X^0) \times c_{\min}(-X^0) \subset i\mathfrak{t}^c$$

and

$$\tilde{c}_{\max}(X^0, -X^0) = c_{\max}(X^0) \times c_{\max}(-X^0) \subset i\mathfrak{t}^c.$$

Note that both cones are $-\tau$-invariant. It follows directly from these definitions and the embedding of $\mathfrak{q} = i\mathfrak{l} \to i\mathfrak{l} \times i\mathfrak{l} = i\mathfrak{g}^c$ that

$$I_{\mathfrak{a}}^{i\mathfrak{l}^c}(\tilde{C}_{\min}) = C_{\min}, \quad I_{\mathfrak{a}}^{i\mathfrak{l}^c}(\tilde{C}_{\max}) = C_{\max}.$$

In particular, we see that the conclusions of Proposition 4.2.12 stay valid.

Let $L_{\mathbb{C}}$ be a simply connected complex Lie group with Lie algebra $\mathfrak{l}_{\mathbb{C}}$ and L the analytic subgroup of $L_{\mathbb{C}}$ with Lie algebra $\mathfrak{l}$. Then the involution σ^c integrates to an involution of $G_{\mathbb{C}} = L_{\mathbb{C}} \times L_{\mathbb{C}}$, again denoted by σ^c, whose group of fixed points is $G^c := L \times L$. Now also the equations (4.24) and (4.25) for the minimal and maximal cones in $i\mathfrak{g}^c$ make sense and yield

$$\tilde{C}_{\min}(X^0, -X^0) = C_{\min}(X^0) \times C_{\min}(-X^0)$$

and

$$\tilde{C}_{\max}(X^0, -X^0) = C_{\max}(X^0) \times C_{\max}(-X^0).$$

Note that both cones are invariant under $-\tau$ and

$$I_{\mathfrak{q}}^{i\mathfrak{g}^c}(\tilde{C}_{\min}) = C_{\min}, \quad I_{\mathfrak{q}}^{i\mathfrak{g}^c}(\tilde{C}_{\max}) = C_{\max}.$$

□

Lemma 4.2.15 *Let $\mathcal{M} = G/H$ be a noncompactly causal symmetric space and $X \in \tilde{C}_{\max}^o$.*

1) *X is semisimple and* $\operatorname{ad} X$ *has real eigenvalues.*

2) *If $G_{\mathbb{C}}/G^c$ is any symmetric space corresponding to $(\mathfrak{g}_{\mathbb{C}}, \sigma^c)$, then there exists a $g \in G_o^c$ such that* $\operatorname{Ad}(g)X \in \tilde{c}_{\max}^o$.

3) *If $X \in \mathfrak{q}$, then there exists an $h \in H_o$ such that* $\operatorname{Ad}(h)X \in c_{\max}$.

Proof: By Theorem 2.1.13 the centralizer of X in G_o^c is compact. Since every compact subgroup in G_o^c is conjugate to one contained in K_o^c, we may assume that $Z_{G_o^c}(X) \subset K_o^c$. But then $\mathfrak{z}_{\mathfrak{g}^c}(iX) \subset \mathfrak{k}^c$. As $iX \in \mathfrak{z}_{\mathfrak{g}^c}(iX)$, it follows that iX is semisimple with purely imaginary eigenvalues. Hence X is semisimple with purely real eigenvalues. Note that $\mathfrak{a}_h \oplus i\mathfrak{a}$ is a Cartan subalgebra of $\mathfrak{k}^c$, so there is a $k \in K_o^c$ such that $\operatorname{Ad}(k)(iX) \in \mathfrak{a}_h \oplus i\mathfrak{a}$. This proves 1) and 2).

Now assume that $X \in \mathfrak{q}$. According to Theorem 1.4.1, we can find an $h \in H_o$ and a θ-stable A-subspace $\mathfrak{b} = \mathfrak{b}_k \oplus \mathfrak{b}_p$ in $\mathfrak{q}$ such that $\operatorname{Ad}(h)X \in \mathfrak{b}$ and $\mathfrak{b}_p \subset \mathfrak{a}$. Proposition 4.2.11 implies that

$$\operatorname{pr}_{i\mathfrak{l}^c}(\operatorname{Ad}(h)X) \in \tilde{c}_{\max}^o.$$

Then Proposition 4.2.12 shows that $\operatorname{pr}_{\mathfrak{a}}(\operatorname{Ad}(h)X) \in c_{\max}^o$ and hence Lemma 4.2.13 implies that $\mathfrak{b}_k = \{0\}$. Thus $\operatorname{Ad}(h)X$ is actually contained in $c_{\max}^o$.
□

Theorem 4.2.16 (Extension of $C_{\min}$ and $C_{\max}$) *Let $\mathcal{M} = G/H$ be a noncompactly causal symmetric space. Then $\tilde{C}_{\min}$ and $\tilde{C}_{\max}$ are $-\tau$-stable and satisfy*

$$I_{\mathfrak{q}}^{i\mathfrak{g}^c}(\tilde{C}_{\min}) = C_{\min}, \quad I_{\mathfrak{q}}^{i\mathfrak{g}^c}(\tilde{C}_{\max}) = C_{\max}.$$

Proof: We may assume that $\mathfrak{g}$ carries no complex structure, since the case of complex $\mathfrak{g}$ was already treated in Remark 4.2.14.

Let $G_{\mathbb{C}}/G^c$ be any symmetric space corresponding to $(\mathfrak{g}_{\mathbb{C}}, \sigma^c)$. Then $G_{\mathbb{C}}/G^c$ is noncompactly causal. Let H^c be the analytic subgroup of $G_{\mathbb{C}}$ with Lie algebra $\mathfrak{h}$ and X^0 the cone-generating element in $c_{\min}$. Then $H^c \subset G^c$ and $C_{\min} \subset \tilde{C}_{\min}$, since they are the H^c-invariant, respectively G^c-invariant, cones generated by X^0. But then

$$C_{\min} \subset I_{\mathfrak{q}}^{i\mathfrak{g}^c}(\tilde{C}_{\min}) \subset P_{\mathfrak{q}}^{i\mathfrak{g}^c}(\tilde{C}_{\min}).$$

Let now $X \in I_{\mathfrak{q}}^{i\mathfrak{g}^c}(\tilde{C}_{\min}^o)$. By Lemma 4.2.15 we can find an $h \in H^c$ such that $\mathrm{Ad}(h)X \in \mathfrak{a}$. But then, by Proposition 4.2.11, and Proposition 4.2.12:

$$\mathrm{Ad}(h)X \in \mathfrak{a} \cap \tilde{C}_{\min} \subset \mathfrak{a} \cap (it^c \cap \tilde{C}_{\min}) = \mathfrak{a} \cap \tilde{c}_{\min} = c_{\min}.$$

Consequently, the H^c-invariance of $C_{\min}$ proves $I_{\mathfrak{q}}^{i\mathfrak{g}^c}(\tilde{C}_{\min}) = C_{\min}$. By duality we get $P_{\mathfrak{q}}^{i\mathfrak{g}^c}(\tilde{C}_{\max}) = C_{\max}$. Now Lemma 2.1.9 implies the claim. □

4.3 The Linear Convexity Theorem

This section is devoted to the proof of the following *convexity theorem*, which generalizes the convexity theorem of Paneitz [147] and which is an important technical tool in the study of H-invariant cones in $\mathfrak{q}$.

Theorem 4.3.1 (The Linear Convexity Theorem) *Let $\mathcal{M} = G/H$ be a noncompactly causal symmetric space and $\mathfrak{a}$ a maximal abelian subspace of $\mathfrak{q}_p$. Further, let $\mathrm{pr}\colon \mathfrak{q} \to \mathfrak{a}$ be the orthogonal projection, $X \in c_{\max}$ and $h \in H_o$. Then*

$$\mathrm{pr}\,(\mathrm{Ad}(h)X) \in \mathrm{conv}(W_0 \cdot X) + c_{\min}.$$

Lemma 4.3.2 *Let $(\mathfrak{g}, \tau)$ be a noncompactly causal symmetric pair. Then*

1) $\mathfrak{h}_p = \mathrm{Im}(\mathrm{id} + \tau)|_{\mathfrak{n}_+}$ *and* $\mathfrak{q}_k = \mathrm{Im}(\mathrm{id} - \tau)|_{\mathfrak{n}_+}$.

2) $\dim \mathfrak{h}_p = \dim \mathfrak{n}_+ = \dim \mathfrak{q}_k$.

Proof: 1) Recall from Lemma 4.1.1 that $\tau = -\theta$ on $\mathfrak{n}_+$. Thus $X + \tau(X) \in \mathfrak{h}_p$ for $X \in \mathfrak{n}_+$. Conversely, let $X \in \mathfrak{h}_p$. Then we can find $L_\alpha \in \mathfrak{g}_\alpha, \alpha \in \Delta^+$ such that $X = \sum_\alpha [L_\alpha + \tau(L_\alpha)]$. In order to show that $L_\alpha \in \mathfrak{n}_+$ for all $\alpha \in \Delta^+$, we only have to show that $L_\alpha \notin \mathfrak{h}^a = \mathfrak{g}(0, X^0)$, where X^0 is a cone generating element in $\mathfrak{q}_p$. But for $L_\alpha \in \mathfrak{h}^a$ we have $L_\alpha + \tau(L_\alpha) \in \mathfrak{h} \cap \mathfrak{k}$, so $X \in \mathfrak{p}$ implies $L_\alpha + \tau(L_\alpha) = 0$. Therefore these L_α can be omitted in the representation of X.

The second statement is proved in the same way.

2) $\ker(\mathrm{id} + \tau) = \mathfrak{q}$ and $\mathfrak{q} \cap \mathfrak{n}_+ = \{0\}$, since no element in $\mathfrak{q}$ can be an eigenvector of $\mathrm{ad}(X^0) \in \mathfrak{q}$. Similarly, $\ker(\mathrm{id} - \tau) = \mathfrak{h}$ and $\mathfrak{h} \cap \mathfrak{n}_+ = \{0\}$. Now the claim follows from 1). □

Lemma 4.3.3 *Let $L \in C^o_{\max}$ and $X \in \mathrm{Ad}(H_o)L$. If $X \notin \mathfrak{q}_p$, then there is a $Z \in \mathrm{Ad}(H_o)L$ such that*

1) $|\mathrm{pr}_{\mathfrak{q}_k} Z| < |\mathrm{pr}_{\mathfrak{q}_k} X|$,

2) $\mathrm{pr}(X) \in \mathrm{conv}(W_0 \cdot \mathrm{pr}(Z)) + c_{\min}$.

Proof: Assume that $X = \mathrm{Ad}(h)L$ and let

$$Y := \mathrm{pr}(X) = \mathrm{pr}(\mathrm{Ad}(h)L) \in \mathrm{pr}(C^o_{\max}) = c^o_{\max}.$$

Assume for the moment that $Y = \mathrm{pr}_{\mathfrak{q}_p}(X)$, i.e., $X \in Y + \mathfrak{q}_k$. Proposition 4.2.7 shows that $\alpha(Y) > 0$ for all $\alpha \in \Delta_+$. By Lemma 4.3.2 we may write X as a linear combination,

$$X = Y + \sum_{\alpha \in \Delta_+} [Y_\alpha - \tau(Y_\alpha)],$$

with $Y_\alpha \in \mathfrak{g}_\alpha$. As $X \neq Y$, there is a $\beta \in \Delta_+$ such that $Y_\beta \neq 0$. Define

$$W := -\frac{1}{\beta(Y)}(Y_\beta + \tau Y_\beta) \in \mathfrak{h}_p$$

and

$$W_1 := \sum_{\alpha \neq \beta} (Y_\alpha - \tau(Y_\alpha)) \in \mathfrak{q}_k.$$

Now a simple calculation gives

$$\begin{aligned} [W, Y] &= -\frac{1}{\beta(Y)}([Y_\beta, Y] + [\tau(Y_\beta), Y]) \\ &= Y_\beta - \tau(Y_\beta) \in \mathfrak{q}_k. \end{aligned}$$

From Lemma 4.1.1 and Lemma 4.1.2 we derive

$$\begin{aligned}[W, Y_\beta - \tau(Y_\beta)] &= -\frac{1}{\beta(Y)}([Y_\beta + \tau(Y_\beta), Y_\beta - \tau(Y_\beta)]) \\ &= \lambda X^\beta,\end{aligned}$$

where $\lambda = |Y_\beta|^2|\beta|^2/\beta(Y) > 0$. Furthermore, $W_2 := [W, W_1] \in \mathfrak{q}_p \cap \mathfrak{a}^\perp$. Let $Z_t = \mathrm{Ad}(\exp tW)X \in \mathrm{Ad}(H_o)L$. It follows from the above calculations that

$$\begin{aligned}Z_t &= X - t[W, X] + O(t^2) \\ &= (Y - t\lambda H_\beta) + (1-t)(Y_\beta - \tau Y_\beta) + W_1 - tW_2 + O(t^2).\end{aligned}$$

Thus

$$\begin{aligned}|\mathrm{pr}_{\mathfrak{q}_k}(Z_t)|^2 &\leq (1-t)^2|Y_\beta - \tau Y_\beta|^2 + |W_1|^2 + \mu t^2 \\ &= |\mathrm{pr}_{\mathfrak{q}_k} X|^2 - t\left((2-t)\,|Y_\beta - \tau Y_\beta|^2 - \mu t\right) \\ &< |\mathrm{pr}_{\mathfrak{q}_k} X|^2\end{aligned}$$

for $t > 0$ sufficiently small.

We claim that for $t > 0$ small enough,

$$Y - \mathrm{pr}\, Z_t = t\lambda H_\beta + O(t^2) \in c_{\min} = c^*_{\max}\,.$$

To see this, let $V \in c_{\max}$. Then $V = \gamma H_\beta + L$, with $\beta(L) = 0$ and $\gamma > 0$. Thus

$$(Y - \mathrm{pr}(Z_t)|V) = t\lambda\gamma|H_\beta|^2 + O(t^2),$$

and this is positive for small t. This proves the lemma if $\mathrm{pr}_{\mathfrak{q}_p} X \in \mathfrak{a}$.

Assume now that $X_p := \mathrm{pr}_{\mathfrak{q}_p}(X) \neq Y$. There exists a $k \in H_o \cap K$ such that $\mathrm{Ad}(k)X_p \in \mathfrak{a}$, since $\mathfrak{a}$ is maximal abelian in $\mathfrak{q}_p$. On the other hand, $\mathrm{Ad}(K \cap H_o)$ is a group of isometries commuting with $\mathrm{pr}_{\mathfrak{q}_p}$ and $\mathrm{pr}_{\mathfrak{q}_k}$. Therefore we get

$$\mathrm{pr}_{\mathfrak{q}_p}(\mathrm{Ad}(k)X) = \mathrm{Ad}(k)X_p$$

and

$$|\mathrm{pr}_{\mathfrak{q}_k}(\mathrm{Ad}(k)X)| = |\mathrm{Ad}(k)\,\mathrm{pr}_{\mathfrak{q}_k}(X)| = |\mathrm{pr}_{\mathfrak{q}_k}(X)|\,.$$

In particular, $\mathrm{Ad}(k)X \notin \mathfrak{q}_p$. By the first part of the proof we may find a $Z \in \mathrm{Ad}(H_o)\,\mathrm{Ad}(k)X = \mathrm{Ad}(H_o)X$ such that

$$|\mathrm{pr}_{\mathfrak{q}_k}(Z)| < |\mathrm{pr}_{\mathfrak{q}_k}(\mathrm{Ad}(k)X)| = |\mathrm{pr}_{\mathfrak{q}_k}(X)|\,.$$

Now Proposition 4.2.9 shows that

$$\begin{aligned} Y &= \operatorname{pr}(\operatorname{Ad}(k^{-1})\operatorname{Ad}(k)X_p) \\ &\in \operatorname{conv}[W_0 \cdot \operatorname{Ad}(k)X_p] \\ &\subset \operatorname{conv}[W_0 \cdot (\operatorname{pr}(Z) + c_{\min})] \\ &= \operatorname{conv}[\operatorname{pr}(Z)] + c_{\min}, \end{aligned}$$

and this implies the claim. □

Recall that adjoint orbits of semisimple elements in semisimple Lie algebras are closed according to a well-known theorem of Borel and Harish-Chandra (cf. [168], p. 106). The following lemma, taken from [20], p. 58, is a generalization of this fact.

Lemma 4.3.4 *Let G/H be a symmetric space with G semisimple. If $X \in \mathfrak{q}$ is semisimple, then the orbit $\operatorname{Ad}(H)X$ is closed in $\mathfrak{q}$.* □

Define a relation $\preceq$ on $\mathfrak{q}$ via

$$X \preceq Y :\Longleftrightarrow \begin{cases} |\operatorname{pr}_{\mathfrak{q}_k}(Y)| \le |\operatorname{pr}_{\mathfrak{q}_k}(X)| \\ \operatorname{pr}(X) \in \operatorname{conv}[W_0 \cdot \operatorname{pr}(Y)] + c_{\min} \end{cases}$$

Lemma 4.3.5 *Let $X \in \mathfrak{q}$. Then the set $\{Y \in \mathfrak{q} \mid X \preceq Y\}$ is $(H_o \cap K)$-invariant and closed in $\mathfrak{q}$.*

Proof: The $(H_o \cap K)$-invariance follows from Proposition 4.2.9 and the $(H_o \cap K)$-equivariance of $\operatorname{pr}_{\mathfrak{q}_k}$. Now assume that $Y_j \in \{Y \in \mathfrak{q} \mid X \preceq Y\}$, $j \in \mathbb{N}$, and that $Y_j \to Y_0 \in \mathfrak{q}$. As $|\operatorname{pr}_{\mathfrak{q}_k}(Y_j)| \le |\operatorname{pr}_{\mathfrak{q}_k}(X)|$, it follows that $|\operatorname{pr}_{\mathfrak{q}_k}(Y_0)| \le |\operatorname{pr}_{\mathfrak{q}_k}(X)|$. Furthermore, there are $Z_j \in \operatorname{conv}[W_0 \cdot \operatorname{pr}(Y_j)]$ and $L_j \in c_{\min}$ such that $\operatorname{pr} X = Z_j + L_j$. But the union of $\operatorname{conv}[W_0 \cdot \operatorname{pr}(Y_j)]$, $j \ge 0$ is bounded, so $\{Z_j\}$ has a convergent subsequence and one easily sees that the limit point is in $\operatorname{conv}(W_0 \cdot Y)$. Thus we can assume that $\{Z_j\}$ converges to $Z_0 \in \operatorname{conv}(W_0 Y)$. Therefore $L_j = \operatorname{pr}(X) - Z_j \to \operatorname{pr}(X) - Z^0 \in c_{\min}$, since $c_{\min}$ is closed. □

Lemma 4.3.6 *Let $X \in c_{\max}^o$ and let $L \in \operatorname{Ad}(H_o)X$. Then the set*

$$M_X(L) := \{Y \in \operatorname{Ad}(H_o)X \mid L \preceq Y\}$$

is compact.

Proof: By Lemma 4.3.4 and Lemma 4.3.5 it follows that $M(L) = M_X(L)$ is closed. Thus we only have to show that $M(L)$ is also bounded. Let $\operatorname{Ad}(h)X \in M(L)$. Write $h = k \exp Z$ with $k \in K \cap H_o$ and $Z \in \mathfrak{h}_p$.

As $\mathrm{Ad}(k)$ is an isometry, $|\mathrm{Ad}(h)X| = |\mathrm{Ad}(\exp Z)X|$. Thus we may as well assume that $h = \exp Z$. Since $\mathrm{ad}_{\mathfrak{q}}(Z)$ is symmetric, we may write $X = \sum_{\lambda\in\mathbb{R}} X_\lambda$, with $X_\lambda \in \mathfrak{q}(\lambda, Z)$. Thus

$$\mathrm{Ad}(h)X = \sum_{\lambda} e^{\lambda} X_\lambda .$$

Now $\theta(\mathfrak{q}(\lambda, Z)) = \mathfrak{q}(-\lambda, Z)$ as $Z \in \mathfrak{h}_p$. From $\theta(X) = -X$ we get

$$X = X_0 + \sum_{\lambda>0} [X_\lambda + X_{-\lambda}] = X_0 + \sum_{\lambda>0} [X_\lambda - \theta(X_\lambda)] , \tag{4.26}$$

with $X_0 \in \mathfrak{q}_p$. Thus

$$\begin{aligned} \mathrm{Ad}(h)X &= X_0 + \sum_{\lambda>0} \left[e^{\lambda} X_\lambda - e^{-\lambda}\theta(X_\lambda)\right] \\ &= X_0 + \sum_{\lambda>0} \sinh(\lambda)\,[X_\lambda + \theta(X_\lambda)] + \sum_{\lambda>0} \cosh(\lambda)\,[X_\lambda - \theta(X_\lambda)] . \end{aligned}$$

In particular,

$$\mathrm{pr}_{\mathfrak{q}_k}(\mathrm{Ad}(h)X) = \sum_{\lambda>0} \sinh(\lambda)\,(X_\lambda + \theta(X_\lambda))$$

and

$$\mathrm{pr}_{\mathfrak{q}_p}(\mathrm{Ad}(h)X) = X_0 + \sum_{\lambda>0} \cosh(\lambda)\,(X_\lambda - \theta(X_\lambda)) .$$

The eigenspaces are orthogonal to each other and $\mathrm{Ad}(h)X \in M(L)$, so we find

$$\begin{aligned} \sum_{\lambda>0} \sinh(\lambda)^2 |X_\lambda + \theta(X_\lambda)|^2 &= |\mathrm{pr}_{\mathfrak{q}_k}(\mathrm{Ad}(h)X)|^2 \\ &\le |\mathrm{pr}_{\mathfrak{q}_k}(L)|^2 \\ &\le |L|^2 . \end{aligned}$$

Furthermore, $|X_\lambda \pm \theta(X_\lambda)|^2 = |X_\lambda|^2 + |\theta(X_\lambda)|^2$. Hence $|X_\lambda - \theta(X_\lambda)|^2 = |X_\lambda + \theta(X_\lambda)|^2$. From $\cosh(t)^2 = 1 + \sinh(t)^2$ we now obtain

$$\begin{aligned} |\mathrm{pr}_{\mathfrak{q}_p}(\mathrm{Ad}(h)X)|^2 &= |X_0|^2 + \sum_{\lambda>0} \cosh(\lambda)^2 |X_\lambda + \theta(X_\lambda)|^2 \\ &= |X_0|^2 + \sum_{\lambda>0} |X_\lambda - \theta(X_\lambda)|^2 + \sum_{\lambda>0} \sinh(\lambda)^2 |X_\lambda + \theta(X_\lambda)|^2 \\ &= |X|^2 + |\mathrm{pr}_{\mathfrak{q}_k}(\mathrm{Ad}(h)X)|^2 \\ &\le |X|^2 + |L|^2 . \end{aligned}$$

Thus $|\operatorname{Ad}(h)X|^2 \leq 2|L|^2 + |X|^2$, which proves the claim. □

Now we are ready to prove Theorem 4.3.1: Let $X \in c_{\max}$. As $\operatorname{conv}(W_0 \cdot X) + c_{\min}$ is closed, we may assume that $X \in c_{\max}^o$. Let $L = \operatorname{Ad}(h)X \in \operatorname{Ad}(H)X$. Since H is essentially connected, we have $\operatorname{Ad}(H)X = \operatorname{Ad}(H_o)X$, so we may assume that $h \in H_o$. Since $M_X(L)$ is compact, the map

$$M_X(L) \ni Y \mapsto |\operatorname{pr}_{\mathfrak{q}_k}(Y)|^2 \in \mathbb{R}$$

attains its minimum at a point $Y = \operatorname{Ad}(a)L$, $a \in H$. By Lemma 4.3.3 we must have $Y \in \mathfrak{q}_p$. Moreover,

$$\operatorname{pr}(L) \in \operatorname{conv}(W_0 \cdot \operatorname{pr}(Y)) + c_{\min} .$$

Because of $Y \in \mathfrak{q}_p$, we can find a $k \in K \cap H_o$ such that $\operatorname{Ad}(k)Y = \operatorname{Ad}(kah)X \in \mathfrak{a}$. By Lemma 4.2.13, part 2), it follows that $kah \in K \cap H_o$. But then $ah \in K \cap H$. Hence Proposition 4.2.9 shows that

$$\operatorname{pr}(Y) \in \operatorname{conv}(W_0 \cdot X),$$

which in turn yields

$$\begin{aligned} \operatorname{pr}(\operatorname{Ad}(h)X) = \operatorname{pr}(L) &\in \operatorname{conv}[W_0 \cdot \operatorname{pr}(Y)] + c_{\min} \\ &= \operatorname{conv}[W_0 \cdot (\operatorname{conv} W_0 \cdot X)] + c_{\min} \\ &= \operatorname{conv}(W_0 \cdot X) + c_{\min} \end{aligned}$$

and therefore proves the theorem. □

4.4 The Classification

Let $\mathcal{M} = G/H$ be a noncompactly causal symmetric space and $\mathfrak{a}$ a maximal abelian subspace of $\mathfrak{q}_p$. Recall the orthogonal projection $\operatorname{pr}\colon \mathfrak{q} \to \mathfrak{a}$ and the corresponding intersection and projection operations for cones. Theorem 4.1.6 implies that any $C \in \operatorname{Cone}_H(\mathfrak{q})$ contains one of the two minimal cones and consequently is contained in the corresponding maximal cone. Then Proposition 4.2.11 implies that

$$c_{\min} \subset I(C) \subset P(C) \subset c_{\max}.$$

Clearly $I(C)$ is W_0-invariant, but Theorem 4.3.1 shows that $P(C)$ is also W_0-invariant. The goal of this section is to show that any W_0-invariant cone between $c_{\min}$ and $c_{\max}$ arises as $I(C)$ for some $C \in \operatorname{Cone}_H(\mathfrak{q})$ and that C is uniquely determined by $I(C)$. We start with a closer examination of the projection pr.

Let $\mathfrak{t}^c$ be a Cartan subalgebra of $\mathfrak{k}^c$ containing $i\mathfrak{a}$. Then $\mathfrak{t}^c$ is a Cartan subalgebra of $\mathfrak{g}^c$ since the simple factors of $\mathfrak{g}^c$ are Hermitian and the analytic subgroup T^c of $e^{\mathrm{ad}(\mathfrak{g}^c)}$ with Lie algebra $\mathrm{ad}(\mathfrak{t}^c)$ is compact. Consider the closed subgroup $\{\varphi \in T^c \mid \tau\varphi\tau = \varphi^{-1}\}$ of T^c. Its connected component $T_a := e^{\mathrm{ad}(i\mathfrak{a})}$ then is a compact connected subgroup of T^c with Lie algebra $\mathrm{ad}(i\mathfrak{a})$. We normalize the Haar measure on T_a in such a way that it has total mass of 1.

Lemma 4.4.1 *Let* $X \in \mathfrak{q}$. *Then* $\mathrm{pr}(X) = \int_{T_a} \varphi(X)\, d\varphi$.

Proof: Write $X = \mathrm{pr}(X) + \sum_{\alpha\in\Delta^+} [L_\alpha - \tau(L_\alpha)]$, with $L_\alpha \in \mathfrak{g}_\alpha$. Then

$$\varphi(X) = \mathrm{pr}(X) + \sum_{\alpha\in\Delta^+} [\varphi^\alpha(L_\alpha) - \varphi^{-\alpha}\tau(L_\alpha)] \ ,$$

where $(e^{\mathrm{ad}\, Y})^\alpha = e^{\alpha(Y)}$ for $Y \in \mathfrak{a}_{\mathbb{C}}$. As $T_a \ni \varphi \mapsto \varphi^\alpha \in \mathbb{C}^*$ is a unitary character, it follows that $\int_{T_a} \varphi^\alpha\, d\varphi = 0$. Hence

$$\begin{aligned}\int_{T_a} \varphi(X) d\varphi &= \mathrm{pr}(X) + \sum_{\alpha\in\Delta^+} \left[\left(\int_{T_a} \varphi^\alpha d\varphi\right) L_\alpha - \left(\int_{T_a} \varphi^{-\alpha} d\varphi\right) \tau(L_\alpha)\right] \\ &= \mathrm{pr}(X)\,.\end{aligned}$$

This implies the claim. □

For $g \in G$ define the linear map $\Phi_g : \mathfrak{g}_{\mathbb{C}} \to \mathfrak{g}_{\mathbb{C}}$ by

$$\Phi_g(X) := \int_{T_a} \varphi\, \mathrm{Ad}(g)\varphi^{-1}(X)\, d\varphi \qquad (4.27)$$

and set $\mathcal{H} := \{\Phi_h \in \mathrm{End}(\mathfrak{a}) \mid h \in H_o\}$. Then Lemma 4.4.1 implies that

$$\mathcal{H} = \{\Phi \in \mathrm{End}(\mathfrak{a}) \mid \exists h \in H_o \,:\, \Phi = \mathrm{pr} \circ \mathrm{Ad}(h)\}\,. \qquad (4.28)$$

Lemma 4.4.2 *1) Let* $Y \in \mathfrak{a}$ *and* $g \in G$. *Then* $\Phi_g \circ \mathrm{ad}\, Y = \mathrm{ad}\, Y \circ \Phi_g$.

2) $\tau \circ \Phi_g = \Phi_{\tau(g)} \circ \tau$ *for all* $g \in G$. *In particular,* $\tau \circ \Phi_h = \Phi_h \circ \tau$ *for all* $h \in H$.

3) $\Phi_b(\mathfrak{z}_\mathfrak{g}(\mathfrak{a})) \subset \mathfrak{z}_\mathfrak{g}(\mathfrak{a})$. *If* $h \in H$ *and* $X \in \mathfrak{a}$, *then* $\Phi_h(X) = \mathrm{pr}(\mathrm{Ad}(h)X)$.

4) Let $k \in N_{K\cap H}(\mathfrak{a})$ *and* $h \in H$. *Then* $\mathrm{Ad}(k)\Phi_h(X) = \Phi_{kh}(X)$.

Proof: 1) Let $Y \in \mathfrak{a}$. Then $e^{\mathrm{ad}\, tiY} \in T_a$ for all $t \in \mathbb{R}$. Hence $\Phi_g e^{ti\, \mathrm{ad}\, Y} = e^{ti\, \mathrm{ad}\, Y}\Phi_g$. Differentiating at $t = 0$ yields $\Phi_g \circ \mathrm{ad}\, Y = \mathrm{ad}\, Y \circ \Phi_g$.

2) Since $\tau\varphi\tau = \varphi^{-1}$ for all $\varphi \in T_a$ and T_a is unimodular, it follows that

$$\begin{aligned}\tau\Phi_g\tau(X) &:= \int_{T_a} \tau\varphi\, \mathrm{Ad}(g)\varphi^{-1}\tau(X)\, d\varphi \\ &= \int_{T_a} \varphi^{-1}\, \mathrm{Ad}(\tau(g))\varphi(X)\, d\varphi \\ &= \Phi_{\tau(g)}(X).\end{aligned}$$

3) From part 1) we get $\Phi_g(\mathfrak{z}_\mathfrak{s}(\mathfrak{a})) \subset \mathfrak{z}_\mathfrak{s}(\mathfrak{a})$. If $X \in \mathfrak{a}$, then $\varphi(X) = X$ for all $\varphi \in T_a$. Hence

$$\Phi_h(X) = \int_T \varphi(\mathrm{Ad}(h)X)\, d\varphi.$$

and this equals $\mathrm{pr}(\mathrm{Ad}(h)X)$ by Lemma 4.4.1.

4) Note first that pr is $N_{K\cap H}(\mathfrak{a})$-equivariant by Lemma 4.2.6. Now the claim follows from part 3). □

Lemma 4.4.3 *Let $h \in H$. Then $\Phi_h^* = \Phi_{\theta(h^{-1})}$. In particular, $\mathcal{H}^* = \mathcal{H}$.*

Proof: Note that $\mathrm{Ad}(g)^* = \mathrm{Ad}(\theta(g)^{-1})$ with the usual inner product on $\mathfrak{g}$. Thus for $X, Y \in \mathfrak{a}$:

$$\begin{aligned}(\Phi_h(X)|Y) &= (\mathrm{pr}(\mathrm{Ad}(h)X)|Y) \\ &= (\mathrm{Ad}(h)X|Y) \\ &= (X|\,\mathrm{Ad}(\theta(h^{-1}))Y) \\ &= (X|\,\mathrm{pr}\,\mathrm{Ad}(\theta(h^{-1}))Y) \\ &= (X|\Phi_{\theta(h^{-1})}(Y)).\end{aligned}$$

From this the lemma follows. □

Remark 4.4.4 Theorem 4.3.1 shows that a convex cone c with $c_{\min} \subset c \subset c_{\max}$ is W_0-invariant if and only if it is $\mathcal{H}$-invariant. □

Lemma 4.4.5 *Let c be an $\mathcal{H}$-invariant cone in $\mathfrak{a}$. Then*

1) *c^* is $\mathcal{H}$-invariant.*

2) *If c is closed and regular, then we can choose a cone-generating element $X^0 \in c$ so that*

$$c_{\min}(X^0) \subset c \subset c_{\max}(X^0).$$

Proof: Let $X \in c \setminus \{0\}$. Let $\tilde{X} := [1/\#W_0] \sum_{w \in W_0} w \cdot X$. By Remark 4.4.4, $\tilde{X} \in c$. Since c is proper, we see that $X \neq 0$. The W_0-invariance of $\tilde{X}$ shows that all $\alpha \in \Delta_0$ vanish on $\tilde{X}$ and hence $\tilde{X} \in \mathfrak{z}(\mathfrak{h}^a)$. Thus we may choose a cone-generating element $X^0 \in c$. Set $c_{\min} := c_{\min}(X^0)$ etc. Then the $\mathcal{H}$-invariance of c shows that

$$c_{\min} = \mathrm{pr}(C_{\min}) = \mathrm{pr}\left(\overline{\mathrm{conv}\left[\mathrm{Ad}(H_o)\left(\mathbb{R}^+ X^0\right)\right]}\right) \subset \mathcal{H} \cdot c = c.$$

The regularity of c is equivalent to the regularity of c^*. But c^* is $\mathcal{H}$-invariant by Lemma 4.4.3. Applying what we have already proved to c^* implies that $c_{\min} \subset c^*$ hence by duality $c \subset c_{\max}$. □

We now turn to the question of which cones in $\mathfrak{a}$ occur as intersections $I(C)$ with $C \in \mathrm{Cone}_H(\mathfrak{q})$. Recall the extension operators for cones from Remark 2.1.12. We set $E := E_{\mathfrak{a}}^{\mathfrak{q},H}$. Then we have

$$C_{\min}(X^0) = E(\mathbb{R}^+ X^0)$$

and

$$P\left(E(U)\right) = \overline{\mathrm{cone}\left\{\mathcal{H}(U)\right\}}. \tag{4.29}$$

Theorem 4.4.6 (Extension, Intersection and Projection) *Let $\mathcal{M} = G/H$ be a noncompactly causal symmetric space and $\mathfrak{a}$ a maximal abelian subspace of $\mathfrak{q}_p$. Then for any $c \in \mathrm{Cone}_{W_0}(\mathfrak{a})$ we have $E(c) \in \mathrm{Cone}_H(\mathfrak{q})$ and*

$$c = P(E(c)) = I(E(c)).$$

Proof: It follows from Lemma 4.4.5 that there exists a cone-generating element $X^0 \in c$. As $C_{\min} := C_{\min}(X^0)$ is generated by the H-orbit of $X^0 \in c_{\min} := c_{\min}(X^0)$ (cf. Lemma 4.2.8), it follows that $C_{\min} \subset E(c_{\min})$. On the other hand, $c_{\min} \subset C_{\min}$ so that $E(c_{\min}) \subset C_{\min}$ and hence $E(c_{\min}) = C_{\min}$. Now $c_{\max} \subset C_{\max}$ (cf. Proposition 4.2.11) implies, $E(c_{\max}) \subset C_{\max}$.

So far we know $C_{\min} \subset E(c) \subset C_{\max}$. But $E(c)$ is generated by $\mathrm{Ad}(H)c = \mathrm{Ad}(H_o)c$ (cf. Theorem 3.1.18), so $E(c) \in \mathrm{Cone}_H(\mathfrak{q})$.

Now let $X \in c$ and $h \in H$. Then Lemma 4.4.2, part 3, implies

$$\mathrm{pr}(\mathrm{Ad}(h)X) = \Phi_h(X) \in \mathcal{H} \cdot X \subset c.$$

Moreover, $P(E(c)) \subset c$, since $E(c)$ is generated by $\mathrm{Ad}(H)c = \mathrm{Ad}(H_o)c$ and c is $\mathcal{H}$-invariant by Remark 4.4.4. But clearly $c \subset E(c)$, which implies $c \subset P(E(c))$. Thus $P(E(c)) = c$. Finally, $c \subset I(E(c)) \subset P(E(c)) = c$ proves $I(E(c)) = c$. □

Theorem 4.4.7 *Let $\mathcal{M} = G/H$ be a noncompactly causal symmetric space and $\mathfrak{a}$ a maximal abelian subspace of $\mathfrak{q}_p$. Then for a closed cone c in $\mathfrak{a}$ the following conditions are equivalent:*

1) *c is W_0-invariant and $c_{\min} \subset c \subset c_{\max}$ for a suitable chosen minimal cone.*

2) *c is regular and $\mathcal{H}$-invariant.*

3) *There exists a cone $C \in \mathrm{Cone}_H(\mathfrak{q})$ such that $P(C) = c$.*

4) *There exists a cone $C \in \mathrm{Cone}_H(\mathfrak{q})$ such that $I(C) = c$.*

Proof: If 1) holds, then by the convexity theorem 4.3.1, $\Phi_h(X) \in c$ for all $h \in H_o$ and $X \in c$. Thus 2) follows. 3) and 4) follow from 2) by Theorem 4.4.6. If 3) holds, then $c_{\min} \subset I(C) \subset c \subset c_{\max}$, which implies 2) and 1). Similarly, 4) implies 2) and 1). □

The last step in our classification program is to show that H-invariant regular cones in $\mathfrak{q}$ are indeed completely determined by their intersections with any maximal abelian subspace of $\mathfrak{q}_p$. In order to do this we again have to use the structure theory provided by the fact that $\mathfrak{g}^c$ has Hermitian simple factors.

Definition 4.4.8 Let $\mathcal{M} = G/H$ be a noncompactly causal symmetric space and $G_{\mathbb{C}}/G^c$ be any symmetric space corresponding to $(\mathfrak{g}_{\mathbb{C}}, \sigma^c)$. A cone $C \in \mathrm{Cone}_H(\mathfrak{q})$ is called *G^c-extendable* if there exists a cone $\tilde{C} \in \mathrm{Cone}_{G^c}(i\mathfrak{g}^c)$ such that $C = I_{\mathfrak{q}}^{i\mathfrak{g}^c}(\tilde{C})$. □

If $C \in \mathrm{Cone}_H(\mathfrak{q})$ is G^c-extendable, we can find $\tilde{C} \in \mathrm{Cone}_{G^c}(i\mathfrak{g}^c)$ such that $C = I_{\mathfrak{q}}^{i\mathfrak{g}^c}(\tilde{C})$ and $-\tau(\tilde{C}) = \tilde{C}$. In fact, if $\tilde{C}_1 \in \mathrm{Cone}_{G^c}(i\mathfrak{g}^c)$ satisfies $C = I_{\mathfrak{q}}^{i\mathfrak{g}^c}(\tilde{C}_1)$, we simply set $\tilde{C} := \tilde{C}_1 \cap \big(-\tau(\tilde{C}_1)\big)$.

We have seen in Theorem 4.2.16 that the minimal and the maximal cones in $\mathfrak{q}$ are G^c-extendable for any symmetric space $G_{\mathbb{C}}/G^c$ corresponding to $(\mathfrak{g}_{\mathbb{C}}, \sigma^c)$. We will see in the next section that this is indeed true for all cones in $\mathrm{Cone}_H(\mathfrak{q})$. What we need now is a much weaker statement.

Lemma 4.4.9 *Let $C \in \mathrm{Cone}_H(\mathfrak{q})$. Then $E_{\mathfrak{q}}^{i\mathfrak{g}^c, G^c}(C) \in \mathrm{Cone}_{G^c}(i\mathfrak{g}^c)$.*

Proof: As $C_{\min} \subset C$ for a suitable cone-generating element, it follows that

$$\tilde{C}_{\min} = E_{\mathfrak{q}}^{i\mathfrak{g}^c, G^c}(C_{\min}) \subset E_{\mathfrak{q}}^{i\mathfrak{g}^c, G^c}(C).$$

But $\tilde{C}_{\max}$ is G^c-invariant, so $C \subset C_{\max} \subset \tilde{C}_{\max}$ implies

$$E_{\mathfrak{q}}^{i\mathfrak{g}^c, G^c}(C) \subset \tilde{C}_{\max}.$$

Therefore $E_{\mathfrak{q}}^{i\mathfrak{g}^c, G^c}(C)$ is regular. □

Theorem 4.4.10 *Let $\mathcal{M} = G/H$ be a noncompactly causal symmetric space and $C \in \mathrm{Cone}_H(\mathfrak{q})$. If $X \in C^o$, then X is semisimple and* $\mathrm{ad}\,X$ *has real eigenvalues and there exists an element $h \in H_o$ such that* $\mathrm{Ad}(h)X \in I(C^o)$.

Proof: According to Lemma 4.4.9, we know that $C \subset \tilde{C}_{\max}$ for a suitable choice of a cone-generating element X^0. Since X^0 is contained in the interiors of C and $\tilde{C}_{\max}$, we see that $C^o \subset \tilde{C}^o_{\max}$. But then Lemma 4.2.15 shows that there exists an $h \in H_o$ such that $\mathrm{Ad}(h)X \in \mathfrak{a} \cap C^o = I(C^o)$. □

From Theorem 4.4.10 we immediately obtain the following theorem, which completes our classification program:

Theorem 4.4.11 (Reconstruction of Cones) *Let $\mathcal{M} = G/H$ be a noncompactly causal symmetric space and $C \in \mathrm{Cone}_H(\mathfrak{q})$. Then*

$$C^o = \mathrm{Ad}(H)I(C^o)$$

and

$$C = E^{\mathfrak{q},H}_{\mathfrak{a},\{1\}}\left(I^{\mathfrak{a}}_{\mathfrak{q}}(C^o)\right) = \overline{\mathrm{Ad}(H)I(C^o)}. \qquad \square$$

Corollary 4.4.12 *Let $\mathcal{M} = G/H$ be a noncompactly causal symmetric space and $C \in \mathrm{Cone}_{H_o}(\mathfrak{q})$. Then $C \in \mathrm{Cone}_H(\mathfrak{q})$.*

Proof: According to Theorem 3.1.18, we have

$$H = H_o Z_{H\cap K}(\mathfrak{a})$$

and $Z_{H\cap K}(\mathfrak{a})$ acts trivially on $I(C^o)$. Therefore the claim follows from Theorem 4.4.11. □

4.5 Extension of Cones

Assume that $\mathcal{M}$ is a noncompactly causal symmetric space and $G_{\mathbb{C}}/G^c$ any symmetric space corresponding to $(\mathfrak{g}^c, \sigma^c)$. The goal of this section is to show that any $C \in \mathrm{Cone}_H(\mathfrak{q})$ is G^c-extendable. We note first that Remark 4.2.14 shows that this is the case if $\mathfrak{g}$ carries a complex structure, since then $\mathfrak{g}^c$ is Hermitian. So we may assume that $G_{\mathbb{C}}/G^c$ is noncompactly causal and by Corollary 4.4.12 we only have to show G^c_o-extendability. In particular, we may assume that $G_{\mathbb{C}}$ is simply connected and that G^c and G are the analytic subgroups of $G_{\mathbb{C}}$ with Lie algebras $\mathfrak{g}^c$ and $\mathfrak{g}$.

Fix a Cartan subalgebra $\mathfrak{t}^c$ of $\mathfrak{k}^c$ as before and consider

$$\tilde{W}_0 = N_{K^c}(\mathfrak{t}^c)/Z_{K^c}(\mathfrak{t}^c),$$

where K^c is the analytic subgroup of G^c corresponding to the Lie algebra $\mathfrak{k}^c$. Then K^c is compact and $\tilde{W}_0$ is the Weyl group for the pair $(\mathfrak{k}^c, \mathfrak{t}^c)$. We choose a cone-generating element $X^0 \in \mathfrak{q}_p \subset i\mathfrak{h}_k + \mathfrak{q}_p$ and a positive system $\tilde{\Delta}^+$ for $\tilde{\Delta} = \Delta(\mathfrak{g}_\mathbb{C}, \mathfrak{t}^c)$ as in (4.12). Let

$$\mathcal{C}(\Delta_0^+) := \{X \in \mathfrak{a} \mid \forall \alpha \in \Delta_0^+ : \alpha(X) > 0\}$$

be the corresponding open *Weyl chamber* in $\mathfrak{a}$. Similarly, let $\tilde{\mathcal{C}}(\tilde{\Delta}_0^+)$ be the chamber in $i\mathfrak{t}^c$. We define $C := \overline{\mathcal{C}(\Delta_0^+)}$ and $\tilde{C} := \overline{\tilde{\mathcal{C}}(\tilde{\Delta}_0^+)}$.

Remark 4.5.1 Let C' be the closure of the Weyl chamber in $\mathfrak{a}' := \mathfrak{a} \cap [\mathfrak{h}^a, \mathfrak{h}^a]$ corresponding to Δ_0^+. Then $C = C' \oplus \mathbb{R}X^0$ and $C^* = (C')^* \subset \mathfrak{a}'$, where $(C')^*$ is the dual of C' in $\mathfrak{a}'$. If $X = X_1 + X_2 \in C$ with $X_1 \in C'$ and $X_2 \in \mathbb{R}X^0$, then

$$s(X) = s(X_1) + X_2$$

for all $s \in W_0$. Similar things hold for $\tilde{C}$ and $\tilde{W}_0$. □

Lemma 4.5.2 *1) Let $\alpha \in \tilde{\Delta}_0$ be such that $\alpha|_\mathfrak{a} \neq 0$. Then α is positive if and only if $-\tau(\alpha)$ is positive.*

2) $C = \tilde{C} \cap \mathfrak{a} = \mathrm{pr}(\tilde{C})$, where $\mathrm{pr}\colon i\mathfrak{t}^c \to \mathfrak{a}$ is the orthogonal projection.

3) $C^* = \mathrm{pr}(\tilde{C}^*)$.

Proof. 1) This follows from (4.12).

2) Obviously $C \subset \tilde{C} \cap \mathfrak{a} \subset \mathrm{pr}(\tilde{C})$. Thus we only have to show that $\mathrm{pr}(\tilde{C}) \subset C$. Let $X \in \tilde{C}$. Then

$$\mathrm{pr}(X) = \frac{1}{2}\left[X - \tau(X)\right].$$

Let $\alpha \in \Delta_0^+$ and choose $\beta \in \tilde{\Delta}_0^+$ such that $\beta|_\mathfrak{a} = \alpha$. Then

$$\alpha = \frac{1}{2}(\beta - \tau\beta).$$

Then 1) shows $-\tau\beta \in \tilde{\Delta}_0^+$. Thus

$$\begin{aligned} \alpha(\mathrm{pr}(X)) &= \beta\left(\frac{1}{2}[X - \tau(X)]\right) \\ &= \frac{1}{2}(\beta(X) + [-\tau\beta(X)]) \geq 0. \end{aligned}$$

Thus $\mathrm{pr}(X) \in C$ as claimed.

3) This follows from 2) and Lemma 2.1.8. □

Consider the following groups (cf. Theorem 4.2.6):

$$\tilde{W}_0(\tau) := \{w \in \tilde{W}_0 \mid \tau \circ w = w \circ \tau\} \tag{4.30}$$
$$\tilde{W}_0^{\mathfrak{a}} := \{w \in \tilde{W}_0 \mid w|_{\mathfrak{a}} = \mathrm{id}\}. \tag{4.31}$$

Lemma 4.5.3 *The restriction to $\mathfrak{a}$ induces an exact sequence*

$$\{1\} \longrightarrow \tilde{W}_0(\tau) \cap \tilde{W}_0^{\mathfrak{a}} \longrightarrow \tilde{W}_0(\tau) \longrightarrow W_0 \longrightarrow \{1\}.$$

Proof. We only have to show that the restriction to $\mathfrak{a}$ induces a surjective map $\tilde{W}_0(\tau) \to W_0$. So let $w \in W_0$ and $k \in N_{K\cap H}(\mathfrak{a})$ be such that $\mathrm{Ad}(k)|_{\mathfrak{a}} = w$. Note that $\mathrm{Ad}(k)(i\mathfrak{t}^c \cap i\mathfrak{k})$ is a maximal abelian subspace of $i\mathfrak{m} = i\mathfrak{z}_{\mathfrak{k}}(\mathfrak{a})$. Therefore there exists an $h \in M_o := Z_K(\mathfrak{a})_o = Z_{K\cap H}(\mathfrak{a})_o$ such that $\mathrm{Ad}(h)\,[\mathrm{Ad}(k)(i\mathfrak{t}^c \cap i\mathfrak{m})] = i\mathfrak{t}^c \cap i\mathfrak{m}$. But then $hk \in N_{K\cap H}(\mathfrak{a}) \cap N_{K\cap H}(i\mathfrak{t}^c \cap i\mathfrak{m}) \subset N_{K\cap H}(i\mathfrak{t}^c)$ and therefore hk corresponds to an element $\tilde{w}$ of $\tilde{W}_0$. As $\tilde{w}$ leaves $\mathfrak{a}$ and $\tilde{\mathfrak{a}} \cap i\mathfrak{m}$ invariant, it commutes with τ. Thus $\tilde{w} \in \tilde{W}_0(\tau)$. Finally, we note that

$$\tilde{w} \cdot X = \mathrm{Ad}(hk)(X) = \mathrm{Ad}(h)\,\mathrm{Ad}(k)(X) = \mathrm{Ad}(h)(w \cdot X) = w \cdot X$$

for $X \in \mathfrak{a}$, which implies the claim. □

Remark 4.5.4 Let $\tilde{w} \in \tilde{W}_0(\tau)$ and $w = \tilde{w}|_{\mathfrak{a}}$. Then we have $w \circ \mathrm{pr}_{\mathfrak{a}} = \mathrm{pr}_{\mathfrak{a}} \circ \tilde{w}$. □

In view of Remark 4.5.1, the following lemma is a consequence of Lemma 8.3 in [45], p. 459:

Lemma 4.5.5 *1) Let $X \in \tilde{C}$. Then*

$$\mathrm{conv}(\tilde{W}_0 \cdot X) = \bigcup_{w\in \tilde{W}_0} w\left[\tilde{C} \cap (X - \tilde{C}^*)\right] = \bigcap_{w\in W_0} w\left(X - \tilde{C}^*\right).$$

2) Let $X \in C$. Then

$$\mathrm{conv}(W_0 \cdot X) = \bigcup_{w\in W_0} w\,[C \cap (X - C^*)] = \bigcap_{w\in \tilde{W}_0} w\,(X - C^*).$$ □

Theorem 4.5.6 *Let $(\mathfrak{g}, \tau)$ be a noncompactly causal pair, $\mathfrak{a}$ a maximal abelian subspace of $\mathfrak{q}_{\mathfrak{p}}$, and $\mathfrak{t}^c$ a Cartan subalgebra of $\mathfrak{k}^c$ containing $i\mathfrak{a}$. Further, let W_0 and $\tilde{W}_0$ be the Weyl groups of $(\mathfrak{g}, \mathfrak{a})$ and $(\mathfrak{k}^c, \mathfrak{t}^c)$, respectively. Then $\mathrm{pr}(\mathrm{conv}\,\tilde{W}_0 \cdot X) = \mathrm{conv}(W_0 \cdot X)$ for all $X \in \mathfrak{a}$, where $\mathrm{pr}\colon i\mathfrak{t}^c \to \mathfrak{a}$ is the orthogonal projection.*

Proof: Set $\tilde{L} = \text{conv}(\tilde{W}_0 \cdot) \subset i\mathfrak{t}^c$, $L = \text{conv}(W_0 \cdot X) \subset \mathfrak{a}$ and $F := \text{pr}(\tilde{L})$. Fix $w \in W_0$. According to Lemma 4.5.3 we can choose a $\tilde{w} \in \tilde{W}_0(\tau)$ such that $\tilde{w}|_{\mathfrak{a}} = w$. Then Remark 4.5.4 and the $\tilde{W}_0$-invariance of $\tilde{L}$ imply that

$$w(F) = w \cdot [\text{pr}(\tilde{L})] = \text{pr}(\tilde{w}\tilde{L}) = F.$$

Therefore F is convex and W_0-invariant. As $X \in F$, it follows that $L \subset F$. To show the converse, we choose $w \in W_0$ such that

$$w(X) \in C \subset \tilde{C}.$$

Choose $\tilde{w} \in \tilde{W}_0$ with $\tilde{w}|_{\mathfrak{a}} = w$. Using Lemma 4.5.5 and Lemma 4.5.2, part 3), we find

$$F \subset \text{pr}(\tilde{w}(X) - \tilde{C}^*) = w(X) - C^*.$$

This together with Lemma 4.5.5, part 2), shows

$$F \cap C \subset [w(X) - C^*] \cap C \subset \text{conv}[W_0 \cdot w(X)] = \text{conv}(W_0 \cdot X)$$

and hence the claim. □

Corollary 4.5.7 *Let c be a W_0-invariant cone contained in $\mathfrak{a}$. Then $\tilde{c} := \text{conv}\{\tilde{W}_0(c)\}$ is a $\tilde{W}_0$-invariant cone in $\tilde{\mathfrak{a}}$ with $\text{pr}(\tilde{c}) = c = \tilde{c} \cap \mathfrak{a}$.*

Proof: We obviously have $c \subset \tilde{c} \cap \mathfrak{a} \subset \text{pr}(\tilde{c})$. Let $X \in c$. Then

$$\text{pr}(\text{conv}\, \tilde{W}_0(X)) = \text{conv}\, W_0 \cdot X \subset c$$

and hence also $\text{pr}(\tilde{c}) \subset c$. □

Theorem 4.5.8 (Extension of Cones) *Suppose that $\mathcal{M} = G/H$ is a non-compactly causal symmetric space. Let $G_{\mathbb{C}}/G^c$ be any noncompactly causal symmetric space corresponding to $(\mathfrak{g}_{\mathbb{C}}, \sigma^c)$. If $C \in \text{Cone}_H(\mathfrak{q})$, then*

1) $E_{\mathfrak{q}}^{i\mathfrak{g}^c, G^c}(C) \in \text{conv}_{G^c}(i\mathfrak{g}^c)$,

2) $-\tau(E_{\mathfrak{q}}^{i\mathfrak{g}^c, G^c}(C)) = E_{\mathfrak{q}}^{i\mathfrak{g}^c, G^c}(C)$, *and*

3) $E_{\mathfrak{q}}^{i\mathfrak{g}^c, G^c}(C) \cap \mathfrak{q} = \text{pr}_{\mathfrak{q}}(E_{\mathfrak{q}}^{i\mathfrak{g}^c, G^c}(C)) = C$.

In particular, every cone in $\text{Cone}_H(\mathfrak{q})$ is G^c-extendable.

Proof: We may assume that $\mathfrak{g}$ carries no complex structure and all groups are contained in the simply connected group $G_{\mathbb{C}}$.

2) is obvious, as $-\tau(C) = C$. We prove 1) and 3) together. Fix a Cartan subalgebra $\mathfrak{t}^c$ of $\mathfrak{k}^c$ containing a maximal abelian subspace $\mathfrak{a}$ of $\mathfrak{q}_p$.

Let $C \in \mathrm{Cone}_H(\mathfrak{q})$. We may assume that $c_{\min} \subset C \cap \mathfrak{a} \subset c_{\max}$. Now we apply Corollary 4.5.7 to find a $\tilde{W}_0$-invariant cone $\tilde{c}$ in $i\mathfrak{t}^c$ such that $\mathrm{pr}(\tilde{c}) = C \cap \mathfrak{a} = \tilde{c} \cap \mathfrak{a}$, where $\mathrm{pr}: i\mathfrak{t}^c \to \mathfrak{a}$ is the orthogonal projection. Replacing $\tilde{c}$ by $\tilde{c} + \tilde{c}_{\min}$ if necessary we may assume that $\tilde{c}_{\min} \subset \tilde{c} \subset \tilde{c}_{\max}$. By Theorem 4.4.7 applied to $G_{\mathbb{C}}/G^c$ there is a cone $\tilde{C} \in \mathrm{conv}_{G^c}(i\mathfrak{g}^c)$ such that $\tilde{C} \cap i\mathfrak{t}^c = P_{\mathfrak{a}}^{i\mathfrak{t}^c}(\tilde{C}) = \tilde{c}$. In particular, we have $\tilde{C} \cap \mathfrak{a} = C \cap \mathfrak{a}$. Thus $\tilde{C} \cap \mathfrak{q}$ and C are both cones in $\mathrm{Cone}_{H_o}(\mathfrak{q})$, and their intersections with $\mathfrak{a}$ agree. Thus Theorem 4.4.11 proves that $\tilde{C} \cap \mathfrak{q} = C$. Since $\tilde{C}$ is G^c-invariant and contains C, it also contains $E_{\mathfrak{q}}^{i\mathfrak{g}^c, G^c}(C)$. Therefore $C = E_{\mathfrak{q}}^{i\mathfrak{g}^c, G^c}(C) \cap \mathfrak{q}$ and then Theorem 4.4.11 implies the claim, since $E_{\mathfrak{q}}^{i\mathfrak{g}^c, G^c}(C)$ is regular by Lemma 4.4.9. □

Notes for Chapter 4

The material in this chapter is taken mainly from Chapter 5 in [129]. The relation between the strongly orthogonal roots in Section 4.1 is from [130]. A more algebraic proof of Lemma 4.1.9 can be found in [133]. The convexity theorem in the group case was proved by Paneitz in [147]. The linear convexity theorem, which was proved in [129], can also be viewed as an infinitesimal version of the convexity theorem of Neeb [116]. It can be derived from general symplectic convexity theorems applied to suitable coadjoint orbits (cf. [62]). The proof presented here is based on the proof of the convexity theorem of Paneitz by Spindler [158]. The classification for simple groups is due to Ol'shanskii [139], Paneitz [147, 148], and Vinberg [165]. Their results were generalized to arbitrary Lie groups by Hilgert, Hofmann, and Lawson in [50]. The extension theorem for invariant cones was proved in [48] for the classical spaces and in [129] for the general case using the classification. The idea of the proof given here is due to Neeb [116]. The invariant cones in the group case have been described quite explicitly by Paneitz in [147] for the classical groups. Thus Theorem 4.5.8 can also be used to obtain explicit descriptions in the general case.

Chapter 5

The Geometry of Noncompactly Causal Symmetric Spaces

If $\mathcal{M} = G_{\mathbb{C}}/G$ is a causal symmetric space, then G/K is a bounded symmetric domain. It is well known that in this case G/K can be realized via the Harish-Chandra embedding as a complex symmetric domain in $\mathfrak{p}^-$, which in our notation for noncompactly causal symmetric spaces corresponds to $\mathfrak{n}_-$. We generalize this embedding to the general case in the first section. More precisely, we show that if G/H is a noncompactly causal symmetric space, then $H/H \cap K$ is a real symmetric domain Ω_- in $\mathfrak{n}_-$ which can also be realized as an open subset $\mathcal{O}$ in a certain minimal flag manifold $\mathcal{X} = G/P_{\max}$ of G. The importance of this observation lies in the fact that the semigroup $S(C)$ associated to the causal orientation of $\mathcal{M} = G/H$ is essentially equal to the semigroup $S(G^\tau, P_{\max})$ of compressions of $\mathcal{O}$. This semigroup consists of all elements in G mapping $\mathcal{O} \subset \mathcal{X}$ into itself. We show that $S(C) = (\exp C)H$ is homeomorphic to $C \times H$. With this information at hand, one easily sees that noncompactly causal symmetric spaces are always ordered and have good control over the geometric structure of the positive domain $\mathcal{M}_+$ of $\mathcal{M}$ which consists of the elements greater than the base point with respect to the causal ordering. In particular, we prove that intervals in this order are compact. Finally, we give a proof of Neeb's non-linear analog of Theorem 4.3.1 which turns out to be extremely useful in the harmonic analysis of noncompactly causal symmetric spaces.

5.1 The Bounded Realization of $H/H \cap K$

In this section we fix a noncompactly causal symmetric space $\mathcal{M} = G/H$ and assume that G is contained in a simply connected complex group $G_{\mathbb{C}}$ with Lie algebra $\mathfrak{g}_{\mathbb{C}}$. Recall the abelian subalgebras $\mathfrak{n}_{\pm} \subset \mathfrak{g}$ which form the irreducible pieces of the $\mathfrak{h}^a$-module $\mathfrak{q} = \mathfrak{n}_- + \mathfrak{n}_+$ (cf. Remark 3.1.17 and Lemma 1.3.4) and let $N_{\pm} = \exp \mathfrak{n}_{\pm}$ be the corresponding analytic subgroups of G which are automatically closed. Similarly, we have a closed nilpotent subgroup $N_0 = \exp \mathfrak{n}_0$ in G. Since H^a centralizes $X^0 \in \mathfrak{q}_p$, it also normalizes $\mathfrak{n}_{\pm}$ and $N_{\pm}$. Therefore

$$P_{\max} = H^a N_+ = H^a N \tag{5.1}$$

defines a maximal parabolic subgroup of G (cf. Appendix A.1 and Lemma 1.3.4).

Consider the involutive anti-automorphism

$$^\sharp : G \to G, \quad g \mapsto \tau(g)^{-1} \tag{5.2}$$

and denote its derivative at 1 also by $^\sharp$. From $\tau^a|_{\mathfrak{a}} = \mathrm{id}$ we obtain the following.

Lemma 5.1.1 *Both $P_{\min}$ and $P_{\max}$ are τ^a-stable. Furthermore,*

$$\theta(P_{\min}) = \tau(P_{\min}) = MAN^\sharp$$

and

$$\theta(P_{\max}) = \tau(P_{\max}) = H^a N_+^\sharp = H^a N_- = H^a N^\sharp. \qquad \square$$

Recall the maximal set Γ of strongly orthogonal roots in $\Delta_+ = \Delta(\mathfrak{n}_+, \mathfrak{a})$ from Remark 4.1.10 and the corresponding maximal abelian subspace $\mathfrak{a}_h^c = \sum_{\gamma \in \Gamma} \mathbb{R} Y^\gamma$ of $\mathfrak{h}_p$. We set

$$A_h^c = \exp \mathfrak{a}_h^c. \tag{5.3}$$

Lemma 5.1.2 *Let the notation be as above. Then $HP_{\min} = HP_{\max} = G^\tau P_{\max} = (G^\tau)_o P_{\max}$ and $H \cap P_{\max} = H \cap K$. Furthermore, $HP_{\max}$ is open in G.*

Proof: Lemma 3.1.22 implies that

$$H^a = (H \cap K) MAN_0$$

so $H = (\exp \mathfrak{h}_p)(H \cap K)$ (cf. (1.8)) proves the first part.

Let $h \in H \cap P_{\max}$. Write $h = k_1 a k_2$ with $k_j \in K \cap H$ and $a \in A_h^c$. From $K \cap H \subset P_{\max}$ it follows that $a \in H \cap P_{\max}$. But then $\tau(a) = a^{-1} \in P_{\max} \cap \overline{P}_{\max} = MA$. But $A_h^c \cap MA = \{1\}$, which implies that $a = 1$.

A simple calculation shows that the differential of

$$G^\tau \times A \times N \ni (h, a, n) \mapsto han \in G$$

is bijective everywhere. Thus $G^\tau AN = G^\tau P_{\max} = HP_{\max}$ is open in G, cf. [99]. □

The generalized Bruhat decomposition (cf. [168], p.76) shows that both $N_- P_{\max}$ and $N^\sharp P_{\min}$ are open and dense in G and the maps

$$N_- \times H^a \times N_+ \ni (n_-, g, n_+) \mapsto n_- g n_+ \in N_- P_{\max}$$

and

$$N^\sharp \times M \times A \times N \ni (\theta n_1, m, a, n_2) \mapsto \theta n_1 m a n_2 \in G$$

are diffeomorphisms onto their images.

Example 5.1.3 Recall the situation of the Examples 1.3.12 and 3.1.13, i.e., $G = \mathrm{SL}(2, \mathbb{R})$. Then $M = \{\pm 1\}$ and

$$\begin{pmatrix} a & b \\ c & d \end{pmatrix}^\sharp = \begin{pmatrix} a & -c \\ -b & d \end{pmatrix}.$$

Moreover, we have

$$N^\sharp = N_- = \left\{ \begin{pmatrix} 1 & 0 \\ y & 1 \end{pmatrix} \middle| y \in \mathbb{R} \right\},$$

$$P_{\min} = P_{\max} = \left\{ \begin{pmatrix} a & r \\ 0 & a^{-1} \end{pmatrix} \middle| r \in \mathbb{R}, a \in \mathbb{R} \setminus \{0\} \right\},$$

and

$$N_- P_{\max} = \left\{ \begin{pmatrix} a & b \\ c & d \end{pmatrix} \middle| a \neq 0 \right\}. \tag{5.4}$$

If $\begin{pmatrix} a & b \\ c & d \end{pmatrix} = \begin{pmatrix} 1 & 0 \\ y & 1 \end{pmatrix} \begin{pmatrix} e & 0 \\ 0 & 1/e \end{pmatrix} \begin{pmatrix} 1 & x \\ 0 & 1 \end{pmatrix} \in N_- P_{\max}$, then

$$e = a \quad y = c/e \quad \text{and} \quad x = b/e.$$ □

Lemma 5.1.2 implies that the set $\mathcal{O} = (G^\tau)_o \cdot \mathbf{o}_\mathcal{X}$ in the real flag manifold $\mathcal{X} = G/P_{\max}$, where $\mathbf{o}_\mathcal{X} := 1P_{\max}/P_{\max}$, is open. Moreover, we have

$$\begin{aligned} H/H \cap K &\simeq HP_{\max}/P_{\max} = G^\tau P_{\max}/P_{\max} = \\ &= (G^\tau)_o P_{\max}/P_{\max} = (G^\tau)_o . \mathbf{o}_\mathcal{X} = \mathcal{O}. \end{aligned} \tag{5.5}$$

We will now describe $\mathcal{O}$ in more detail using the symmetric $\mathrm{SL}(2, \mathbb{R})$-reduction from Section 4.1.

Lemma 5.1.4 *$H \subset N_-P_{\max}$. Furthermore,*

$$\exp\sum_{\gamma\in\Gamma} t_\gamma Y^\gamma$$
$$= \left(\exp\sum_{\gamma\in\Gamma}\tanh t_\gamma Y_{-\gamma}\right)\left(\exp\sum_{\gamma\in\Gamma}\log\cosh t_j X^\gamma\right)\left(\exp\sum_{\gamma\in\Gamma}\tanh t_\gamma Y_\gamma\right)$$

Proof: The polar decomposition for H gives $H = (H\cap K)A_h^c(H\cap K)$. As $H\cap K \subset G^{\tau^a}$ and $H \cap K$ normalizes N_-, we only have to show that $A_h^c \subset N_-G^{\tau^a}$. But Example 5.1.3 implies that

$$\exp\sum_{\gamma\in\Gamma} t_\gamma Y^\gamma = \left(\exp\sum_{\gamma\in\Gamma}\tanh t_\gamma Y_{-\gamma}\right)\left(\exp\sum_{\gamma\in\Gamma}\log\cosh t_\gamma X^\gamma\right) \cdot \left(\exp\sum_{\gamma\in\Gamma}\tanh t_\gamma Y_\gamma\right).$$

This shows that $A_h^c \subset N_-P_{\max}$ and the lemma follows. □

Define a map $\kappa : \mathfrak{n}_- \to \mathcal{X}$ by

$$\kappa(X) = (\exp X)\cdot P_{\max}. \tag{5.6}$$

Using $H/H\cap K = HP_{\max}/P_{\max} \subset N_-P_{\max}/P_{\max} \simeq N_- \simeq \mathfrak{n}_-$ we see that κ is injective and find for $h \in H$:

$$hP_{\max} = \exp(\kappa^{-1}(h(H\cap K))P_{\max})P_{max}.$$

Let

$$\Omega_- := \kappa^{-1}(\mathcal{O}) \subset \mathfrak{n}_-. \tag{5.7}$$

By Lemma 5.1.4 we have the following.

Lemma 5.1.5 *Let $a = \exp\sum_{\gamma\in\Gamma} t_\gamma Y^\gamma \in A_h^c$. Then*

$$\kappa^{-1}(aP_{\max}) = \sum_{\gamma\in\Gamma}\tanh t_\gamma Y_{-\gamma}.$$ □

Lemma 5.1.6 *Let $h\in H$ and $X \in \Omega_-$. Then $h\cdot X = \mathrm{Ad}(h)X$.*

Proof: Let $X\in\Omega_-$ and $h \in H$. Then $h\exp X = [\exp \mathrm{Ad}(h)X]h$ and the claim follows from $\mathrm{Ad}(h)X \in \mathfrak{n}_-$. □

Lemma 5.1.7 *Let $Y \in \mathfrak{n}_-$. Then there exists a $k \in K \cap H_o$ such that* $\mathrm{Ad}(k)Y \in \sum_{j=1}^{r} \mathbb{R}Y_{-\gamma}$.

Proof: $\tau = -\theta$ on $\mathfrak{n}_-$ (cf. Lemma 4.1.1) implies that $Y + \tau(Y) \in \mathfrak{h}_p$. Hence there exists a $k \in K \cap H_o$ such that

$$\mathrm{Ad}(k)[Y + \tau(Y)] = \sum_{\gamma\in\Gamma} t_\gamma(Y_\gamma + Y_{-\gamma}) \in \mathfrak{a}_h .$$

But then $\mathrm{Ad}(k)Y = \sum_{\gamma\in\Gamma} t_\gamma Y_{-\gamma}$ and the claim follows. □

Theorem 5.1.8 *Let the notation be as above. Then*

$$\Omega_- = \mathrm{Ad}(K \cap H)\left\{ \sum_{\gamma\in\Gamma} y_\gamma Y_{-\gamma} \,\middle|\, -1 < y_\gamma < 1,\ \gamma \in \Gamma \right\} .$$

Proof: Let $h = k_1 a k_2 \in H$ with $k_j \in K \cap H$ and $a = \exp \sum_{\gamma\in\Gamma} t_\gamma Y^\gamma \in A_h^c$. Then

$$\begin{aligned} \kappa^{-1}(h(H \cap K)) &= \mathrm{Ad}(k_1)\kappa^{-1}\big(\exp \sum_{\gamma} t_\gamma Y^\gamma (P_{\max})\big) \\ &= \mathrm{Ad}(k_1) \sum_{\gamma\in\Gamma} \tanh t_\gamma Y_{-\gamma} \in \Omega_- . \end{aligned}$$

On the other hand let $X = \mathrm{Ad}(k) \sum_{\gamma\in\Gamma} y_\gamma Y_{-\gamma}$ be in Ω_- and $t_\gamma = \operatorname{arctanh}(y_\gamma)$. Then

$$\kappa(X) = \left(k \exp \sum_{\gamma\in\Gamma} t_\gamma Y^\gamma \right) P_{\max} .$$

This proves the theorem. □

Remark 5.1.9 As τ interchanges N_- and N_+ in such a way that $\tau(Y_{-\gamma}) = Y_\gamma$ and $Y^\gamma \in \mathfrak{h}_p$ we see that for $\Omega_+ := \tau(\Omega) \subset \mathfrak{n}_+$ we have

$$H\tau(P_{max})/\tau(P_{max}) \simeq \mathrm{Ad}(K \cap H)\left\{ \sum_{\gamma\in\Gamma} y_\gamma Y_\gamma \,\middle|\, -1 < y_\gamma < 1,\ \gamma \in \Gamma \right\} = \Omega_+$$

□

Example 5.1.10 Let G be a Hermitian Lie group, i.e., G is semisimple and G/K is a bounded symmetric domain. Then $G_\mathbb{C}/G$ is a noncompactly causal symmetric space. In this case $\mathfrak{a} = i\mathfrak{t}$, where $\mathfrak{t} \subset \mathfrak{k}$ is a Cartan subalgebra of $\mathfrak{k}$ and of $\mathfrak{g}$. Furthermore, $\Delta_+ = \Delta_n$ is the set of noncompact roots, $\Delta_0 = \Delta_k$ the set of compact roots, and $\mathfrak{n}_\pm = \mathfrak{p}^\pm$ in the notation of Appendix A.3. Hence the above result states that $G \subset P^-K_\mathbb{C}P^+$ and G/K can be realized as an bounded symmetric domain in $\mathfrak{p}^-$. In other words, Theorem 5.1.8 is a generalization of the Borel–Harish-Chandra realization of Hermitian symmetric spaces as bounded domains in $\mathfrak{p}^-$. □

Let σ and η be the involutions on $G_\mathbb{C}$ with fixed point groups G and G^c (cf. Section 1.1). Then G^c/K^c is a bounded symmetric domain. We want to relate Ω_- and $\mathcal{O}$ to the Borel–Harish-Chandra realization of G^c/K^c.

Recall that $\mathfrak{g}^c = \mathfrak{h} + i\mathfrak{q}$ is a Hermitian Lie algebra. Moreover, let $\mathfrak{t}^c = \mathfrak{t}_m + i\mathfrak{a}$ be a compactly embedded Cartan algebra in $\mathfrak{g}^c$ and $\mathfrak{k}^c = \mathfrak{h}_k + i\mathfrak{q}_p$ the uniquely determined maximal compactly embedded subalgebra of $\mathfrak{g}^c$ containing $\mathfrak{t}^c$ (cf. [50], Theorem A.2.40, and Section 4.1). As before, we write $\tilde{\Delta} = \Delta(\mathfrak{g}_\mathbb{C}, \mathfrak{t}^c_\mathbb{C})$ for the corresponding set of roots. Then $\tilde{\Delta}_0$ denotes the roots of $\mathfrak{k}^c_\mathbb{C}$ which we call *compact* and $\tilde{\Delta}_+$ the corresponding set of noncompact roots. The Cartan decomposition of $\mathfrak{g}^c$ with respect to $\theta\sigma$ is $\mathfrak{g}^c = \mathfrak{k}^c + \mathfrak{p}^c$ with $\mathfrak{p}^c = \mathfrak{h}_p + i\mathfrak{q}_k$. Note that $(\mathfrak{k}^c)_\mathbb{C} = (\mathfrak{h}^a)_\mathbb{C}$ and $(\mathfrak{p}^c)_\mathbb{C} = \sum_{\alpha\in\tilde{\Delta}_p} \mathfrak{g}^\alpha_\mathbb{C}$.

For

$$(\mathfrak{p}^c)^\pm = \sum_{\alpha\in\tilde{\Delta}_p^\pm} \mathfrak{g}^\alpha_\mathbb{C}$$

we have

$$(\mathfrak{p}_{\max})_\mathbb{C} = (\mathfrak{k}^c)_\mathbb{C} + (\mathfrak{p}^c)^+_\mathbb{C}\,.$$

The theory of Hermitian symmetric spaces (cf. Appendix A.3) says that G^c/K^c embeds as an open G^c-orbit $\mathcal{O}_\mathbb{C}$ into the complex flag manifold $\mathcal{X}_\mathbb{C} = G_\mathbb{C}/(P_{\max})_\mathbb{C}$ and then as a bounded symmetric domain $(\Omega_-)_\mathbb{C}$ in $(\mathfrak{p}^c)^-$ (and $(\Omega_+)_\mathbb{C}$ in $(\mathfrak{p}^c)^+$).

The complex parabolic $(P_{\max})_\mathbb{C}$ is stable under the conjugation σ. Hence σ yields a complex conjugation on $G_\mathbb{C}/(P_{\max})_\mathbb{C}$ which we still denote by σ. We write $\left(G_\mathbb{C}/(P_{\max})_\mathbb{C}\right)^\sigma$ for the set of σ-fixed points.

Lemma 5.1.11 $\mathcal{X}^\sigma_\mathbb{C} = \mathcal{X}$.

Proof: $(N_-)_\mathbb{C}.\mathbf{o}_\mathcal{X}$ is open dense in $\mathcal{X}_\mathbb{C}$ and invariant under σ. If $x \in \mathcal{X}^\sigma_\mathbb{C}$, then there exists a sequence of $n_j \in (N_-)_\mathbb{C}$ such that $n_j.\mathbf{o}_\mathcal{X}$ converges to x. But then $\sigma(n_j).\mathbf{o}_\mathcal{X}$ converges to x as well, whence $n_j\sigma(n_j)^{-1}.\mathbf{o}_\mathcal{X}$ converges to $\mathbf{o}_\mathcal{X}$. Thus $n_j\sigma(n_j)^{-1}$ converges to 1 in $(N_-)_\mathbb{C}$. Now, the fact that $(N_-)_\mathbb{C}$ is a vector group shows that the imaginary part of n_j converges to zero (identifying $(\mathfrak{n}_-)_\mathbb{C}$ and $(N_-)_\mathbb{C}$) and we can replace n_j by its real part without changing the limit. This proves $x \in \mathcal{X}$ and hence the claim.□

Lemma 5.1.12 *1)* $\mathcal{O}_{\mathbb{C}} \cap \mathcal{X} = \mathcal{O}$.

2) $(\Omega_-)_{\mathbb{C}} \cap \mathfrak{n}_- = \Omega_-$.

Proof: 1) Let $g \in G^c$ with $\sigma(g.o) = g.o$. Since $K^c \subset (P_{\max})_{\mathbb{C}}$, we may w.l.o.g. assume that $g = \exp X$ with $X \in \mathfrak{p}^c$. Then the hypothesis implies that $\exp\left(\sigma(X)\right) \in \exp(X)K^c$, so that the Cartan decomposition of G^c yields that $\sigma(X) = X$. Therefore $X \in \mathfrak{h}_p$ and consequently $g \in G^\tau$.

2) follows immediately from 1). □

Example 5.1.13 We continue the SL(2, ℝ)-Example 5.1.3. In that context we have $\mathcal{X}_{\mathbb{C}} = \mathbb{CP}^1$ and $\mathcal{X} = \mathbb{RP}^1$. Under the Harish-Chandra embedding we find

$$\mathcal{O}_{\mathbb{C}} = (\Omega_-)_{\mathbb{C}} = \left\{ \begin{pmatrix} 0 & 0 \\ z & 0 \end{pmatrix} \middle| \, |z| < 1, z \in \mathbb{C} \right\}$$

and

$$\mathcal{O} = \Omega_- = \left\{ \begin{pmatrix} 0 & 0 \\ r & 0 \end{pmatrix} \middle| \, |r| < 1, z \in \mathbb{R} \right\}.$$

The action of $G_{\mathbb{C}}$ on $(\Omega_-)_{\mathbb{C}}$ is given by

$$\begin{pmatrix} a & b \\ c & d \end{pmatrix} \cdot \begin{pmatrix} 0 & 0 \\ z & 0 \end{pmatrix} = \begin{pmatrix} 0 & 0 \\ \frac{dz+c}{bz+a} & 0 \end{pmatrix}.$$ □

5.2 The Semigroup $S(C)$

If G is an arbitrary Lie group, then the differential of the exponential map at the point $X \in \mathfrak{g}$ is given by

$$d_X \exp = (d_1 \ell_{\exp X}) \frac{1 - e^{-\operatorname{ad} X}}{\operatorname{ad} X} = (d_1 \ell_{\exp X}) f(\operatorname{ad} X)$$

where $f(t) = \sum_{n=0}^{\infty} (-1)^n t^n/(n+1)!$ and $\ell_g : G \to G$ denotes left multiplication by g. We derive similar formulas for arbitary symmetric spaces G/H. Define

$$f_h(t) := \frac{1 - \cosh t}{t} = -\sum_{n=1}^{\infty} \frac{t^{2n-1}}{(2n)!} \tag{5.8}$$

$$f_q(t) := \frac{\sinh t}{t} = \sum_{n=0}^{\infty} \frac{t^{2n}}{(2n+1)!}. \tag{5.9}$$

Lemma 5.2.1 *1) The functions f_q and f_h are analytic in the complex plane.*

2) $f_h^{-1}(0) = 2\pi i\mathbb{Z} \setminus \{0\}$.

3) $f_q^{-1}(0) = \pi i\mathbb{Z} \setminus \{0\}$.

4) $f = f_h + f_q$. □

Lemma 5.2.2 *Let the notation be as above and $X, Y \in \mathfrak{q}$. Then*

$$f(\operatorname{ad} X)Y = f_h(\operatorname{ad} X)Y + f_q(\operatorname{ad} X)Y$$

and $f_h(\operatorname{ad} X)Y \in \mathfrak{h}$, $f_q(\operatorname{ad} X)Y \in \mathfrak{q}$.

Proof: This follows immediately from Lemma 5.2.1 and $\operatorname{ad}(X)^k\mathfrak{q} \subset \mathfrak{h}$ if k is odd and $\operatorname{ad}(X)^k\mathfrak{q} \subset \mathfrak{q}$ if k is even. □

Lemma 5.2.3 *Define $\varphi : \mathfrak{q} \times H \to G$ by $\varphi(X, h) = (\exp X)h$. Then for all $X, Y \in \mathfrak{q}$, $Z \in \mathfrak{h}$, and $h \in H$ the following holds:*

$$d_{(X,h)}\varphi(Y, Z) = \left[Z + \operatorname{Ad}(h^{-1})(f_h(\operatorname{ad} X)Y)\right] + \operatorname{Ad}(h^{-1})\left[f_q(\operatorname{ad} X)Y\right].$$

Here we identify T_hH with $\mathfrak{h}$ and T_gG with $\mathfrak{g}$ via the left multiplication.

Proof: It is clear that $d_{(X,h)}\varphi(0, Z) = Z$. Let $F \in C^\infty(G)$. Let $a(h, X) := \exp Xh[\exp(-\operatorname{Ad}(h^{-1})X]$. Then

$$\begin{aligned}
\frac{d}{dt}F(\exp(X + tY)h)_{t=0} &= \frac{d}{dt}F(\exp Xh[h^{-1}\exp(-X)\exp(X + tY)h])_{t=0} \\
&= \frac{d}{dt}F(a(h, X)\exp(\operatorname{Ad}(h^{-1})X + t\operatorname{Ad}(h^{-1})Y))_{t=0}
\end{aligned}$$

from which the lemma follows. □

If we identify $T_o(\mathcal{M})$ with $\mathfrak{q}$, then the exponential map $\operatorname{Exp} : \mathfrak{q} \to G/H$ is given by $\operatorname{Exp} X = (\exp X)H = \pi(\varphi(X, 1))$, where $\pi : G \to \mathcal{M}$ is the canonical projection. Identify $T_X(\mathfrak{q})$ with the vector space $\mathfrak{q}$ in the usual way. Then, using that $d_1\ell_a : T_1G \to T_aG$ is an isomorphism, we have

$$d_X \operatorname{Exp} = d_1\ell_{\exp X} \circ f_q(\operatorname{ad} X), \quad X \in \mathfrak{q}. \tag{5.10}$$

Hence Exp is a local diffeomorphism for all X such that $\operatorname{spec}(\operatorname{ad} X) \cap (\pi i\mathbb{Z} \setminus \{0\}) = \emptyset$. We will actually need more than this. Define for $\lambda > 0$

$$U(\lambda) := \{X \in \mathfrak{q} \mid \max_{\mu \in \operatorname{spec}(\operatorname{ad} X)} |\operatorname{Im}\mu| < \lambda\} \tag{5.11}$$

$$V(\lambda) := \operatorname{Exp}(U(\lambda)) \tag{5.12}$$

$$W(\lambda) := \{\exp(X)h \mid X \in U(\lambda), h \in H\} = \pi^{-1}(V(\lambda)) \tag{5.13}$$

Let H act on $V(\lambda) \times H$ by $h \cdot (X, k) = (\operatorname{Ad}(h)X, hkh^{-1})$.

Lemma 5.2.4 *Let the notation be as above. Then the following hold:*

1) *$U(\lambda)$ is an open H-invariant 0-neighborhood in $\mathfrak{q}$.*

2) *If $\lambda < \frac{\pi}{2}$, then $W(\lambda)$ is an open H-invariant 1-neighborhood in G and $\varphi : V(\lambda) \times H \to W(\lambda)$ is an H-equivariant diffeomorphism, where H acts on G by conjugation.*

3) *If $\lambda < \frac{\pi}{2}$, then $V(\lambda)$ is an open H-invariant o-neighborhood in $\mathcal{M}$ and $\operatorname{Exp} : U(\lambda) \to V(\lambda)$ is an H-equivariant diffeomorphism.*

Proof: The first part is obvious. To prove 2) assume that $\lambda < \frac{\pi}{2}$. We first show that φ is a local diffeomorphism. This will imply that $W(\lambda)$ is open. By Lemma 5.2.3 we have

$$d_{(X,h)}\varphi(Y,Z) = \left[Z + \operatorname{Ad}(h^{-1})(f_{\mathfrak{h}}(\operatorname{ad} X)Y)\right] + \operatorname{Ad}(h^{-1})[f_{\mathfrak{q}}(\operatorname{ad} X)Y].$$

If $d_{(X,h)}\varphi(Y,Z) = 0$, then

$$Z + \operatorname{Ad}(h^{-1})\,(f_{\mathfrak{h}}(\operatorname{ad} X)Y)) = 0 \quad \text{and} \quad \operatorname{Ad}(h^{-1})(f_{\mathfrak{q}}(\operatorname{ad} X)Y) = 0$$

according to Lemma 5.2.2. But then $Y = 0$ as $\varphi_{\mathfrak{q}}(\operatorname{ad} X) : \mathfrak{q} \to \mathfrak{q}$ is an isomorphism for $X \in U(\lambda)$. Therefore $Z = 0$, too and it follows that $d_{(X,h)}\varphi$ is an isomorphism. Thus – by the implicit function theorem – φ is a local diffeomorphism. Now we only have to show that φ is injective. Assume that

$$g = \exp(X)h = \exp(Y)k, \quad X, Y \in U(\lambda),\ h, k \in H.$$

Then $g\tau(g)^{-1} = \exp 2X = \exp 2Y$. By [163], p. 193, it follows that $X = Y$. But then also $h = k$. The H-equivariance follows from

$$\varphi(\operatorname{Ad}(k)X, khk^{-1}) = \exp(\operatorname{Ad}(k)X)khk^{-1} = k\varphi(X,h)k^{-1}.$$

3) As $d_X \operatorname{Exp}$ is a local diffeomorphism for $X \in U(\lambda)$, we only have to show that Exp is bijective. Assume that $\operatorname{Exp}(X) = \operatorname{Exp}(Y)$ for $X, Y \in U(\lambda)$. Then $\exp X = (\exp Y)h$ for some $h \in H$. By (2) this shows that $X = Y$. □

Theorem V.4.57 in [50] says

Theorem 5.2.5 *Let G/H be a symmetric space. Let $C \subset \mathfrak{q}$ be a regular H-invariant cone in $\mathfrak{q}$ such that $\operatorname{spec} \operatorname{ad}(X) \subset \mathbb{R}$ for all $X \in C$. If $(\exp C)H$ is closed in G, then $S(C) := (\exp C)H$ is a semigroup in G with $\mathbf{L}(S(C)) = C + \mathfrak{h}$.* □

From now on we will always assume that $\mathcal{M}$ is a noncompactly causal irreducible semisimple symmetric space.

Theorem 5.2.6 *Let $\mathcal{M} = G/H$ be a noncompactly causal semisimple symmetric space. Let $C \in \mathrm{Cone}_H(\mathfrak{q})$. Define*

$$S = S(C) = (\exp C)H = \varphi(C, H). \tag{5.14}$$

Then S is a closed semigroup in G. Furthermore, the following hold:

1) $S \cap S^{-1} = H$.

2) $C \times H \ni (X, h) \to \exp(X)h \in S$ is a homeomorphism.

3) $S^o = \exp(C^o)H$ and $C^o \times H \ni (X, h) \to \exp(X)h \in S^o$ is a diffeomorphism.

4) $S = H(S \cap A)H$.

Proof. As $C \subset U(\lambda)$ for all $\lambda > 0$, 2) and 3) follow from Lemma 5.2.4. Assume that $s = \exp(X)h \in S \cap S^{-1}$. Then $s^{-1} = (\exp Y)k$ for some $Y \in C$ and $k \in H$. Hence

$$(\exp Y)k = h^{-1}\exp(-X) = \exp(-\mathrm{Ad}(h^{-1})X)h^{-1}$$

As $-\mathrm{Ad}(h^{-1})X \in U(\lambda)$, it follows that $Y = -\mathrm{Ad}(h^{-1})X \in C \cap -C = \{0\}$. Hence $Y = 0$ and $s \in H$. This implies 1).

As $C \times H$ is closed in $U(\lambda)$, it follows that $(\exp C)S$ is closed. Now Theorem 5.2.5 shows that $S(C)$ is a semigroup.

The last assertion now follows from the reconstruction theorem 4.4.11. □

The cone C defines a G-invariant topological causal orientation $\preceq$ on $\mathcal{M}$. From Theorem 2.3.3 we obtain

Theorem 5.2.7 *Let $\mathcal{M}$ be a noncompactly causal symmetric space, $C \in \mathrm{Cone}_H(\mathfrak{q})$ and $\preceq$ the corresponding causal orientation on $\mathcal{M}$. Then $\preceq$ is antisymmetric and*

$$S(C) = \{s \in G \mid \mathrm{o} \preceq s \cdot \mathrm{o}\},$$

i.e., $S(C)$ is the causal semigroup of $\mathcal{M}$. □

In particular, Theorem 5.2.7 shows that $\preceq$ and the order $\leq_{S(C)}$ defined in Section 2.3 agree (cf. Remark 2.3.2), so that the positive cone is simply the $S(C)$-orbit of o:

$$\mathcal{M}_+ = S(C) \cdot \mathrm{o}. \tag{5.15}$$

We conclude this section with the observation that one can view the noncompactly causal space $\check{\mathcal{M}}$ as a subspace of $G_{\mathbb{C}}/\check{G}$ (cf. Section 1.1).

Proposition 5.2.8 *$\mathcal{N} := G_{\mathbb{C}}/\check{G}^c$ is a causal symmetric space and the canonical map*

$$\pi : \check{\mathcal{M}} \to \mathcal{N}, \quad g(G_{\mathbb{C}}^{\sigma} \cap G_{\mathbb{C}}^{\tau}) \mapsto g\check{G}^c$$

is a $\check{G}$-equivariant homeomorphism onto its (closed) image and preserves the causal orientation.

Proof: It follows directly from the definitions that the map is well defined, $\check{G}$-equivariant, and injective. Let $C \in \mathrm{Cone}_H(\mathfrak{q})$ and $\tilde{C}$ be the minimal G^c-invariant extension to $i\mathfrak{g}^c$, cf. Section 4.5. As C is H-stable, it follows that $\tilde{C}$ is in $\mathrm{Cone}_{G^c}(i\mathfrak{g}^c)$. In particular, $\mathcal{N}$ is causal and π is a causal map. Theorem 5.2.6 implies that $\check{\mathcal{M}}_+$ and $\mathcal{N}_+$ is homeomorphic to C, respectively $\tilde{C}$. But then homogeneity and G-equivariance show that π is a proper map. In particular, it is closed, which implies the claim. □

5.3 The Causal Intervals

In this section we will show that the causal intervals $[x, y]$, $x, y \in \mathcal{M}$ are compact. Fix x and choose $g \in G$ such that $g \cdot x = \mathbf{o}$. Since $\ell_g : \mathcal{M} \to \mathcal{M}$ is an order-preserving diffeomorphism, it follows that

$$\ell_g([x, y]) = [\mathbf{o}, g \cdot y]$$

and $[x, y]$ is compact if and only if $[\mathbf{o}, g \cdot y]$ is compact. Thus we may assume that $x = \mathbf{o}$ and $y \in [\mathbf{o}, \infty) = S(C) \cdot \mathbf{o}$.

Let $\lambda < \frac{\pi}{2}$; then $\mathrm{Exp} : U(\lambda) \to V(\lambda)$ is a diffeomorphism. In particular, we may define $\mathrm{Log} : V(\lambda) \to U(\lambda)$ to be the inverse of Exp.

Theorem 5.3.1 $\mathrm{Log} : [\mathbf{o}, \infty) \to C$ *is order-preserving.*

We prove this theorem in several steps. Consider the function

$$\varphi(x) = \frac{x}{\sinh x}.$$

Then $\varphi = 1/f_q$, with f_q as in (5.9).

Lemma 5.3.2 *Let φ be as above. Then*

$$\varphi(x) = \frac{\pi}{4} \int_{-\infty}^{\infty} e^{ixy} \frac{1}{\cosh^2(\pi y/2)}\, dy.$$

In particular, φ is positive definite.

Proof: It is well known that

$$\pi \tanh \pi y = \int_0^\infty \frac{\sin yx}{\sinh(x/2)}\, dx.$$

As sinh is an odd function, the integral on the right-hand side equals

$$\frac{1}{2i}\int_{-\infty}^{\infty} e^{ixy}\frac{1}{\sinh(x/2)}\, dx.$$

Differentiating with respect to y gives

$$\frac{\pi^2}{\cosh^2(\pi y)} = \frac{1}{2}\int_{-\infty}^{\infty} e^{ixy}\frac{x}{\sinh(x/2)}\, dx\,.$$

Taking the inverse Fourier transforms now yields:

$$\frac{x}{\sinh(x/2)} = \pi\int_{-\infty}^{\infty} e^{-ixy}\frac{1}{\cosh^2(\pi y)} dy\,.$$

Finally, we replace x by $2x$ to obtain

$$\frac{x}{\sinh x} = \frac{\pi}{4}\int_{-\infty}^{\infty} e^{-ixy}\frac{1}{\cosh^2(\pi y/2)}\, dy\,. \qquad \square$$

Lemma 5.3.3 *Let* $C \in \mathrm{Cone}_H(\mathfrak{q})$. *Then*

$$\frac{\operatorname{ad} X}{\sinh X} C \subset C$$

for every $X \in C$.

Proof: Let X and C be as in the lemma. We may assume that $X \in C^o$ as φ is continuous. Then $\operatorname{ad} X$ has only real eigenvalues. Let $G_{\mathbb{C}}$ be the complex Lie group generated by $\exp(\operatorname{ad} X)$, $X \in \mathfrak{g}_{\mathbb{C}}$. Further, let G^c be the closed subgroup of $G_{\mathbb{C}}$ generated by $\exp(\operatorname{ad} X)$, $X \in \mathfrak{g}^c = \mathfrak{h} \oplus i\mathfrak{q}$. Consider the minimal extension D of C to a G^c-invariant cone in $i\mathfrak{g}^c = i\mathfrak{h} \oplus \mathfrak{q}$ (cf. Section 4.5). Then $D \cap \mathfrak{q} = \mathrm{pr}_{\mathfrak{q}} D = C$. As $iX \in \mathfrak{g}^c$, it follows that

$$e^{iy(\operatorname{ad} X)} C \subset D$$

for all $y \in \mathbb{R}$. But $1/\cosh^2((\pi y)/2) > 0$ for all $y \in \mathbb{R}$, so

$$e^{iy \operatorname{ad} X}\frac{1}{\cosh(\pi y/2)} C \subset D$$

for all $y \in \mathbb{R}$ and Lemma 5.3.2 shows that $\varphi(\operatorname{ad} X)C \subset D$. But obviously $\varphi(\operatorname{ad} X)C \subset \mathfrak{g}$. Hence

$$\varphi(\operatorname{ad} X)C \subset D \cap \mathfrak{g} = D \cap \mathfrak{q} = C,$$

which proves the lemma. □

Proof of Theorem 5.3.1 : We have to show that $d_{\operatorname{Exp} X} \operatorname{Log}(C_{\operatorname{Exp} X}) \subset C$ for all $X \in C$. Using (5.10) we calculate

$$\begin{aligned} d_{\operatorname{Exp} X} \operatorname{Log} &= (d_X \operatorname{Exp})^{-1} \\ &= (d_1 \ell_{\exp X} \circ f_{\mathfrak{q}}(\operatorname{ad} X))^{-1} \\ &= \varphi(\operatorname{ad} X) \circ (d_1 \ell_{\exp X})^{-1}. \end{aligned}$$

The claim now follows from Lemma 5.3.3, since $C_{\operatorname{Exp} X} = d_1 \ell_{\exp X}(C)$. □

Definition 5.3.4 Let G/H be an ordered symmetric space. Then G/H is called *globally hyperbolic* if all the intervals $[m, n]$, $m, n \in G/H$ are compact. □

Theorem 5.3.5 *Let $\mathcal{M}$ be a noncompactly causal symmetric space. Then $\mathcal{M}$ is globally hyperbolic.*

Proof: Let $\gamma : [0, a] \to \mathcal{M}$ be a causal curve with $\gamma(0) = \mathbf{o}$. By Theorem 5.3.1, $\operatorname{Log} \circ \gamma$ is a causal curve in $\mathfrak{q}$ with $\operatorname{Log} \circ \gamma(0) = 0$. In particular, $\operatorname{Log}(\gamma(t)) \in C \cap [C - \operatorname{Log}(\gamma(a))]$ for all $t \in [0, a]$. It follows that

$$\operatorname{Log}([\mathbf{o}, \operatorname{Exp} X]) \subset C \cap (C - X),$$

As $C \cap (C - X)$ is compact and Log is a homeomorphism, it follows that $[\mathbf{o}, \operatorname{Exp} X]$ is compact. □

5.4 Compression Semigroups

In this section we show how closely related the semigroups of type $S(C)$ are to the semigroup of self-maps of the open domain $\mathcal{O}$ in $\mathcal{X}$. Recall that we assumed G to be contained in a simply connected complexification $G_{\mathbb{C}}$.

Lemma 5.4.1 *Let $F_\tau \subset F \subset M$ be a set of representatives for $(H_o \cap F)\backslash F$. Then the group multiplication gives a diffeomorphism*

$$H_o \times F_\tau \times A \times N \to H_o P_{\max}$$

Proof: Lemma 3.1.22 implies that multiplication gives a diffeomorphism $H_o \times F_\tau \to G^\tau$, since F is a finite subgroup of G^τ normalizing H_o. According to Lemma 5.1.2 we have

$$G^\tau \cap AN \subset K \cap AN = \{1\}.$$

This shows that the map is injective. The surjectivity is also a consequence of Lemma 5.1.2. Finally, we recall that the bijectivity of the differential has already been observed in the proof of Lemma 5.1.2. □

Recall the cones $c_{\max}, \tilde{c}_{\max}$, and $\tilde{C}_{\max}$ from (4.21), (4.23), and (4.25). We know that

$$\tilde{C}_{\max} = \overline{\mathrm{Ad}(G^c)\tilde{c}_{\max}} = E^{i\mathfrak{g}^c, G^c}_{\tilde{\mathfrak{a}},\{1\}}(\tilde{c}_{\max})$$

is a closed convex G^c-invariant cone in $i\mathfrak{g}^c$ whose intersection with and projection to $\tilde{\mathfrak{a}}$ is $\tilde{c}_{\max}$ (cf. Theorem 4.4.6). One associates the Ol'shanskii semigroup

$$S(\tilde{C}_{\max}) := G^c \exp(\tilde{C}_{\max}) \tag{5.16}$$

with $\tilde{C}_{\max}$ and observes that it is closed and maximal in $G_{\mathbb{C}}$ (cf. [52], Corollaries 7.36 and 8.53).

Definition 5.4.2 Let $\mathcal{X}$ be a locally compact space on which a locally compact group G acts continuously. Further, let $\mathcal{O}$ be a nonempty subset of $\mathcal{X}$. Then $S(\mathcal{O})$ is defined by

$$S(\mathcal{O}) := \{g \in G \mid g \cdot \mathcal{O} \subset \mathcal{O}\}.$$ □

From this definition and Proposition B.1.3 we immediately obtain the following.

Lemma 5.4.3 *$S(\mathcal{O})$ is a subsemigroup of G. If G acts transitively on $\mathcal{X}$, then the interior $S(\mathcal{O})^o$ of $S(\mathcal{O})$ is given by*

$$S(\mathcal{O})^o = \{g \in G \mid g \cdot \overline{\mathcal{O}} \subset \mathcal{O}\}.$$ □

Recall the special situation from Section 5.1. If $g \in S(\mathcal{O})$ and $X \in \Omega_-$, then we define $g \cdot X$ by

$$g \cdot X = \kappa^{-1}(g\kappa(X)). \tag{5.17}$$

This turns κ into a $S(\mathcal{O})$-equivariant map. We note that $H \subset S(\mathcal{O})$.

For any group G and any pair of closed subgroups L, Q of G we write

$$S(L,Q) := \{g \in G : gL \subset LQ\} = \{g \in G : gLQ \subset LQ\} = S(LQ/Q)$$

and called it the *compression semigroup* in G of the L-orbit in G/Q. Then it follows from [52], Proposition 8.45, that $S(\tilde{C}_{\max})$ coincides with the subsemigroup $S(G^c, B)$, where B is the Borel subgroup belonging to $\mathfrak{h}_{\mathbb{C}}$ and $\tilde{\Delta}^+$. It is important for the cases we are interested in to observe that [52], Lemma 8.41, implies $S(G^c, B) = S(G^c, (P_{\max})_{\mathbb{C}})$ (this also follows from Lemma 5.1.2 applied to $G_{\mathbb{C}}/G^c$). Thus we have $S(\tilde{C}_{\max}) = S(G^c, (P_{\max})_{\mathbb{C}})$.

Lemma 5.4.4 *Fix any parabolic subgroup Q between $P_{\min}$ and $P_{\max}$. Then*

$$S(G^\tau, Q) = S(H_o, Q) = S(H_o, P_{\min}) = S(G^\tau, P_{\min}).$$

Proof: This follows from Lemma 3.1.22, Lemma 5.1.2, and the observation that

$$S(G^\tau, P_{\min}) \subset S(G^\tau, Q) \subset S(G^\tau, P_{\max})$$

as well as the corresponding relation for H_o. □

Consider $C_{\max} = \tilde{C}_{\max} \cap \mathfrak{q}$ and $S(C_{\max}) = H \exp(C_{\max})$.

Remark 5.4.5 The cone $C_{\max}$ is G^τ-invariant because

$$G^\tau = G_{\mathbb{C}}^{\tau c} \cap G_{\mathbb{C}}^{\sigma} = G_{\mathbb{C}}^{\tau c} \cap G_{\mathbb{C}}^{\sigma\tau c} = G_{\mathbb{C}}^{\tau c} \cap G^c = (G^c)^\sigma$$

and $\tilde{C}_{\max}$ is G^c-invariant. Note also that

$$\begin{aligned} G \cap S(\tilde{C}_{\max}) &= G_{\mathbb{C}}^{\sigma} \cap S(\tilde{C}_{\max}) \\ &= (G^c)^\sigma \exp(C_{\max}) \\ &= G^\tau \exp(C_{\max}) = FS(C_{\max}). \end{aligned}$$

This is sometimes helpful to reduce questions concerning the semigroups $H \exp(C_{\max})$ to similar problems for $S(\tilde{C}_{\max})$. □

The closed subsemigroup $S(C_{\max})$ is called the *real maximal Ol'shanskii semigroup*.

Lemma 5.4.6 $S(C_{\max}) \subset S(G^\tau, P_{\max})$.

Proof: Let $g \in S(C_{\max})$ and $x \in \mathcal{O} = G^\tau \cdot \mathrm{o}_{\mathcal{X}} \subset \mathcal{X} \subset \mathcal{X}_{\mathbb{C}}$. Then

$$S(C_{\max}) \subset S(\tilde{C}_{\max}) = S(G^c, (P_{\max})_{\mathbb{C}})$$

implies

$$g.\mathrm{o}_{\mathcal{X}} \in G^c \cdot \mathrm{o}_{\mathcal{X}} \cap \mathcal{X}_{\mathbb{C}}^{\sigma} = \mathcal{O} = G^\tau \cdot \mathrm{o}_{\mathcal{X}}$$

(Lemma 5.1.11 and Lemma 5.1.12). This shows that

$$S(C_{\max}) = G \cap S(\tilde{C}_{\max}) \subset S(G^\tau, P_{\max}).$$ □

Theorem 5.4.7 *1)* $S(C) \subset HAN \cap N^\sharp AH$.

2) $S(C) = H(S(C) \cap AN) = (S(C) \cap N^\sharp A)$.

3) $G^\tau \exp(C) \cap AN = S(C) \cap AN = S(C)_o \cap AN$ *is connected.*

Proof: 1) Lemma 5.4.6 shows that $S(C) \subset HP_{\max}$. In particular, we have $S(C)_o \subset (HP_{\max})_o$, which is equal to $(G^\tau)_o AN = H_o AN$ by Lemma 5.4.1. But then $S(C) = HS(C)_o \subset HAN$. Since $S(C) = \tau(S(C)^{-1})$ we also have $S(C) \subset \tau(N)AH$.

2) This follows from 1) in view of $H \subset S(C)$.

3) Since $H_o \subset S(C)_o$, Lemma 5.4.1 shows that $S(C)_o \cap AN$ is connected. Now the claim follows from $G^\tau \exp(C) = G^\tau S(C)_o$, $S(C) = HS(C)_o$, and $G^\tau \cap AN = \{1\}$. □

Theorem 5.4.8 *1)* $S(C_{\max}) \cap B^\sharp$ *is a generating Lie semigroup in* $B^\sharp$ *with the pointed generating tangent cone*

$$\mathbf{L}(S(C_{\max}) \cap B^\sharp) = (C_{\max} + \mathfrak{h}) \cap (\mathfrak{n}^\sharp + \mathfrak{a}) \supset c_{\max}.$$

2) $S(C_{\max}) \cap B^\sharp \subset N^\sharp \exp(c_{\max})$.

Proof: 1) Let $G_1 := N^\sharp A \times H$ act on G by $(t,h) \cdot g = tgh^{-1}$. Then the orbit of 1 is the open subset $N^\sharp AH$ of G. Moreover, the stabilizer of 1 is trivial. We define the field Θ of cones on $N^\sharp AH$ by

$$\Theta(g) := d_1\lambda_g(C_{\max} + \mathfrak{h}) \qquad \forall g \in N^\sharp AH,$$

where $\lambda_g : G \to G, x \mapsto gx$ denotes left multiplication by g.

We claim that Θ is invariant under the action of G_1, i.e., that

$$d_1\mu_{(t,h)}\Theta(1) = \Theta(th^{-1}),$$

where $\mu_{(t,h)} : N^\sharp AH \to N^\sharp AH, g \mapsto tgh^{-1}$. To see this, we first note that

$$\mu_{(t,h)} = \lambda_t \circ \rho_{h^{-1}} = \lambda_{th^{-1}} \circ I_h,$$

where $I_h : g \mapsto hgh^{-1}$. Therefore

$$d\mu_{(t,h)}(1)\Theta(1) = d\lambda_{th^{-1}}(1)\,\mathrm{Ad}(h)(C_{\max} + \mathfrak{h}) = \Theta(th^{-1})$$

is a consequence of $\mathrm{Ad}(H)(C_{\max} + \mathfrak{h}) = C_{\max} + \mathfrak{h}$.

The semigroup $S(C_{\max})_o$ is the set of all points in $N^\sharp AH_o$ for which there exists a $\Theta(1)$-causal curve. Since this set is closed, the inverse image of $S(C_{\max})_o$ under the orbit mapping

$$\Phi : G_1 \to G, \quad (t,h) \mapsto th^{-1}$$

is a Lie semigroup whose tangent wedge agrees with

$$d\Phi^{-1}(1)\,(\Theta(1)) = d\Phi^{-1}(1)(C_{\max}+\mathfrak{h}) = \mathfrak{h} + (C_{\max}+\mathfrak{h}) \cap (\mathfrak{a}+\mathfrak{n}^\sharp)$$

([114], p. 471).

We know that $S(C_{\max})_o = (S(C_{\max}) \cap B^\sharp)H_o$ and therefore

$$\Phi^{-1}(S(C_{\max})_o) = (S(C_{\max})_o \cap B^\sharp) \times H_o$$

is a Lie semigroup. We conclude with Theorem 5.4.7 that $S(C_{\max} \cap B^\sharp) = S(C_{\max})_o \cap B^\sharp$ is a Lie semigroup with $\mathbf{L}(S(C_{\max} \cap B^\sharp)) = \mathbf{L}(S(C_{\max})) \cap (\mathfrak{a}+\mathfrak{n}^\sharp) = (C_{\max}+\mathfrak{h}) \cap (\mathfrak{a}+\mathfrak{n}^\sharp)$.

2) The mapping $p\colon B^\sharp \to A, n^\sharp a \mapsto a$ is a group homomorphism because $N^\sharp$ is a normal subgroup of $B^\sharp$. Therefore $p(S(C_{\max}) \cap B^\sharp)$ is a subsemigroup of A which is contained in the Lie semigroup generated by

$$\begin{aligned} V := dp(1)\mathbf{L}(S(C_{\max}) \cap B^\sharp) &= \left[(C_{\max}+\mathfrak{h}) \cap (\mathfrak{a}+\mathfrak{n}^\sharp) + \mathfrak{n}^\sharp\right] \cap \mathfrak{a} \\ &= (C_{\max}+\mathfrak{h}+\mathfrak{n}^\sharp) \cap \mathfrak{a}. \end{aligned}$$

This cone is the projection of $C_{\max}$ along $\mathfrak{h}+\mathfrak{n}^\sharp$ onto $\mathfrak{a}$. Let $w \in C_{\max} \subset \mathfrak{q}$. Then there exists $X \in \mathfrak{a}$ and $Y \in \mathfrak{n}^\sharp$ such that $w = X + Y - \tau(Y)$. Hence

$$w \in X + 2Y - [\tau(Y)+Y] \in X + \mathfrak{n}^\sharp + \mathfrak{h}.$$

Therefore V is the orthogonal projection of $C_{\max}$ in $\mathfrak{a}$. Thus $V = \mathfrak{a} \cap C_{\max} = c_{\max}$ and hence $p(S(C_{\max}) \cap B^\sharp) = \exp(c_{\max})$. Consequently, we find $S(C_{\max}) \cap B^\sharp \subset N^\sharp \exp(c_{\max})$. □

Lemma 5.4.9 $S(G^\tau, P_{\max}) \cap \exp \mathfrak{a} = \exp c_{\max}$.

Proof: The inclusion $\exp c_{\max} \subset S(G^\tau, P_{\max})$ is clear. To show the converse direction, we let $X \in \mathfrak{a} \setminus c_{\max}$. Then there exists an $\alpha \in \Delta_+$ such that $\alpha(X) < 0$. Consider the subalgebra $\mathfrak{s}_\alpha = \varphi_\alpha(\mathfrak{sl}(2,\mathbb{R}))$ described in (4.8). Then $\exp(\mathbb{R}Y^\alpha) \subset (G^\tau)_o$, and it suffices to show that $\exp(X)\exp(\mathbb{R}Y^\alpha)\cdot o$ cannot be contained in $\mathcal{O}$.

Lemma 5.4.1 shows that the map

$$a_H\colon G^\tau AN = G^\tau P_{\max} \to \mathfrak{a}, \quad han \mapsto \log(a) \tag{5.18}$$

is well defined and analytic. We call it the *causal Iwasawa projection*. When we restrict this map to the group generated by $\exp(\mathfrak{s}_\alpha)$, a simple $\mathrm{SL}(2,\mathbb{R})$-calculation shows that

$$a_H\left(\exp(tX^\alpha)\exp(sY^\alpha)\right) = \left[t + \frac{1}{2}\log(1+(1-e^{-2t})\sinh^2(\frac{s}{2}))\right] X^\alpha.$$

Now choose $X' := X - t_o X^\alpha$, where $t_o = \alpha(X)$. Then X' commutes with $\mathfrak{s}_\alpha$ and hence we calculate

$$\begin{aligned} a_H(\exp(X)\exp(sY^\alpha)) &= a_H\left(\exp(t_o X^\alpha)\exp(sY\alpha)\exp(X')\right) \\ &= a_H(\exp(t_o X^\alpha)\exp(sY^\alpha)) + X' \\ &= X' + \left[t_o + \frac{1}{2}\log\left(1 + \frac{(e^{2t_o}-1)\sinh^2(\frac{s}{2})}{e^{2t_o}}\right)\right] X^\alpha. \end{aligned}$$

But this function in the parameter s is not extendable to all of $\mathbb{R}$ so that $\exp(X)\exp(sY^\alpha)\cdot o$ cannot be contained in $\mathcal{O}$ for all $s \in \mathbb{R}$. □

So far we know that the semigroup $S(G^\tau, P_{\max})$ contains the Ol'shanskii semigroup $S(C_{\max})$ and that the intersection of A with $S(G^\tau, P_{\max})$ is not bigger than the intersection with the Ol'shanskii semigroup. The remainder of this section will be devoted to the proof of the equality $S(G^\tau, P_{\max}) = S(C_{\max})$.

We start with a description of the open H-orbits in the flag manifolds $G/P_{\min}$.

Lemma 5.4.10 *Let $\mathfrak{a}', \mathfrak{a}'' \subset \mathfrak{p}$ be τ-invariant maximal abelian subspaces such that $\mathfrak{a}' \cap \mathfrak{q}_p$ and $\mathfrak{a}'' \cap \mathfrak{q}_p$ are maximal abelian in $\mathfrak{q}_p$. Then there exists $k \in (K^\tau)_o$ such that* $\mathrm{Ad}(k)\mathfrak{a}' = \mathfrak{a}''$.

Proof: (cf. also Lemma 7 in [99], p. 341.) Since the maximal abelian subspaces of $\mathfrak{q}_p$ are conjugate under K_0^τ ([44], p.247), we even may assume that $\mathfrak{a}' \cap \mathfrak{q}_p = \mathfrak{a}'' \cap \mathfrak{q}_p$.

Set $\mathfrak{g}^0 := \mathfrak{z}_\mathfrak{g}(\mathfrak{a}'')$. We consider the symmetric Lie algebra $(\mathfrak{g}^0, \tau)$ which is invariant under θ and therefore reductive ([168], Corollary 1.1.5.4). Then $\mathfrak{g}^0 \cap \mathfrak{q}_p = \mathfrak{a}''$ is central in $\mathfrak{g}^0$ and $\mathfrak{a}' = (\mathfrak{a}'' \cap \mathfrak{q}_p) \oplus (\mathfrak{a}' \cap \mathfrak{h}_p)$. Hence $\mathfrak{a}' \cap \mathfrak{h}_p \subset \mathfrak{g}^0 \cap \mathfrak{h}_p$ is maximal abelian in $\mathfrak{h}_p \cap \mathfrak{g}^0$. The same holds for $\mathfrak{a}'' \cap \mathfrak{h} \subset \mathfrak{h}_p \cap \mathfrak{g}^0$. The pair $(\mathfrak{h} \cap \mathfrak{g}^0, \theta)$ is Riemannian symmetric, hence $\mathfrak{a}'' \cap \mathfrak{h}_p$ and $\mathfrak{a}' \cap \mathfrak{h}_p$ are conjugate under $\exp(\mathfrak{h}_k \cap \mathfrak{g}^0)$ ([44], p.247). We conclude that $\mathfrak{a}'$ is conjugate to $\mathfrak{a}''$ under $(K^\tau)_o$. □

Note that H-orbits in $G/P_{\min}$ correspond to H-conjugacy classes of minimal parabolic subalgebras of $\mathfrak{g}$. According to [99], p. 331, each minimal parabolic subalgebra of $\mathfrak{g}$ is $(G^\tau)_o$-conjugate to one of the form $\mathfrak{m} + \mathfrak{a} + \mathfrak{n}$ for some τ-invariant maximal abelian subspace $\mathfrak{a}$ of $\mathfrak{p}$ and some positive system in $\Delta(\mathfrak{g}, \mathfrak{a})$. Let $\mathfrak{a}_1, \mathfrak{a}_2, \ldots, \mathfrak{a}_s$ be a set of representatives of the K^τ–conjugacy classes of maximal abelian τ-invariant subspaces of $\mathfrak{p}$. [99], §3, Proposition 1 (cf. also [156], Proposition 7.1.8), among other things, says:

Lemma 5.4.11 *Let Δ_j^+ be a positive system for $\Delta_j := \Delta(\mathfrak{g}, \mathfrak{a}_j)$. Denote the corresponding minimal parabolic subalgebra by $\mathfrak{p}(\mathfrak{a}_j, \Delta_j^+)$. Then the H-*

conjugacy class of $\mathfrak{p}(\mathfrak{a}_j, \Delta_j^+)$ *corresponds to an open* H*-orbit in* $G/P_{\min}$ *if and only if the following conditions are satisfied:*

1) $\mathfrak{a}_j \cap \mathfrak{q}$ *is maximal abelian in* $\mathfrak{q}_p$.

2) Δ_j^+ *is* $\mathfrak{q}$*-compatible, i.e., the set* $\Delta_j^+ \setminus (\mathfrak{a}_j \cap \mathfrak{q})^\perp$ *is* $-\tau$*-invariant.* □

We call a τ-invariant maximal abelian subspace $\mathfrak{a}'$ of $\mathfrak{p}$ a $\mathfrak{q}$-*maximal* subspace if $\mathfrak{a}' \cap \mathfrak{q}$ is maximal abelian in $\mathfrak{q}_p$ (cf. [156], p. 118). Note that according to Lemma 5.4.10, $\mathfrak{q}$-maximal maximal abelian subspaces of $\mathfrak{p}$ are conjugate under $(K^\tau)_o$. This shows that only one of the $\mathfrak{a}_j$ can be $\mathfrak{q}$-maximal; i.e., condition 1) can be satisfied only by one of the $\mathfrak{a}_j$.

The part of [99], §3, Proposition 1 we have not yet stated here concerns the number of open H-orbits. For a fixed τ-invariant maximal abelian subspace $\mathfrak{a}$ of $\mathfrak{p}$ we consider the Weyl groups

$$\begin{aligned} W(\mathfrak{a}) &= N_K(\mathfrak{a})/Z_K(\mathfrak{a}), \\ W_\tau(\mathfrak{a}) &:= \{s \in W(\mathfrak{a}) \mid s(\mathfrak{a} \cap \mathfrak{h}) = \mathfrak{a} \cap \mathfrak{h}\}, \\ W_0(\mathfrak{a}) &:= N_{K\cap H}(\mathfrak{a})/Z_{K\cap H}(\mathfrak{a}). \end{aligned}$$

Then

$$W_0(\mathfrak{a}) \subset W_\tau(\mathfrak{a}) \subset W(\mathfrak{a}),$$

and [99], §3, Proposition 1 says that the number of open H-orbits in $G/P_{\min}$ is the number of cosets in $W_\tau(\mathfrak{a})/W_0(\mathfrak{a})$.

Remark 5.4.12 Let $\mathfrak{a}$ be a τ-invariant $\mathfrak{q}$-maximal abelian subspace of $\mathfrak{p}$ and Δ^+ a positive system for $\Delta(\mathfrak{g}, \mathfrak{a})$. Then Δ^+ is $\mathfrak{q}$-compatible if and only $\mu|_{\mathfrak{a}\cap\mathfrak{q}} \neq 0$ implies $-\tau(\mu) \in \Delta^+$ for all $\mu \in \Delta^+$ (cf. [156], p. 120, and [99], p. 355). □

Lemma 5.4.13 *1) Let* $\mathfrak{a}$ *be a* τ*-invariant* $\mathfrak{q}$*-maximal abelian subspace of* $\mathfrak{p}$ *and* Δ_1^+, Δ_2^+ *two* $\mathfrak{q}$*-compatible positive systems for* $\Delta(\mathfrak{g}, \mathfrak{a})$*. Then the corresponding minimal parabolic subalgebras belong to the same open* H*-orbit if and only if there exists a* $s \in W_0(\mathfrak{a})$ *such that* $s \cdot \Delta_1^+ = \Delta_2^+$.

2) Let $\mathfrak{a}$ *be a* τ*-invariant* $\mathfrak{q}$*-maximal abelian subspace of* $\mathfrak{p}$*. Assume that* $\Delta^+(\mathfrak{g}, \mathfrak{a})$ *is* $\mathfrak{q}$*-compatible. Then the open* H*-orbits in* G/P *are precisely the* $HsP_{\min}/P_{\min}$ *with* $s \in W_\tau(\mathfrak{a})$*. The orbits* $Hs_1P_{\min}/P_{\min}$ *and* $Hs_2P_{\min}/P_{\min}$ *agree if and only if there exists an* $s \in W_0(\mathfrak{a})$ *with* $ss_1 = s_2$*. In particular,* $HgP_{\min}/P_{\min}$ *is open if and only if* $g \in HW_\tau(\mathfrak{a})P_{\min}$*. The union of open* H*-orbits is dense in* $G/P_{\min}$.

Proof. 1) The Weyl group $W_\tau(\mathfrak{a})$ acts simply transitively on the set of $\mathfrak{q}$-compatible positive systems (cf. [156], Proposition 7.1.7]). Clearly, two $\mathfrak{q}$-compatible positive systems belong to the same H-orbit if they are conjugate under $W_0(\mathfrak{a})$. Now the formula for the number of open orbits implies the claim.

2) Only the last claim remains to be shown. But that follows from the fact that there are only finitely many H-orbits (cf. [99], Theorem 3, and [52], Proposition 8.10(ii)), since each orbit is an immersed manifold. □

Lemma 5.4.14 *If $P_1 \subset P_2$ are parabolic subgroups of G, $\mathcal{F}_i = G/P_i$ the corresponding flag manifolds, $x_i \in \mathcal{F}_i$, and $\pi: \mathcal{F}_1 \to \mathcal{F}_2$ the natural projection, then the following assertions hold:*

1) *If $H \cdot x_1$ is open in $\mathcal{F}_1$, then $\pi(H \cdot x_1) = H \cdot \pi(x_1)$ is open in $\mathcal{F}_2$.*

2) *If $H \cdot x_2$ is open in $\mathcal{F}_2$, then $\pi^{-1}(H \cdot x_2)$ contains an open H-orbit in $\mathcal{F}_1$.*

Proof. 1) follows from the fact that π is open and G-equivariant.

2) $\pi^{-1}(H \cdot x_2)$ is open by continuity. Suppose for a moment that P_1 is a minimal parabolic. Then Lemma 5.4.13.2) says that the union of open H-orbits is dense in $\mathcal{F}_1$ and therefore $\pi^{-1}(H \cdot x_2)$ intersects, hence contains, an open H-orbit. If we now apply 1), this argument shows that for *any* flag manifold the union of the open H-orbits is dense, and we can prove our claim for arbitrary P_1. □

Lemma 5.4.15 *Let P' be an arbitrary parabolic, $x \in G/P'$, and P'_x the stabilizer of x in G. Then the following statements are equivalent:*

1) *$H \cdot x$ is open in G/P'.*

2) *There exist a $\mathfrak{q}$-maximal maximal abelian subspace $\mathfrak{a}_\sharp$ of $\mathfrak{p}$ and a $\mathfrak{q}$-compatible positive system $\Delta_\sharp^+$ of $\Delta(\mathfrak{g}, \mathfrak{a}_\sharp)$ such that P'_x is H-conjugate to a standard parabolic associated to $\Delta_\sharp^+$.*

Proof. In the case where P' is a minimal parabolic, our claim is just Proposition 5.4.11.

"1) ⇒ 2)": For the general case recall that G/P' can be identified with the set of parabolic subgroups of G conjugate to P' and the natural projection $\pi: G/P_{\min} \to G/P'$ maps a conjugate $gP_{\min}g^{-1}$ of $P_{\min}$ to $gP'g^{-1}$. Identifying x and P'_x, one has that $\pi^{-1}(x)$ consists of all minimal parabolics contained in P'_x. Lemma 5.4.14 says that there is one such minimal parabolic $P_\sharp$ that lies in an open H-orbit. But then Proposition 5.4.11 shows that $\mathfrak{p}_\sharp$ is of the form $\mathfrak{p}(\mathfrak{a}_\sharp, \Delta_\sharp^+)$ for suitable $\mathfrak{a}_\sharp$ and $\Delta_\sharp^+$, and thus P'_x must have the right form since it contains $P_\sharp$.

"2) $\Rightarrow$ 1)": For the converse we invoke Lemma 5.4.14.1) to see that the H-orbits of the parabolic group associated to a $\mathfrak{q}$-compatible system $\Delta_\sharp^+$ are always open. $\square$

Lemma 5.4.16 *Recall the flag manifold $\mathcal{X} := G/P_{\max}$ with base point $o_\mathcal{X}$. Then the following assertions hold:*

1) The open orbit $H \cdot o_\mathcal{X}$ is contained in the open Bruhat cell $N_- \cdot o_\mathcal{X}$.

2) Every other open H-orbit is not entirely contained in $N_- \cdot o_\mathcal{X}$.

3) $H \cdot o_\mathcal{X}$ is the largest open subset of the open cell which is H-invariant.

Proof: 1) This is a consequence of Lemma 5.1.12.

2) Let $y \in \mathcal{X}$ and suppose that the H-orbit of y is open and different from the H-orbit of the base point. Then it follows from Proposition 5.4.15 that there exists a point in this H-orbit which is fixed by the subgroup A. Since the base point is the only A-fixed point in the open Bruhat–cell, we conclude that $H \cdot y$ cannot be contained in the open cell.

3) Since the set of all elements in $\mathcal{X}$ whose H-orbit is open is dense ([52], Proposition 8.10(ii)), this follows from the fact that $H \cdot o_\mathcal{X}$ is the interior of its closure. $\square$

Lemma 5.4.17 *The following assertions hold:*

1) $HP_{\max} = \exp(\Omega_-)H^a N_+$ is the largest open H-left-invariant subset of $N_- H^a N_+ = N_- P_{\max}$.

2) There exists an open bounded subset $\Omega_+ \subset \mathfrak{n}_+$ such that $P_{\max}^\sharp H = N_- H^a \exp(\Omega_+)$. This set is the largest open H-right-invariant subset of $N_- H^a N_+$.

3) Every H-biinvariant open subset of $N_- H^a N_+$ is contained in the open set $\exp(\Omega_-)H^a \exp(\Omega_+)$.

Proof: 1) Lemma 5.1.12 shows $HP_{\max} = \exp(\Omega_-)H^a N_+$ and $H^a N_+ = P_{\max}$ follows from Lemma 3.1.22. Now the claim follows from Lemma 5.4.16.3).

2) This follows from 1) by applying the automorphism τ.

3) This is a consequence of 1) and 2). $\square$

Lemma 5.4.18 *$S(G^\tau, P_{\max})^o$ is the largest open H-biinvariant subset of $N_- H^a N_+$.*

Proof: It follows from Lemma 5.4.16.1) that $S(G^\tau, P_{\max}) \subset HP_{\max} \subset N_- H^a N_+$. Moreover, for every $s \in S(G^\tau, P_{\max})$ the double coset HsH is contained in $S(G^\tau, P_{\max})$ and therefore in $N_- H^a N_+$. This shows that

$S(G^\tau, P_{\max})^o$ is an open H-biinvariant subset of $N_- H^a N_+$. Now suppose that $E \subset N_- H^a N_+$ is an open H-biinvariant set. Then we first use Lemma 5.4.17 to see that

$$EH \subset E \subset \exp(\Omega_-) H^a \exp(\Omega_+) \subset \exp(\Omega_-) P_{\max} = HP_{\max}.$$

It follows in particular that $E \subset S(G^\tau, P_{\max})$. Thus $S(G^\tau, P_{\max})$ is the unique maximal open H-biinvariant subset of $N_- H^a N_+$. □

Corollary 5.4.19 *$S(G^\tau, P_{\max})$ is invariant under the involution $s \mapsto s^\sharp$.*

Proof: This is a consequence of Lemma 5.4.18 because the set $N_- H^a N_+$ is invariant under this involution and therefore the same is true for the maximal H-biinvariant subset of this set. □

Theorem 5.4.20 *Let $\mathcal{M} = G/H$ be a noncompactly causal symmetric space and τ the corresponding involution on G. Assume that G embeds into a simply connected complex group $G_\mathbb{C}$ and let Q be a parabolic subgroup between $P_{\min}$ and $P_{\max} = H^a N_+$. Then*

$$S(H, Q) = G^\tau \exp(C_{\max}) = S(G^\tau, P_{\max}).$$

Proof: It follows from Lemmas 5.1.2 and 5.4.4 that we may w.l.o.g. assume that $Q = P_{\max}$ and that $H = G^\tau$. First we apply Theorem 1.4.2 to obtain further information on the semigroup $S(G^\tau, P_{\max})$. Recall the notation from Section 1.4 and let $\mathfrak{a}_q \subset \mathfrak{q}$ be a θ-invariant A-subspace. Suppose that $\pi^{-1}(A_q') \cap S(G^\tau, P_{\max})^o \neq \emptyset$. In view of Corollary 5.4.19, the semigroup $S(G^\tau, P_{\max})$ is invariant under the mapping π and therefore we find $s \in S_\tau^o \cap A_q'$. Next we recall that $A_q = (A_q \cap K)A_q^p$, where $A_q^p := \exp(\mathfrak{a}_q \cap \mathfrak{q}_p)$. We consider the semigroup $S_A := S(G^\tau, P_{\max})^o \cap A_q$. Then the semigroup $S_A A_q^p / A_q^p$ is an open subsemigroup of a compact group, so that it must contain the identity element (cf. [52], Corollary 1.21). We conclude that S_A intersects the subgroup A_q^p. This subgroup is conjugate to a subgroup of A (Lemma 5.4.10). Suppose that $A_q \cap K$ is nontrivial. Then $\mathfrak{a}_q \cap \mathfrak{p}$ is not maximal abelian in $\mathfrak{q}_p$ and the description of the $W_0(\mathfrak{a})$-conjugacy classes of A-subspaces given in [143], p. 413 shows that the conjugate of A_q^p in A lies in the exponential image of the set $\bigcup_{\alpha \in \Delta_p^+} \ker \alpha$. It follows that this set contains interior points of $S(G^\tau, P_{\max})$. On the other hand, we know already that $S(G^\tau, P_{\max}) \cap A = \exp(c_{\max})$ (Lemma 5.4.9). This is a contradiction because every element in $c_{\max}$ which is in the kernel of a noncompact root lies on the boundary. Thus we have shown that the only A-subspace A_q for which the open subset $H\phi^{-1}(A_q)$ intersects $S(G^\tau, P_{\max})^o$ is $\tilde{A} := Z_{\phi(G)}(\mathfrak{a})$. Let $s \in S(G^\tau, P_{\max}) \cap H\phi^{-1}(\tilde{A})$. Then

$$\pi(s) = ss^\sharp \in \tilde{A}.$$

We have to get a better picture of the set $\tilde{A}$. So we first remark that $\tilde{A} = Z_{\phi(G)}(A) = (M \cap \phi(K))A$ (Theorem 1.4.2). Let $k \in \phi(K) \cap M$. Then $k^\sharp = k$ and on the other hand $\tau(k) = k$ by Lemma 3.1.22. Therefore $k = k^{-1}$, i.e., $k^2 = 1$. Moreover, the surjectivity of the exponential function of the Riemannian symmetric space $K/K^\tau \cong \phi(K)$ yields that $\pi(K) = \exp(\mathfrak{q}_k)$.

We write $ss^\sharp = k\exp(Z)$ with $k \in \phi(K) \cap M$ and $Z \in \mathfrak{a}$. Then we we find $Y \in \mathfrak{q}_k$ with $k = \exp(2Y)$. We set $k' := \exp Y$. We claim that $\mathrm{Ad}(k')\mathfrak{a} \subset \mathfrak{q}_p$. To see this, pick $X \in \mathfrak{a}$. Then

$$\tau\left(\mathrm{Ad}(k')X\right) = \mathrm{Ad}(k')^{-1}\tau(X) = -\mathrm{Ad}(k')^{-1}X = -\mathrm{Ad}(k')X$$

and similarly

$$\theta\left(\mathrm{Ad}(k')X\right) = \mathrm{Ad}(k')\theta(X) = -\mathrm{Ad}(k')X.$$

This proves our claim. Now we find that

$$\begin{aligned}\left[k'\exp(\tfrac{1}{2}e^{\mathrm{ad}\,Y}Z)\right]\left[k'\exp(\tfrac{1}{2}e^{\mathrm{ad}\,Y}Z)\right]^\sharp &= k'\exp(e^{\mathrm{ad}\,Y}Z)k' \\ &= k\exp(Z) \\ &= ss^\sharp.\end{aligned}$$

We conclude that

$$\begin{aligned}k\exp(\tfrac{1}{2}Z)(k')^{-1} &= k'\exp\left(\tfrac{1}{2}e^{\mathrm{ad}\,Y}Z\right) \in \phi^{-1}(s) \\ &= sH \subset S(G^\tau, P_{\max}).\end{aligned}$$

We have already seen that $\mathrm{Ad}(k')\mathfrak{a} \subset \mathfrak{q}_p$. Hence there exists $k'' \in (K^\tau)_0$ such that $\mathrm{Ad}(k'')\,\mathrm{Ad}(k')\mathfrak{a} = \mathfrak{a}$ (Lemma 5.4.10). This means that $k''k' \in N_K(\mathfrak{a})$. Multiplying with k'' on the left, we find that

$$k''k'\exp(Z)\cdot x_0 \in N_K(\mathfrak{a}).x_0,$$

so that this point is an A-fixed point in $G/P_{\max}$. On the other hand, the semigroup $S(G^\tau, P_{\max})$ is contained in the set $N_-P_{\max}$, so x_0 is the only A-fixed point in the set $S(G^\tau, P_{\max})\cdot x_0$. Thus

$$k''k' \in P_{\max} \cap N_K(\mathfrak{a}) \subset H^a \cap K = Z_K(\mathfrak{c}) \subset H \subset G^\tau.$$

It follows that $k''k' \in H$ and therefore that $k' \in H$. Thus $\exp(Z) \in A \cap S(G^\tau, P_{\max}) \subset G^\tau \exp C_{\max}$ and hence $s \in G^\tau \exp C_{\max}$, which finally shows that $S(G^\tau, P_{\max})$ is contained in the semigroup $G^\tau \exp C_{\max}$. □

Corollary 5.4.21 *Recall the open domain $\mathcal{O} = G_o^\tau \cdot o_\mathcal{X}$ in $\mathcal{X} = G/P_{\max}$. Then*

$$S(C_{\max}) = \{s \in G \mid s \cdot \mathcal{O} \subset \mathcal{O}\} \text{ and } S(C_{\max})^o = \{s \in G \mid s \cdot \overline{\mathcal{O}} \subset \mathcal{O}\}. \quad \square$$

Remark 5.4.22 One can show that $G^\tau \exp(C_{\max})$ is actually a maximal subsemigroup of G (cf. [58], Theorem V.4). □

Example 5.4.23 In the situation of SL(2, $\mathbb{R}$)-Example 5.1.13 we have

$$S(C_{\max}) \cap B^\sharp = \left\{ \begin{pmatrix} a & 0 \\ c & a^{-1} \end{pmatrix} \middle| \forall |r| < 1 : \left| \frac{r}{a^2} + \frac{c}{a} \right| < 1 \right\}.$$

An elementary argument shows that the condition on c and a can be reformulated as

$$S(C_{\max}) \cap B^\sharp = \left\{ \begin{pmatrix} a & 0 \\ c & a^{-1} \end{pmatrix} \middle| |c| < a - a^{-1}; 0 < a \right\}. \quad \square$$

5.5 The Nonlinear Convexity Theorem

In this section we again consider a noncompactly causal symmetric space $\mathcal{M} = G/H$ such that G is contained in a simply connected complexification $G_\mathbb{C}$. We will prove Neeb's nonlinear convexity theorem, which says that

$$a_H(aH) = \operatorname{conv}(W_0 \cdot \log a) + c_{\min}$$

for all $a \in \exp(c_{\max})$. This will be done first for the special case $\mathcal{N} = G_\mathbb{C}/\check{G}^c$ (cf. Section 1.1). Then the general result can be obtained via the suitable intersections with smaller spaces. Note that in our situation $G = \check{G}$. We write G^c for $\check{G}^c$.

Recall the situation described in Section 4.5. In particular, let $\mathfrak{t}^c$ be a Cartan subalgebra of $\mathfrak{k}^c$ containing $i\mathfrak{a}$ and $\tilde{\Delta} = \Delta(\mathfrak{g}_\mathbb{C}, \mathfrak{t}^c_\mathbb{C})$. We set

$$\tilde{\mathfrak{n}} := \sum_{\tilde{\alpha} \in \tilde{\Delta}^+} (\mathfrak{g}_\mathbb{C})_{\tilde{\alpha}}, \tag{5.19}$$

where $\tilde{\Delta}^+$ is chosen as on p. 95. Further we set $\tilde{\mathfrak{a}} := i\mathfrak{t}^c$, $\tilde{A} := \exp \tilde{\mathfrak{a}}$ and $\tilde{N} := \exp \tilde{\mathfrak{n}}$. Then (5.18) yields a causal Iwasawa projection $a_{G^c}\colon G^c\tilde{A}\tilde{N} \to \tilde{\mathfrak{a}}$. The derivative $d_1 a_{G^c}\colon \mathfrak{g}_\mathbb{C} \to \tilde{\mathfrak{a}}$ is simply the projection along $\mathfrak{g}^c + \tilde{\mathfrak{n}}$.

Lemma 5.5.1 *Let $\tilde{p} : \mathfrak{g}_\mathbb{C} \to \mathfrak{t}^c_\mathbb{C}$ denote the projection along the sum of the root spaces. Then $\tilde{p}|_{i\mathfrak{g}^c} = d_1 a_{G^c}|_{i\mathfrak{g}^c}$.*

Proof: Let $X \in i\mathfrak{g}^c$. Then we can write $X = Y + Z - \overline{Z}$, where $Y \in \tilde{\mathfrak{a}}$ and $Z \in \tilde{\mathfrak{n}}$. Therefore $\tilde{p}(X) = \tilde{p}(Y + Z - \sigma^c(Z)) = Y$ and

$$\begin{aligned} d_1 a_{G^c}(X) &= d_1 a_{G^c}(Y + Z - \sigma^c(Z)) \\ &= Y - d_1 a_{G^c}(\sigma^c(Z)) \\ &= Y - d_1 a_{G^c}(Z + \sigma^c(Z)) = Y. \end{aligned}$$

From this the lemma follows. □

We note that for $g \in G^c$, $a \in \tilde{A}$ and $n \in \tilde{N}$ we have

$$a_{G^c} \circ \lambda_g = a_{G^c} \quad \text{and} \quad a_{G^c} \circ \rho_{an} = a_{G^c} + \log a,$$

where as usual $\lambda_g(x) = gx$ and $\rho_g(x) = xg$ denote left and right multiplication.

We briefly recall the basic definitions concerning homogeneous vector bundles. Let L/U be a homogeneous space of L and V a vector space on which U acts by the representation $\tau: U \to \mathrm{GL}(V)$. Then we obtain an action of U on $L \times V$ via $u.(l,v) := (lu^{-1}, \tau(u).v)$ and the space of U-orbits is denoted $L \times_U V$ and called a *homogeneous vector bundle*. We write $[l,v]$ for the element of $L \times_U V$ which corresponds to the orbit of (l,v) in $L \times V$ and note that L acts from the left on $L \times_U V$ by $l.[l',v] := [ll',v]$. If L is a complex group, U is a complex subgroup, and the representation τ is holomorphic, the corresponding vector bundle is holomorphic.

Fix a linear functional $\omega \in i\tilde{c}_{\max} = i\tilde{c}^*_{\min}$ (cf. (4.22)) such that $i\omega$ integrates to a character χ of $T^c = \exp(\mathfrak{t}^c)$ and $\omega(i[\sigma^c(X), X]) \geq 0$ for $X \in (\mathfrak{g}_{\mathbb{C}})_{\tilde{\alpha}}$ with $\tilde{\alpha} \in \tilde{\Delta}_0^+$. We put $\Sigma := \{\tilde{\alpha} \in \tilde{\Delta} \mid (\forall X \in (\mathfrak{g}_{\mathbb{C}})_{\tilde{\alpha}})\ \omega(i[\sigma^c(X), X]) \geq 0\}$. Then the subalgebra

$$\tilde{\mathfrak{b}} := \mathfrak{t}^c_{\mathbb{C}} \oplus \sum_{\tilde{\alpha} \in \Sigma} (\mathfrak{g}_{\mathbb{C}})_{\tilde{\alpha}}$$

is a (complex) parabolic subalgebra of $\mathfrak{g}_{\mathbb{C}}$. Let $\tilde{B}$ be the corresponding parabolic subgroup of $G_{\mathbb{C}}$ and G^c_ω the stabilizer of ω in G^c w.r.t. the coadjoint action. Then $\tilde{B} \cap G^c = G^c_\omega$ by Theorem I.3 in [56], and we obtain a complex structure on the coadjoint orbit $G^c \cdot \omega \cong G^c/G^c_\omega$ by embedding $G^c \cdot \omega$ as the open orbit $G^c \cdot \mathbf{o}_{\tilde{B}}$ of the base point $\mathbf{o}_{\tilde{B}}$ in the complex homogeneous space $G_{\mathbb{C}}/\tilde{B}$.

We find a holomorphic character $\chi : \tilde{B} \to \mathbb{C}$ with $\chi(\exp X) = e^{i\omega(X)}$ for $X \in \tilde{\mathfrak{b}}$, where we set $\omega(X) = 0$ if X belongs to the sum of root spaces. Thus we obtain two homogeneous holomorphic line bundles: the line bundle $E := G^c \times_{G^c_\omega} \mathbb{C}_\chi$ and the line bundle $E' := G_{\mathbb{C}} \times_{\tilde{B}} \mathbb{C}_\chi$. The bundle E embeds as the open subset $E'|_{G^c \cdot \mathbf{o}_{\tilde{B}}}$ of E'.

Let $q : G^c \times \mathbb{C} \to E$ denote the quotient mapping which identifies the elements (g, z) and $(gh^{-1}, \chi(h)z)$ for $h \in G^c_\omega$. We define a function h on E by $h([g,z]) := |z|^2$ for $g \in G^c$, $z \in \mathbb{C}$.

We have already seen that the bundle E inherits a complex structure by its embedding in the complex bundle E'. We write I for the tensor field denoting multiplication by i in each tangent space. For a 1-form α on a complex manifold $\mathcal{Y}$ we define a 1-form $I\alpha$ by $\langle I\alpha, v\rangle := \langle \alpha, -Iv\rangle$ on each tangent space $T_p(\mathcal{Y})$.

Let $G^{c,\sharp} := G^c \times \mathbb{C}^*$ and $G^\sharp_\mathbb{C} := G_\mathbb{C} \times \mathbb{C}^*$. Then $G^{c,\sharp}$ acts transitively on the complement E_0 of the zero section in E and similarly $G^\sharp_\mathbb{C}$ acts transitively on E'_0 by $(g, \zeta) \cdot [g', v] = [gg', \zeta v]$.

Lemma 5.5.2 *The* 1*-form* $\alpha = I(d \log h) = \frac{1}{h} Idh$ *on* E_0 *is invariant under the action of* $G^{c,\sharp}$.

Proof: Since the action of $G^{c,\sharp}$ on E_0 is holomorphic and G^c preserves the function h, the G^c-invariance is clear. On the other hand, we have for $z \in \mathbb{C}^*$ and $\mu_z([g,x]) = [g, zx]$ that $h \circ \mu_z = |z|^2 h$. Hence $\log(h \circ \mu_z) = \log h + \log |z|^2$. Thus

$$\mu_z^*(d \log h) = d \log(h \circ \mu_z) = d(\log h + \log |z|^2) = d \log h.$$

This proves the assertion. □

To calculate the 1-form α, we have to calculate its pull-back $q^*\alpha$ to the group $G^c \times \mathbb{C}^*$ which is a left-invariant 1-form on this group. Its value in the unit element $(1,1)$ is given by $(q^*\alpha)(1,1) = \alpha([1,1])d_{(1,1)}q = -d(\log h)Id_{(1,1)}q$. To calculate this expression, we have to pass from q to the mapping $q' : G_\mathbb{C} \times \mathbb{C} \to E'$, which restricts to q on $G^c \times \mathbb{C}$. The calculation of $q'^*\alpha$ on $G_\mathbb{C} \times \mathbb{C}^*$ in the unit element is easier since q' is a holomorphic mapping:

$$(q'^*\alpha)(1,1) = -d(\log h)Id_{(1,1)}q' = -d(\log h)d_{(1,1)}q'I = -d_{(1,1)} \log(h \circ q')I.$$

The function $h \circ q'$ is given on the subset $G^c\tilde{A}\tilde{N} \times \mathbb{C}^*$ of $G_\mathbb{C}$ by

$$\begin{aligned} h \circ q'(gan, z) &= h([gan, z]) = h([an, z]) = h([1, \chi(an)z]) \\ &= h([1, \chi(a)z]) = |\chi(a)|^2 |z|^2 \end{aligned}$$

and therefore

$$h \circ q'(s, z) = e^{2\langle i\omega, a_{G^c}(s)\rangle} |z|^2 \tag{5.20}$$

for $s \in G^c\tilde{A}\tilde{N}$. This entails that $\log(h \circ q')(s,z) = 2\langle i\omega, a_{G^c}(s)\rangle + \log |z|^2$ and permits us to compute the differential of $\log(h \circ q')$ in $(1,1)$:

$$d_{(1,1)} \log(h \circ q')(Y, \zeta) = 2i\omega \circ d_1 a_{G^c}(Y) + 2\,\mathrm{Re}(\zeta).$$

Finally, we use Lemma 5.5.1 to calculate the form $q^*\alpha$ in $(1,1)$:

$$\begin{aligned}(q^*\alpha)(1,1)(X,\zeta) &= -2i\omega \circ d_1 a_{G^c}(iX) - 2\operatorname{Re}(i\zeta)\\ &= -2i\omega \circ \tilde{p}(iX) + 2\operatorname{Im}\zeta\\ &= -2i\omega(iX) + 2\operatorname{Im}\zeta\\ &= 2\omega(X) + 2\operatorname{Im}\zeta.\end{aligned}$$

This proves the assertion. □

Lemma 5.5.3 *For $X \in \mathfrak{g}^c$, $\zeta \in \mathbb{C}$, and $z \in \mathbb{C}^*$ we have*

$$\alpha([g,z])d_{(g,z)}q(d_1\rho_g(X), z\zeta) = 2\langle \operatorname{Ad}^*(g).\omega, X\rangle + 2\operatorname{Im}\zeta.$$

Proof. We write $[g,z] = (g,z)\cdot[1,1] = \mu_{(g,z)}([1,1])$. Therefore Lemma 5.5.2 yields that

$$\begin{aligned}\alpha([g,z])d_{(g,z)}q(d_1\rho_g(X), z\zeta) &= (q^*\alpha)(g,z)(d_1\rho_g(X), z\zeta)\\ &= (q^*\alpha)(g,z)(d_1\lambda_g(\operatorname{Ad}(g^{-1})X), z\zeta)\\ &= (q^*\alpha)(1,1)(\operatorname{Ad}(g^{-1})X, \zeta)\\ &= 2\langle \operatorname{Ad}^*(g)\cdot\omega, X\rangle + 2\operatorname{Im}\zeta.\end{aligned}$$

Now the Lemma follows. □

Corollary 5.5.4 *For $X \in i\mathfrak{g}^c$, $g \in G^c$, $\zeta \in \mathbb{C}$, and $z \in \mathbb{C}^*$ we have*

$$-d\log h([g,z])Id_{(g,z)}q(d_1\rho_g(iX), z\zeta) = 2\langle \operatorname{Ad}^*(g)\cdot\omega, iX\rangle + 2\operatorname{Im}\zeta.$$

Proposition 5.5.5 *For $X \in i\mathfrak{g}^c$ let $m_X := \sup\langle G^c\cdot\omega, iX\rangle$ and define the vector field $\dot{\sigma}(iX)$ on E by*

$$\dot{\sigma}(iX)(p) := d/dt \exp(-tiX)\cdot p|_{t=0}$$

for $p \in E$. Then $\langle d_p \log h, I\dot{\sigma}(iX)(p)\rangle \leq 2m_X$ for all $p \in E_0$.

Proof. Let $p = [g,z] \in E_0$. Then

$$\begin{aligned}\dot{\sigma}(iX)([g,z]) &= d/dt \exp(-tiX)\cdot[g,z]|_{t=0}\\ &= d/dt[\exp(-tiX)g, z]|_{t=0}\\ &= d_{(g,z)}q(-d_1\rho_g(iX), 0)\end{aligned}$$

and therefore

$$\begin{aligned}\langle d_p(\log h), I\dot{\sigma}(iX)(p)\rangle &= \langle d_p\log h, Id_{(g,z)}q(-d_1\rho_g(iX), 0)\rangle\\ &= 2\langle \operatorname{Ad}^*(g)\cdot\omega, iX\rangle \leq 2m_X.\end{aligned}$$

From this the proposition follows. □

Consider the compression semigroup $S := S(G^c, \tilde{B})$ in $G_{\mathbb{C}}$. Then S acts holomorphically on the bundle E' and since S leaves $G^c \cdot o_B$ invariant, the action on E' leaves the subbundle E invariant. Note that $S \subset G^c\tilde{B} = G^c\tilde{A}\tilde{N}$ since $i\tilde{\mathfrak{a}} \subset \mathfrak{g}^c$. Therefore we can write each $s \in S$ as $s = gan$ with $g \in G^c$, $a \in \tilde{A}$, and $n \in \tilde{N}$ and we find with (5.20) that

$$\log h([s, 1]) = 2\langle i\omega, a_{G^c}(s)\rangle.$$

It follows in particular that $\log h([a, 1]) = 2\langle i\omega, \log a\rangle$ for $a \in \tilde{A} \cap S = \exp(\tilde{c}_{\max})$ (cf. Lemma 5.4.9).

Fix $g \in G^c$ and $X \in \tilde{c}_{\max} = (\tilde{\Delta}_+)^\star$. We set $F(t) := \log h(\exp(tX)\cdot[g, 1])$. Then $\exp \mathbb{R}^+ X \subset S$ and therefore $\exp tX \cdot [g, 1] = [\exp tXg, 1] \in E_0$ for all $t \geq 0$. Hence we can use Proposition 5.5.5 to see that

$$F'(t) = \langle d(\log h), I\dot{\sigma}(iX)\rangle([\exp tXg, 1]) \leq 2m_X.$$

Therefore

$$2\langle i\omega, a_{G^c}(\exp Xg)\rangle = \log h(\exp X\cdot[g, 1]) = F(1) \leq 2m_X\cdot 1 = 2m_X. \quad (5.21)$$

We want to use the linear convexity theorem (Theorem 4.3.1) to calculate m_X for $X \in \tilde{c}_{\max}$. To this end we recall that our assumptions on ω, say in particular that $\omega \in i\tilde{c}^\star_{\min} = i\tilde{c}_{\max}$, cf. (4.23). Let $\mathrm{pr}: i\mathfrak{g}^c \to \tilde{\mathfrak{a}}$ be the orthogonal projection. Then Theorem 4.3.1 implies that

$$\mathrm{pr}(G^c \cdot [-i\omega]) \subset \mathrm{conv}[\tilde{W}_0 \cdot (-i\omega)] + \tilde{c}_{\min},$$

where $\tilde{W}_0$ is the Weyl group for $(\mathfrak{k}^c, \mathfrak{t}^c)$ (cf. Section 4.5). Since $X \in \tilde{c}_{\max} = \tilde{c}^\star_{\min}$, it follows that

$$m_X = \sup\langle iX, G^c \cdot \omega\rangle = \sup\langle iX, \mathrm{conv}(\tilde{W}_0 \cdot \omega)\rangle. \quad (5.22)$$

Now we obtain with (5.21) and (5.22)

$$\langle i\omega, a_{G^c}(\exp Xg)\rangle \leq \sup\langle X, \tilde{W}_0 \cdot (i\omega)\rangle \quad (5.23)$$

for $X \in \tilde{c}_{\max}$.

Recall the cone $(\tilde{\Delta}_0^+)^* = \tilde{C}$ from Section 4.5 and consider the set

$$\mathcal{R} := \{\omega \in \tilde{\mathfrak{a}}^* \mid \forall \tilde{\alpha} \in \tilde{\Delta}_0^+ : \frac{2(i\omega \mid \alpha)}{(\alpha \mid \alpha)} \in \mathbb{Z}\}$$

of *integral weights* in $\tilde{\mathfrak{a}}^*$. Then

$$\mathcal{R}_+ := \{\omega \in \tilde{\mathfrak{a}}^* \mid \forall \tilde{\alpha} \in \tilde{\Delta}_0^+ : \frac{2(i\omega \mid \alpha)}{(\alpha \mid \alpha)} \in \mathbb{N}_0^+\}$$

is the set of *dominant* integral weights.

Lemma 5.5.6 *The cone $ic_{\max} \cap (-i\tilde{C})$ is generated by its dominant integral elements which integrate to a character of T^c.*

Proof: Let $\Upsilon := \{\tilde{\alpha}_0, \tilde{\alpha}_1, \ldots, \tilde{\alpha}_k\}$ be a basis for $\tilde{\Delta}^+$ such that $\Upsilon_0 := \{\tilde{\alpha}_1, \ldots, \tilde{\alpha}_k\}$ is a basis for $\tilde{\Delta}_0^+$. Recall that there is an element $X^0 \in i\mathfrak{z}(\mathfrak{k}^c)$ such that $\tilde{\alpha}_1(X^0) = 1$. Let $\lambda_0 \in (\mathfrak{t}^c_{\mathbb{C}})^*$ be determined by $\lambda_0(X^0) = 1$ and $\lambda_0(X) = 0$ for all $X \in \mathfrak{t}^c \cap [\mathfrak{k}^c, \mathfrak{k}^c]$. According to [79], p. 85, each dominant integral element of the lattice

$$\mathcal{R}' := \mathcal{R} \cap i(\mathbb{Z}\lambda_0 + \sum_{j=1}^{k} \mathbb{Z}\tilde{\alpha}_j)$$

integrates to a character of T^c. Let d be the maximal distance between elements of $\mathcal{R}'$. Then, given $\epsilon > 0$ and an element ω in the interior of $ic_{\max} \cap (-i\tilde{C})$, we can find an $n \in \mathbb{N}$ and a $\nu \in \mathcal{R}'$ such that $|n\omega - \nu| < d$ and $\frac{d}{n} < \epsilon$. Thus

$$|\omega - \frac{1}{n}\nu| < \epsilon$$

and $\nu \in ic_{\max} \cap (-i\tilde{C})$ for ϵ small enough. Therefore it suffices to show that ν is dominant. But that is clear since $\nu \in -i(\tilde{\Delta}_0^+)^*$. □

Proposition 5.5.7 *Let $X \in (\tilde{\Delta}^+)^*$. Then $a_{G^c}(\exp(X)G^c) \subset X + \tilde{c}_{\min} - \tilde{C}^*$.*

Proof: Let $\omega \in ic_{\max} \cap (-i\tilde{C})$ be dominant integral and such that it integrates to a character of T^c. Then $\tilde{W}_0 \cdot (i\omega) \in i\omega - \tilde{C}^*$ by Lemma 4.5.5, so that

$$\sup\langle X, \tilde{W}_0 \cdot (i\omega)\rangle = i\omega(X). \tag{5.24}$$

Combining this with (5.23) yields $\langle i\omega, X - a_{G^c}(\exp XG^c)\rangle \subset \mathbb{R}^+$. Now Lemma 5.5.6 proves that

$$X - a_{G^c}(\exp(X)G^c) \subset \left[i\tilde{c}^*_{\min} \cap (-i\tilde{C})\right]^* = -\tilde{c}_{\min} + \tilde{C}^*,$$

i.e., $a_{G^c}(\exp XG) \subset X + \tilde{c}_{\min} - \tilde{C}^*$. □

Proposition 5.5.8 *Let $X \in c_{\max}$ and $a = \exp(X)$. Then the set $a_{G^c}(aG^c)$ is invariant under the Weyl group $\tilde{W}_0$. Moreover, if $Y \in a_{G^c}(aG^c)$, then*

$$\operatorname{conv}(\tilde{W}_0 \cdot Y) \subset a_{G^c}(aG^c).$$

Proof: Set $F := a_{G^c}(aG^c)$. The Weyl group $\tilde{W}_0$ is generated by the reflections $s_{\tilde{\alpha}}$, where $\tilde{\alpha}$ is a root contained in the set Υ_0 simple roots in $\tilde{\Delta}_0^+$. We claim that the line segment $\{Y, s_{\tilde\alpha}(Y)\}$ between Y and $s_{\tilde\alpha}(Y)$ is contained in F whenever $Y \in F$ (cf. [45], p. 477). Let $\tilde\alpha \in \Upsilon_0$ and denote the complex image of the homomorphism $\varphi_{\tilde\alpha}: \mathfrak{sl}(2,\mathbb{C}) \to \mathfrak{g}_\mathbb{C}$ by $\mathfrak{s}_\mathbb{C}^{\tilde\alpha}$. Note that $\mathfrak{s}_\mathbb{C}^{\tilde\alpha} \subset \mathfrak{k}_\mathbb{C}^c$. We set

$$\tilde{\mathfrak{n}}' := \sum_{\beta\in\tilde\Delta^+\setminus\{\tilde\alpha\}} (\mathfrak{g}_\mathbb{C})_\beta.$$

Note that $\tilde\alpha \in \Upsilon_0$ implies $s_{\tilde\alpha}(\tilde\Delta^+ \setminus \{\tilde\alpha\}) \subset \tilde\Delta^+$ since $\tilde\Delta_+$ is $\tilde W_0$-invariant. Therefore

$$\tilde{\mathfrak{n}} = \tilde{\mathfrak{n}}' + (\mathfrak{g}_\mathbb{C})_{\tilde\alpha} \quad \text{and} \quad [\mathfrak{s}_\mathbb{C}^{\tilde\alpha}, \tilde{\mathfrak{n}}'] \subset \tilde{\mathfrak{n}}'. \tag{5.25}$$

According to [45], pp. 440, 477, we have the semidirect decomposition $\tilde N = \tilde N' \times \tilde N^{\tilde\alpha}$, where $\tilde N' = \exp\tilde{\mathfrak{n}}'$ and $\tilde N^{\tilde\alpha} = \exp(\mathfrak{g}_\mathbb{C})_{\tilde\alpha}$.

Let $Y \in F$ and $b = \exp(Y)$. Then there exist $g, v \in G^c$ and $n \in \tilde N$ such that $av = gbn$. We decompose $Y = Y_{\tilde\alpha} + Y_{\tilde\alpha}^\perp$, where $Y_{\tilde\alpha} \in \mathbb{R}X_{\tilde\alpha}$ and $Y_{\tilde\alpha}^\perp \in \ker\tilde\alpha$. Then

$$s_{\tilde\alpha}(Y) = s_{\tilde\alpha}(Y_{\tilde\alpha}) + Y_{\tilde\alpha}^\perp = -Y_{\tilde\alpha} + Y_{\tilde\alpha}^\perp$$

and $\{Y, s_{\tilde\alpha}(Y)\} = [-1,1]Y_{\tilde\alpha} + Y_{\tilde\alpha}^\perp$. We put $b_{\tilde\alpha} := \exp(Y_{\tilde\alpha})$, $b_{\tilde\alpha}^\perp := \exp Y_{\tilde\alpha}^\perp$ and write $n = n_{\tilde\alpha}n'$ in accordance with $\tilde N = \tilde N^{\tilde\alpha}\tilde N'$. Then

$$g^{-1}av = bn = b_{\tilde\alpha}b_{\tilde\alpha}^\perp n_{\tilde\alpha}n' = b_{\tilde\alpha}n_{\tilde\alpha}b_{\tilde\alpha}^\perp n'. \tag{5.26}$$

Let $c_{\tilde\alpha} \in \exp([-1,1]Y_{\tilde\alpha})$ and let $S_\mathbb{C}^{\tilde\alpha}$ be the group generated by $\exp\mathfrak{s}_\mathbb{C}^{\tilde\alpha}$. Then $S_\mathbb{C}^{\tilde\alpha} \subset K_\mathbb{C}^c$. Using Lemma 10.7 in [45], p. 476, we find elements $k_{\tilde\alpha}, v_{\tilde\alpha} \in S_\mathbb{C}^{\tilde\alpha} \cap K^c$ and $n_{\tilde\alpha}^0 \in \tilde N^{\tilde\alpha}$ such that

$$k_{\tilde\alpha}b_{\tilde\alpha}n_{\tilde\alpha}v_{\tilde\alpha} = c_{\tilde\alpha}n_{\tilde\alpha}^0. \tag{5.27}$$

Now $[Y_{\tilde\alpha}^\perp \cap \mathfrak{a}, (\mathfrak{g}_\mathbb{C})_{\tilde\alpha}] = \{0\}$ and (5.26) imply that

$$k_{\tilde\alpha}g^{-1}avv_{\tilde\alpha} = c_{\tilde\alpha}n_{\tilde\alpha}^0 v_{\tilde\alpha}^{-1}b_{\tilde\alpha}^\perp n'v_{\tilde\alpha} = c_{\tilde\alpha}b_{\tilde\alpha}^\perp n_{\tilde\alpha}^0 v_{\tilde\alpha}^{-1}n'v_{\tilde\alpha}.$$

We use (5.25) to see that $n_{\tilde\alpha}^0 v_{\tilde\alpha}^{-1}n'v_{\tilde\alpha} \in n_{\tilde\alpha}^0\tilde N' \subset \tilde N$. Thus

$$a_{G^c}(k_{\tilde\alpha}g^{-1}avv_{\tilde\alpha}) = a_{G^c}(avv_{\tilde\alpha}) = \log(c_{\tilde\alpha}b_{\tilde\alpha}^\perp) = \log c_{\tilde\alpha} + Y_{\tilde\alpha}^\perp.$$

Since $c_{\tilde\alpha}$ was arbitrary in $\exp([-1,1]Y_{\tilde\alpha})$, we conclude that

$$\{Y, s_{\tilde\alpha}(Y)\} \subset a_{G^c}(aG^c).$$

This proves the $\tilde W_0$-invariance of $a_{G^c}(aG^c)$ because $\tilde W_0$ is generated by the reflections $s_{\tilde\alpha}$ for $\tilde\alpha$ simple. Let $\tilde\beta \in \tilde\Delta_0$. Then there exists $w \in \tilde W_0$ such that $w \cdot \tilde\beta \in \Upsilon$ and we have for each $Y \in F$ that

$$w\{w^{-1}Y, s_{\tilde\beta}w^{-1}\cdot Y\} = \{Y, ws_{\tilde\beta}w^{-1}\cdot Y\} = \{Y, s_{w\cdot\tilde\beta}\cdot Y\} \subset F.$$

Now Lemma 10.4 in [Hel84, p. 474] implies that $\operatorname{conv}(\tilde{W}_0 \cdot Y) \subset F$ for every element $Y \in F$. □

Proposition 5.5.9 *Let $a \in \exp(c_{\max})$. Then*

$$a_{G^c}(aG^c) \subset \operatorname{conv}(\tilde{W}_0 \cdot \log a) + \tilde{c}_{\min}.$$

Proof: Applying a suitable element of $\tilde{W}_0$, we may assume that $X := \log a \in (\tilde{\Delta}^+)^*$ because the sets on the right- and left-hand sides do not change if we replace X by $w \cdot X$ for $w \in \tilde{W}_0$ (Proposition 5.5.8).

Now Proposition 5.5.7 entails that

$$a_{G^c}(aG^c) \subset X + \tilde{c}_{\min} - \tilde{C}^*$$

and since the set on the left-hand side is invariant under $\tilde{W}_0$, again by Proposition 5.5.8, we conclude with Lemma 4.5.5 that

$$a_{G^c}(aG^c) \subset \bigcap_{w \in \tilde{W}_0} w \cdot \left(X + \tilde{c}_{\min} - \tilde{C}^*\right) = \operatorname{conv}(\tilde{W}_0 \cdot X) + \tilde{c}_{\min}$$

since $\tilde{c}_{\min}$ is $\tilde{W}_0$-invariant. □

Recall that $G^\tau \subset G^c$, $A \subset \tilde{A}$ and $N \subset \tilde{N}$ are the σ^c-fixed points of the respective groups. Therefore a_{G^c} commutes with σ^c and the map $a_{G^\tau}: G^\tau AN \to \mathfrak{a}$ is simply the restriction of a_{G^c} to $G^\tau AN$. In view of Theorem 4.5.6, this implies that

$$a_{G^\tau}(aG^\tau) \subset \mathfrak{a} \cap \left[\operatorname{conv}(\tilde{W}_0 \cdot \log a) + \tilde{c}_{\min}\right] = \operatorname{conv}(W_0 \cdot \log a) + c_{\min} \quad (5.28)$$

for $a \in \exp(c_{\max})$.

In order to prove the converse inclusion we need some additional information. Note first that Proposition 3.2.2 yields the following lemma.

Lemma 5.5.10 *Let Υ be a basis of the system Δ^+ . Then $\Upsilon_0 := \Upsilon \cap \Delta_0^+$ is a basis of Δ_0^+ and Υ contains exactly one root not contained in Δ_0.* □

Lemma 5.5.11 *Let $C = (\Delta_0^+)^*$. Then the highest root γ in Δ^+ satisfies*

$$c_{\min} \subset \mathbb{R}^+\gamma - C^*.$$

Proof: We note first that the considerations in Section 4.1 show that Δ is an irreducible root system. Further, we note that the highest root automatically is contained in Δ_+. Let $\Upsilon = \{\alpha_0, \alpha_1, \ldots, \alpha_k\}$ be the simple system for Δ^+ such that $\Upsilon_0 = \{\alpha_1, \ldots, \alpha_k\}$ is the simple system for Δ_0^+. Now suppose that $\beta = \sum_{j=0}^k m_j\alpha_j \in \Delta_+$ and $\gamma = \sum_{j=0}^k n_j\alpha_j$. Then $n_0 = m_0 = 1$ and $\gamma - \beta = \sum_{j=1}^k (n_j - m_j)\alpha_j \in \Delta_0^+ \subset C^*$. Therefore $\beta \in \gamma - C^*$ for all $\beta \in \Delta_+$ which implies the claim since $c_{\min} = \operatorname{cone}(\Delta_+)$. □

Lemma 5.5.12 *Let* $\log a = X \in c_{\max}$ *and* $\alpha \in \Delta_+$ *be such that* $\alpha(X) > 0$. *Then*

$$X + \mathbb{R}^+ X^\alpha \subset a_{G^\tau}(aG^\tau).$$

Proof: $\mathfrak{sl}(2,\mathbb{R})$-reduction yields

$$a_{G^\tau}(\exp(X)\exp \mathbb{R}Y^\alpha) = X + \mathbb{R}^+ X^\alpha$$

and this implies the claim (cf. Lemma 5.4.9 and its proof). □

Theorem 5.5.13 (The Nonlinear Convexity Theorem) *Let* $\mathcal{M}$ *be a noncompactly causal symmetric space,* $\mathfrak{a} \subset \mathfrak{q}_{\mathfrak{p}}$ *a maximal abelian subspace, and* $a_H : HAN \to \mathfrak{a}$ *the corresponding projection. Then*

$$a_H(aH) = \mathrm{conv}(W_0 \cdot \log a) + c_{\min}$$

for $1 \neq a \in \exp(c_{\max})$.

Proof: Note first that Lemma 3.1.22 implies that we may assume $H = G^\tau$. In view of (5.28), we only have to show the inclusion $\supset$. Replacing $X = \log a$ by a suitable W_0-conjugate, we may also assume that $X \in C = (\Delta_0^+)^*$. Since $X \neq 0$, there exists a $\alpha \in \Delta_+$ such that $\alpha(X) > 0$. Let $\gamma \in \Delta_+$ be the highest root of Δ^+. Then $\gamma(X) \geq \alpha(X) > 0$ and hence Lemma 5.5.12 implies that $X + \mathbb{R}^+ X^\gamma \subset a_H(aH)$. Now Proposition 5.5.8 implies that it suffices to show

$$\mathrm{conv}\left[W_0 \cdot (X + \mathbb{R}^+ X^\gamma)\right] = \mathrm{conv}(W_0 \cdot X) + c_{\min}. \tag{5.29}$$

To do this, note first that $\mathbb{R}^+ X^\gamma = \mathbb{R}^+\gamma \subset c_{\min}$ and that both sides of (5.29) are closed, convex, and W_0-invariant. Thus it remains to verify

$$[\mathrm{conv}(W_0 \cdot X) + c_{\min}] \cap C \subset \mathrm{conv}\left(W_0 \cdot (X + \mathbb{R}^+ X^\gamma)\right) \cap C. \tag{5.30}$$

According to Lemma 5.5.11 and Lemma 4.5.5 we have

$$\mathrm{conv}(W_0 \cdot X) + c_{\min} \subset (X - C^*) + (\mathbb{R}^+\gamma - C^*) = (X + \mathbb{R}^+\gamma) - C^*.$$

Note that $(Y - C^*) \cap C \subset \mathrm{conv}(W_0 \cdot Y)$ for all $Y \in C$ by Lemma 4.5.5. But [8], §1.8, Proposition 25 implies that $\gamma \in C$. Thus for any $r > 0$ we have

$$\left[(x + r\gamma) - C^*\right] \cap C \subset \mathrm{conv}\left[W_0 \cdot (X + r\gamma)\right] \cap C.$$

This implies (5.30) and hence the claim. □

Corollary 5.5.14 *1) Let* $a \in \exp(c_{\max})$ *and* $n^\sharp \in N^\sharp \cap HAN$. *Then* $an^\sharp a^{-1} \subset HAN$ *and*

$$a_H(an^\sharp a^{-1}) - a_H(n^\sharp) \in c_{\min}.$$

2) $a_H(N^\sharp \cap HAN) \subset ic_{\min}$.

Proof: 1) Let $p_H : HAN \to P$ be the projection onto the H-factor. Then

$$a_H(xy) = a_H(xp_H(y)) + a_H(y).$$

Therefore

$$a_H(an^\sharp a^{-1}) = a_H(ap_H(n^\sharp)) + a_H(n^\sharp) - \log a.$$

Now Theorem 5.5.13 shows that

$$a_H(ap_H(n^\sharp)) \in a_H(aH) \subset \log a + c_{\min}$$

and this implies the claim.

2) Let $X \in (c_{\max} \cap C)^o$ and $n^\sharp \in N^\sharp \cap HAN$. Then

$$\lim_{t\to\infty} \exp(tX)n^\sharp \exp(-tX) = 1$$

and therefore

$$-a_H(n^\sharp) = \lim_{t\to\infty} a_H(\exp(tX)n^\sharp \exp(-tX)) - a_H(n^\sharp) \in c_{\min}. \qquad \square$$

Example 5.5.15 For $G = \mathrm{SL}(2,\mathbb{R})$ the nonlinear convexity theorem can made very explicit. In the situation of Example 5.1.3 we have $W_0 = \{1\}$ and $c_{\max} = c_{\min} = \mathbb{R}^+ X^0$. The causal Iwasawa projection is given by

$$a_H\begin{pmatrix} a & b \\ c & d \end{pmatrix} = \sqrt{a^2 - c^2}$$

whenever it is defined. $\square$

5.6 The $B^\sharp$-Order

Let $\mathcal{M} = G/H$ still be a noncompactly causal symmetric space such G is contained in a simply connected complexification $G_\mathbb{C}$. We write S for the maximal real Ol'shanskii semigroup $S(C_{\max})$. Then Theorem 5.2.7 implies that $\mathcal{M}_+ = S\cdot \mathrm{o} = (S \cap B^\sharp)\cdot \mathrm{o}$. This shows that many questions concerning the positive cone of $\mathcal{M}$ can be treated via $B^\sharp$, which has a fairly simple structure.

Remark 5.6.1 Theorem 5.2.7 implies that $S \cap B^\sharp = \{b^\sharp \in B^\sharp \mid \mathrm{o} \preceq b^\sharp \cdot \mathrm{o}\}$. Therefore the restriction of $\leq_S$, cf. (2.14), to $B^\sharp$ defines an order. On the other hand, $S \cap B^\sharp$ defines an order $\leq_{S\cap B^\sharp}$ on $B^\sharp$ via

$$b \leq_{S\cap B^\sharp} b' \quad :\Leftrightarrow \quad b' \in b(S \cap B^\sharp).$$

We claim that the two orders agree. This follows from

$$\downarrow b^\sharp \cap B^\sharp = b^\sharp S^{-1} \cap B^\sharp = b^\sharp \left(S^{-1} \cap B^\sharp\right) = b^\sharp \left(S \cap B^\sharp\right)^{-1}.$$

This means that in particular we can use the notation $\downarrow b^\sharp$ without any ambiguity. □

Proposition 5.6.2 *The map*

$$I_{B^\sharp}: \mathcal{F}_\downarrow(G) \to \mathcal{F}_\downarrow(B), \quad F \mapsto F \cap B^\sharp$$

is $B^\sharp$-equivariant, continuous, and surjective. It is injective on the closed set $\{F \in \mathcal{F}_\downarrow(G) \mid F^o \subset N^\sharp AH\}$.

Proof. The equivariance is obvious. Let $F \in \mathcal{F}_\downarrow(G) \subset \mathcal{F}(G)^H$ (cf. Lemma 2.4.1). Then $I_{B^\sharp}(F) = F \cap B^\sharp$ is closed and for $s \in S \cap B^\sharp$ we have that

$$\left(F \cap B^\sharp\right) s^{-1} \subset F(S \cap B^\sharp)^{-1} \cap B^\sharp = F \cap B^\sharp,$$

whence

$$F \cap B^\sharp \in \mathcal{F}_\downarrow\left(B^\sharp\right).$$

Let $F_n \to F$ in $\mathcal{F}_\downarrow(G)$ and assume that $F_n \cap B^\sharp \to E$. To see that $E = I_{B^\sharp}(F)$, let $e \in E$ and $f_n \in F_n \cap B^\sharp$ with $f_n \to e$ (cf. Lemma B.1.1). Then $e = \lim f_n \in \lim F_n = F$. On the other hand, for $f \in F \cap B^\sharp$ there exists a sequence $f_n \in F_n$ with $f_n \to f$. Since $F_n H = F_n$, we have that $F_n = \left(F_n \cap B^\sharp\right) H$, so we find that $b_n \in F_n \cap B^\sharp$ and $h_n \in H$ with $f_n = b_n h_n$. According to Lemma 5.4.1 we get that $b_n \to f$ and $h_n \to 1$. Thus

$$f = \lim b_n \in \lim \alpha(F_n) = E.$$

It follows that $E = I_{B^\sharp}(F)$.

For $E \in \mathcal{F}_\downarrow\left(B^\sharp\right)$ we set $\beta(E) := \overline{EH}$. Let $s = gh \in S$, where $g \in B^\sharp \cap S$ and $h \in H$. Moreover,

$$Hg^{-1} \subset S^{-1} = \left(S^{-1} \cap B^\sharp\right) H$$

by Theorem 5.4.7. Thus

$$\beta(E)s^{-1} = \overline{EHh^{-1}g^{-1}} = \overline{EHg^{-1}} \subset \overline{E\left(S^{-1} \cap B^\sharp\right) H} = \overline{(\downarrow_{S\cap B^\sharp} E) H} = \overline{EH}$$

shows $\downarrow \beta(E) = \beta(E)$. The inclusion

$$E \subset I_{B^\sharp}\left(\beta(E)\right) = \overline{EH} \cap B^\sharp$$

is clear. Let $b^\sharp \in \overline{EH} \cap B^\sharp$ and $e_n \in E$, $h_n \in H$ with $e_n h_n \to b^\sharp$. Then $e_n \to b$ and $b \in \overline{E} = E$. It follows that $\beta(E) \cap B^\sharp = E$, and hence $I_{B^\sharp}$ is surjective.

If $F \in \mathcal{F}_\downarrow(G)$ with $F^o \subset N^\sharp AH$, then

$$\begin{aligned} F &= \overline{F^o} = \overline{[F^o \cap B^\sharp]\, H} \\ &= \overline{((F \cap B^\sharp)^o)\, H} \\ &= \overline{(F \cap B^\sharp)\, H} = \overline{I_{B^\sharp}(F) H}. \end{aligned}$$

Here we used that $F \cap B^\sharp \in \mathcal{F}_\downarrow\left(B^\sharp\right)$, so $F \cap B^\sharp$ has dense interior by Lemma 2.4.7.

Now we see that $I_{B^\sharp}(F') = I_{B^\sharp}(F)$ and $(F')^o \subset N^\sharp AH$ imply $F' = F$. It remains to show that the set $\{F \in \mathcal{F}_\downarrow(G) \mid F^o \subset N^\sharp AH\}$ is closed. We let $F_n \in \mathcal{F}_\downarrow(G)$ with $F_n \to F$ and $F_n^o \subset B^\sharp H$. We have to show that $F^o \subset B^\sharp H$. For $f \in F^o$ there exist an $f' \in (\uparrow f)^o \cap F^o$ and $n_0 \in \mathbb{N}$ with

$$F_n \cap (\uparrow f)^o \cap F^o \neq \emptyset \quad \forall n \geq n_0.$$

Pick f_n in this set. Then

$$f \in (\downarrow f_n)^o \subset F_n^o \subset B^\sharp H$$

which proves that $F^o \subset B^\sharp H$. □

Lemma 5.6.3 *Consider the order compactification map*

$$\eta_{B^\sharp}: B^\sharp \to \mathcal{F}_\downarrow(B^\sharp), \quad g \mapsto g(S \cap B^\sharp)^{-1}$$

(cf. Lemma 2.4.2*).*

1) If $X \in \left(\mathbf{L}(S \cap B^\sharp)\right)^o$, *then* $\lim_{t\to\infty} \eta_{B^\sharp}(\exp(-tX)) = \emptyset$.

2) $\overline{\eta_{B^\sharp}(B^\sharp)} = \{\emptyset\} \cup B^\sharp \cdot \overline{\eta_{B^\sharp}(S \cap B^\sharp)}$

Proof. 1) We consider the projection

$$p: B^\sharp \cong N^\sharp \rtimes A \to A \tag{5.31}$$

and set $X' := d_1 q(X)$. Note that $\eta_{B^\sharp}(\exp(-tX))$ is decreasing in t, so it has a limit by Lemma B.1.1. Suppose that $g \in \lim \eta_{B^\sharp}(\exp(-tX))$, $t \to \infty$. Then there exist $t_n \in \mathbb{R}$ and $s_n \in S \cap B^\sharp$ with $t_n \to \infty$ and

$g = \lim_{n\to\infty} \exp(-t_n X)s_n^{-1}$. Thus

$$\begin{aligned} p(g) &= \lim_{n\to\infty} \exp(-t_n X')p(s_n)^{-1} \\ &= \lim_{n\to\infty} \exp(-t_n X')\exp(-c_{\max}) \\ &= \exp\left(-\lim[t_n X' + c_{\max}]\right) = \emptyset \end{aligned}$$

because of Theorem 5.4.8.3) and the fact that $t_n X' + c_{\max} \to \emptyset$ whenever $X' \in c^o_{\max} = d_1 p\left(\mathbf{L}(S \cap B^\sharp)^o\right)$ and $t_n \to \infty$.

2) In view of 1), this is just a special case of Lemma 2.4.3.3). □

Lemma 5.6.4 *The restriction of $I_{B^\sharp}$ to $\overline{\eta(B^\sharp)} \subset \mathcal{F}_\downarrow(G)$ yields a homeomorphism*

$$\overline{\eta(B^\sharp)} \to \overline{\eta_{B^\sharp}(B^\sharp)} \subset \mathcal{F}_\downarrow(B^\sharp).$$

Proof: Remark 5.6.1 implies that $I_{B^\sharp}(\eta(g)) = \eta_{B^\sharp}(g)$ for all $g \in B^\sharp$. Using the continuity of $I_{B^\sharp}$ we find that

$$I_{B^\sharp}(\overline{\eta(B^\sharp)}) \subset \overline{I_{B^\sharp}(\eta(B^\sharp))} = \overline{\eta_{B^\sharp}(B^\sharp)}.$$

Since $I_{B^\sharp}$ is $B^\sharp$-equivariant and η is even G-equivariant we have

$$I_{B^\sharp}\left(\eta(B^\sharp)\right) = B^\sharp \cdot \eta_{B^\sharp}(1)$$

and hence $I_{B^\sharp}(\overline{\eta(B^\sharp)})$, which is closed because of compactness, contains the closure of the $B^\sharp$-orbit of $\eta_{B^\sharp}(1)$, i.e., all of $\overline{\eta_{B^\sharp}(B^\sharp)}$.

Recall that for $F \in \eta\left(B^\sharp\right)$ we have

$$F = \downarrow g = gS^{-1} \subset B^\sharp(B^\sharp H) = B^\sharp H$$

and therefore also $F^o \subset B^\sharp H$ since $B^\sharp H$ is open in G. Now again by the continuity of $I_{B^\sharp}$ we get $F^o \subset B^\sharp H$ for all $F \in \overline{\eta(B^\sharp H)}$, and then Proposition 5.6.2 shows that $I_{B^\sharp}$ restricted to $\overline{\eta(B^\sharp)}$ is injective. Finally, compactness yields the claim. □

Lemma 5.6.5 *We have*

$$\overline{\eta(A)} = \emptyset \cup A \cdot \overline{\eta\left(\exp(c_{\max})\right)}.$$

If $\eta(a_n) \to F \neq \emptyset$, then the sequence $a_n \in A$ is bounded from below with respect to the restriction of the ordering $\leq_S$ to A.

Proof. Using Lemma 5.6.4 and Lemma 5.6.3, we see that

$$\begin{aligned}\emptyset &= \left[\overline{\lim_{t\to\infty} \eta_{B^\sharp}(\exp(-tX))\,H}\right] \\ &= \lim_{t\to\infty} \overline{\eta_{B^\sharp}(\exp(-tX))\,H} \\ &= \lim_{t\to\infty} \eta(\exp(-tX))\end{aligned}$$

for all $X \in \mathbf{L}(S \cap B^\sharp)^o$. Suppose that $\eta(a_n) \to F \neq \emptyset$. Then Lemma 5.6.3 and Lemma 5.6.4 yield

$$\eta_{B^\sharp}(a_n) \to F \cap B^\sharp \neq \emptyset.$$

Let $f \in \operatorname{Int}_{B^\sharp} F$ and $a := p(f) \in A$. Then there exists $n_0 \in \mathbb{N}$ with $f \leq_S a_n$ for all $n \geq n_0$. Hence $a = p(f) \leq_S a_n$ for all $n \geq n_0$. Pick t_0 with $\exp(-t_0 Y^0) \leq a$. Then

$$\exp(t_0 Y^0) a_n \in S \cap A = \exp(c_{\max})$$

(cf. Lemma 5.4.9) and

$$\eta(a_n) \to \exp(-t_0 Y^0) \lim_{n\to\infty} \eta\left(\exp(t_0 Y^0) a_n\right) \in A \cdot \overline{\eta\left(\exp(c_{\max})\right)}.$$

This proves the claim. □

Theorem 5.6.6 *1)* $\mathcal{M}^{cpt} = G \cdot \mathcal{M}_+^{cpt} \cup \{\emptyset\}$.

2) $\mathcal{M}^{cpt}$ *has only finitely many G-orbits.*

Proof. 1) follows from Lemma 5.6.5 and Lemma 2.4.3.
2) is a consequence of 1) and (6.8). □

The point of Lemma 5.6.5 and Theorem 5.6.6 is that they will enable us to derive the G-orbit structure of $\mathcal{M}^{cpt}$ from the structure of $\overline{\eta(S \cap A)} \subset \mathcal{M}_+^{cpt}$.

We conclude this section with the useful observation that the projection $p: B^\sharp \cong N^\sharp \rtimes A \to A$ defined in (5.31) is proper. This is an immediate consequence of the following more general lemma.

Lemma 5.6.7 *Let B be a connected Lie group and N a closed normal subgroup such that $A := B/N$ is a vector group. Suppose that $C \subset \mathfrak{b}$ is a pointed closed convex cone such that $C \cap \mathfrak{n} = \{0\}$ and S the closed subsemigroup of B generated by C. Then the homomorphism $\phi: B \to A, b \mapsto bN$ induces a proper semigroup homomorphism $\pi: S \to \phi(S)$.*

Proof: Let $D := d_1\phi(C)$. Then the condition $C \cap \mathfrak{n} = \{0\}$ shows that D is a pointed cone in the abelian Lie algebra $\mathfrak{a}$. Since A is a vector group, we can identify α with A.

Let $\omega \in \mathrm{Int}\{\nu \in \mathfrak{a}^* \mid \forall X \in D : \omega(X) \geq 0\}$. Then ω can be viewed as a function on A and then the function $f := \omega \circ \phi$ satisfies the hypothesis of Theorem 1.32 in [Ne91] because it is a group homomorphism, hence has biinvariant differential. So we find that the order intervals $sS^{-1} \cap S$ in B are compact. Let $K \subset \phi(S)$ be compact and L the maximal value of ω on K. Then $\pi^{-1}(K) \subset f^{-1}([0, L]) \cap S$. Now Theorem 1.32 in [114] implies also that there exists a left invariant Riemannian metric d on B such that the length $L(\gamma)$ of γ is not bigger than L for all curves $\gamma\,[0,T] \to B$ with $\gamma(0) = 1$ and $f(\gamma(T)) \leq L$, which are monotone w.r.t. $\leq_S$. Therefore $d(x, 1) \leq L$ holds for all $x \in \pi^{-1}(K)$. Finally, the theorem of Hopf-Rinow shows that these sets are compact. □

5.7 The Affine Closure of $B^\sharp$

Retain the hypotheses and notation from Section 5.6. In this section we realize $S \cap B^\sharp$ as a semigroup of affine selfmaps and in this way find a suitable compactification which helps us to make the abstract order compactification much more concrete.

Lemma 5.7.1 *1) $B^\sharp$ is a twofold semidirect product $B^\sharp \cong N_- \rtimes (N_0^\sharp \rtimes A)$.*

2) Let $\alpha \in \Delta_+$, $\beta \in \Delta_0^+$, and $X_\beta \in \mathfrak{g}_\beta$. Then $[X_\beta, \mathfrak{g}_{\alpha+n\beta}] \neq \{0\}$ whenever $\alpha + (n+1)\beta \in \Delta$. In particular, if $(\alpha \mid \beta) \neq 0$, we find that $[X_\beta, \sum_{n\in\mathbb{Z}} \mathfrak{g}_{\alpha+n\beta}] \neq \{0\}$.

3) The mapping

$$B^\sharp \to N_- \rtimes \mathrm{Aut}(N_-), \quad (n_-, n_0^\sharp, a) \mapsto (n_-, I_{n_0^\sharp a}),$$

where $I_{n_0^\sharp a}$ denotes the automorphism $n_- \mapsto (n_0^\sharp a)n_-(n_0^\sharp a)^{-1}$, is an injective homomorphism.

Proof: 1) The first assertion follows immediately from the fact that $\mathfrak{n} = \mathfrak{n}_- \rtimes \mathfrak{n}_0^\sharp$, which is a consequence of $\mathfrak{n}_0^\sharp = \mathfrak{z}_{\mathfrak{n}^\sharp}(Y^0)$.

2) Recall the algebra $\mathfrak{s}_\beta \cong \mathfrak{sl}(2, \mathbb{R})$ from Section 4.1. The space

$$V_{\alpha,\beta} := \bigoplus_{n\in\mathbb{Z}} \mathfrak{g}_{\alpha+n\beta}$$

is an $\mathfrak{s}_\beta$-module. The above decomposition of $V_{\alpha,\beta}$ is precisely the H_β-weight decomposition. Suppose that $\alpha + (n+1)\beta \in \Delta$. Then there exists a simple $\mathfrak{s}_\beta$-submodule V of $V_{\alpha,\beta}$ with

$$V \cap \mathfrak{g}_{\alpha+(n+1)\beta} \neq \{0\}.$$

But now the classification of $\mathfrak{sl}(2,\mathbb{R})$ modules says that

$$V = \bigoplus_{m=m_-}^{m_+} V^{\alpha+m\beta},$$

where $V^{\alpha+m\beta} = \mathfrak{g}_{\alpha+m\beta} \cap V$ is one-dimensional. If $\mathfrak{g}_{\alpha+n\beta} \subset \ker \operatorname{ad} X_\beta$, then $n = m_- - 1$, since $\mathfrak{g}_{\alpha+(n+1)\beta} \cap V \neq \{0\}$, and $[X_\beta, V^{\alpha+m\beta}] = V^{\alpha+(m+1)\beta}$ for $m = m_-, m_- + 1, ..., m_+$. Thus

$$(\alpha + n\beta)(H_\beta) < (\alpha + m_-\beta)(H_\beta) < 0.$$

Now [10], Chapter VIII, §7, no. 2, Proposition 3(iii), yields a contradiction to $\mathfrak{g}_{\alpha+n\beta} \subset \ker \operatorname{ad} X_\beta$.

3) It follows from the fact that $c_{\max}$ is pointed and generating that the kernel intersects A trivially. Now 2) shows that the kernel also intersects $N_0^\sharp$ trivially and the assertion follows as $I_{n_0^\sharp a} = I_{n_0^\sharp} I_a$ is the Jordan decomposition when we identify $\mathfrak{n}_-$ with N_-. □

Recall the flag manifold $\mathcal{X}_\mathbb{C} = G_\mathbb{C}/(P_{\max})_\mathbb{C}$ and its base point $\mathbf{o}_\mathcal{X} = 1(P_{\max})_\mathbb{C}$ from Section 5.1.

Lemma 5.7.2 *The mapping $\zeta : \mathfrak{n}_- \to (N_-) \cdot \mathbf{o}_\mathcal{X}, X \mapsto \exp(X) \cdot \mathbf{o}_\mathcal{X}$ is an equivariant mapping of $B^\sharp$-spaces. Here the action of $B^\sharp$ on $\mathfrak{n}_-$ is given by*

$$(N_-)N_0^\sharp A \times \mathfrak{n}_- \to \mathfrak{n}_-, \; (\exp(X)\exp(Y)\exp(Z), E) \mapsto X + e^{\operatorname{ad} Y} e^{\operatorname{ad} Z} E.$$

For $E = \sum_{\alpha\in\Delta_-} E_\alpha$ with $E_\alpha \in \mathfrak{g}_\alpha$ we have that

$$e^{\operatorname{ad} Z} E = \sum_{\alpha\in\Delta_-} e^{\alpha(Z)} E_\alpha, \quad \forall Z \in \mathfrak{a}$$

Proof: Let $X, E \in \mathfrak{n}_-$, $Y \in \mathfrak{n}_0^\sharp$, and $Z \in \mathfrak{a}$. Using that $\mathfrak{n}_-$ is abelian, we have

$$\begin{aligned}
\exp(X) &= \exp(Y)\exp(Z)\exp(E)(P_{\max})_\mathbb{C} \\
&= \exp(X)\exp(Y)\exp(Z)\exp(E)\exp(-Z)\cdot \\
&\quad \cdot \exp(-Y)\exp(Y)\exp(Z)(P_{\max})_\mathbb{C} \\
&= \exp(X + e^{\operatorname{ad} Y} e^{\operatorname{ad} Z} E)\exp(Y)\exp(Z)(P_{\max})_\mathbb{C} \\
&= \exp(X + e^{\operatorname{ad} Y} e^{\operatorname{ad} Z} E)(P_{\max})_\mathbb{C}
\end{aligned}$$

because $\exp(Y)\exp(Z) \in H^a \subset (P_{\max})_{\mathbb{C}}$. □

Recall the domain $\Omega_- \subset \mathfrak{n}_-$ from (5.7) on p. 123. We set

$$\mathrm{Aff}(N_-) := N_- \rtimes \mathrm{End}(N_-) = N_- \rtimes \mathrm{End}(\mathfrak{n}_-) \tag{5.32}$$

and

$$\mathrm{Aff}_{com}(N_-) := \{(n_-,\gamma) \in \mathrm{Aff}(N_-) \mid n_-\gamma(\overline{\Omega}_-) \subset \overline{\Omega}_-\}. \tag{5.33}$$

Here we identify N_- and $\mathfrak{n}_-$ via the exponential function of N_-. We refer to these semigroups as the *affine semigroup* of N_- and the *affine compression semigroup* of $\overline{\Omega}_-$.

Proposition 5.7.3 $\mathrm{Aff}_{com}(N_-) \cap B^\sharp = S \cap B^\sharp$, *where* $B^\sharp$ *is identified with a subgroup of* $\mathrm{Aff}(N_-)$ *via* Lemma 5.7.1.

Proof: In view of Lemma 5.7.2, the claim follows from $S \cap B^\sharp = \{b^\sharp \in B^\sharp \mid b^\sharp \cdot \mathcal{O} \subset \mathcal{O}\}$ and $\mathcal{O} = \exp(\Omega_-)\cdot o_{\mathcal{X}}$. □

Proposition 5.7.4 $\mathrm{Aff}_{com}(N_-)$ *is compact.*

Proof: Note first that $\mathrm{Aff}_{com}(N_-)$ is closed in $\mathrm{Aff}(N_-)$. Now let $(n_-,\gamma) \in \mathrm{Aff}_{com}(N_-)$. Then

$$n_- = (n_-,\gamma)\cdot 1 \in \overline{\Omega}_- \subset N_-$$

so that

$$\gamma(\overline{\Omega}_-) \subset n_-^{-1}\overline{\Omega}_- \subset (\overline{\Omega}_-)^{-1}(\overline{\Omega}_-).$$

Since $\overline{\Omega}_-$ is a compact neighborhood of 0 in $\mathfrak{n}_-$, we can find a norm on $\mathfrak{n}_-$ and a constant $c > o$ such that $||\gamma|| \le c$ for all $(n_-,\gamma) \in \mathrm{Aff}_{com}(N_-)$. In other words,

$$\mathrm{Aff}_{com}(N_-) \subset \{(n_-,\gamma) \in \mathrm{Aff}(N_-) \mid n_{\in}\overline{\Omega}_-, ||\gamma|| \le c\}$$

is relatively compact, hence compact. □

Lemma 5.7.5 *The action of* $\mathrm{Aff}_{com}(N_-)$ *on* $\overline{\Omega}_-$ *extends to a continuous action of* $\mathrm{Aff}_{com}(N_-)$ *on* $\mathcal{F}(\overline{\Omega}_-)$.

Proof: This follows from the more general fact that $\mathrm{End}(N_-)$ acts continuously on $\mathcal{C}(N_-)$, the set of compact subsets of N_- equipped with the Vietoris topology for the one-point compactification of N_-. To see this, let $K_n \to K$ in $\mathcal{C}(N_-)$ and $s_n \to s$ in $\mathrm{End}(N_-)$. Let U be a symmetric neighborhood of 1 in $\theta(N)$ and V another symmetric 1-neighborhood with

$V \subset U$ and $s_n(V) \subset U$ for all $n \in \mathbb{N}$. Take $n_0 \in \mathbb{N}$ such that $K_n \subset KV$ and $s_n(K) \subset s(K)U$ for $n \geq n_0$. (Note that s_n converges uniformly on compact sets). Then (cf. B.1.2)

$$s_n(K_n) \subset s_n(KV) = s_n(K)s_n(V) \subset s(K)U^2.$$

Since U^2 is symmetric, we also conclude that $s(K) \subset s_n(K_n)U^2$. Hence $s_n(K_n) \to s(K)$. □

Let $\overline{B^\sharp}$ denote the closure of $B^\sharp$ in $\mathrm{Aff}(N_-)$ and S_A^{cpt} the closure $\overline{S \cap A} = \overline{\exp c_{\max}} \subset \overline{B^\sharp}$ of $S \cap A$ in $\overline{B^\sharp}$. Then $\overline{B^\sharp}$, $\overline{S \cap B^\sharp}$ and S_A^{cpt} are compact semigroups.

We describe the structure of S_A^{cpt}.

Theorem 5.7.6 (The structure of S_A^{cpt}) *Let the notation be as above. Then the following assertions are true:*

1) $S_A^{cpt} = \overline{\exp(\mathbb{R}^+X_1)} \cdot \ldots \cdot \overline{\exp(\mathbb{R}^+X_n)}$, *where* $c_{\max} = \sum_{i=1}^n \mathbb{R}^+X_i$.

2) *For* $F \in \mathrm{Fa}(-c_{\max}^*) = \mathrm{Fa}(\mathrm{cone}(\Delta_-))$ *we define* $e_F \in \mathrm{End}(\mathfrak{n}_-)$ *by*

$$e_F(X) = \begin{cases} 0, & \text{if } X \in \mathfrak{g}_\alpha, \alpha \notin F \cap \Delta_-, \\ X, & \text{if } X \in \mathfrak{g}_\alpha, \alpha \in F \cap \Delta_-. \end{cases}$$

Then the mapping $F \mapsto e_F$ *defines an isomorphism of* $\mathrm{Fa}(-c_{\max}^*)$ *and the lattice of idempotents* $E(S_A^{cpt})$ *of the compact abelian semigroup* S_A^{cpt}.

3) *For* $X \in c_{\max}$ *we have that*

$$\lim_{t\to\infty} \exp(tX) = e_F, \quad \text{where} \quad F = X^\perp \cap (-c_{\max}^*), \tag{5.34}$$

and conversely,

$$e_F = \lim_{t\to\infty} \exp(tX) \quad \text{for every} \quad X \in \mathrm{Int}_{F^\perp}(c_{\max} \cap F^\perp).$$

4) $S_A^{cpt} = (S \cap A) \cdot E(S_A^{cpt})$.

Proof: 1) is obvious.

2), 3) Let $F \in \mathrm{Fa}(-c_{\max}^*)$. Then there exists an element $X \in c_{\max}$ with $F = X^\perp \cap (-c_{\max}^*)$. The functions $t \mapsto e^{\alpha(tX)}$ are decreasing for all $\alpha \in -c_{\max}^*$. More precisely,

$$\alpha(X) \begin{cases} = 0, & \text{if } \alpha \in F \\ < 0, & \text{if } \alpha \notin F. \end{cases}$$

This shows that

$$\lim_{t\to\infty} \exp(tX) = e_F.$$

It is clear that $e_F^2 = e_F$ and therefore $e_F \in E(S_A^{cpt})$. Let $e \in E(S_A^{cpt})$. Then there exists an element $X \in c_{\max}$ such that $e = \lim_{t\to\infty} \exp(tX)$ because $c_{\max}$ is polyhedral ([154], pp. 11, 26). Thus $e = e_F$ for $F = X^\perp \cap \mathrm{cone}(\Delta_-)$ and with

$$\mathrm{Int}_{F^\perp}(c_{\max} \cap F^\perp) = \{Y \in c_{\max} \mid Y^\perp \cap -c_{\max}^* = F\} \tag{5.35}$$

we find that

$$\mathrm{Int}_{F^\perp}(c_{\max} \cap F^\perp) = \{X \in c_{\max} \mid \lim_{t\to\infty} \exp(tX) = e_F\}.$$

This proves that $F \mapsto e_F$ is a bijection. Since it is clearly order-preserving, the claim follows.

4) Let $s = s_1 \cdot \ldots \cdot s_n \in S_A^{cpt}$ with $s_i \in \overline{\exp(\mathbb{R}^+ X_i)}$. Then either $s_i \in \exp(\mathbb{R}^+ X_i) \subset \exp(c_{\max})$ or $s_i = \lim_{t\to\infty} \exp(tX_i) \in E(S_A^{cpt})$. Thus $s \in \exp(c_{\max}) E(S_A^{cpt})$. □

Example 5.7.7 In the situation of the $\mathrm{SL}(2,\mathbb{R})$ Example 5.4.23,we identify $\mathrm{Aff}(N_-)$ with $\mathbb{R}\times\mathbb{R}$, where the multiplication is

$$(n,\gamma)(n',\gamma') = (n+\gamma n', \gamma\gamma')$$

and the action on $\mathbb{R} = \mathfrak{n}_-$ is given by

$$(n,\gamma)\cdot n' = n + \gamma n'.$$

Then $\mathrm{Aff}_{com}(N_-)$ corresponds to the set

$$\begin{aligned} &\{(n,\gamma) \in \mathbb{R}\times\mathbb{R} \mid \forall |r| < 1 : |n+\gamma r| < 1\} \\ = \;&\{(n,\gamma) \in \mathbb{R}\times\mathbb{R} \mid |\gamma| \le 1, |n| \le 1 - |\gamma|\}. \end{aligned}$$

The embedding of $B^\sharp$ into $\mathrm{Aff}(N_-)$ is

$$\begin{pmatrix} a & 0 \\ c & a^{-1} \end{pmatrix} \mapsto \left(\frac{c}{a}, \frac{1}{a^2}\right)$$

so the image of $B^\sharp$ is simply

$$\{(n,\gamma) \mid \gamma > 0, \nu \in \mathbb{R}\}.$$

Thus we obtain

$$S \cap B^\sharp = \{(n,\gamma) \in \mathbb{R}\times\mathbb{R} \mid 0 < \gamma \le 1, |n| \le 1-\gamma\}$$

and

$$S \cap A = \{(0,\gamma) \in \mathbb{R} \times \mathbb{R} \mid 0 < \gamma \leq 1\}.$$

The faces of $-c^*_{\max} = -c_{\max} = \mathbb{R}^- X^0 = -\mathbb{R}^+\alpha$ are $F_1 = \{0\}$ and $F_2 = -\mathbb{R}^+\alpha$. Therefore we have $e_{F_1} = (0,0)$ and $e_{F_2} = (0,1)$. Thus

$$\begin{aligned} S_A^{cpt} &= \{(0,\gamma) \in \mathbb{R} \times \mathbb{R} \mid 0 \leq \gamma \leq 1\} \\ &= \{(0,0)\} \cup (S \cap A) \\ &= (S \cap A)e_{F_1} \cup (S \cap A)e_{F_2} \end{aligned}$$

by direct calculation. □

Let $\mathfrak{a}^+ := \{X \in \mathfrak{a} \mid \forall \alpha \in \Delta^+ : \alpha(X) \geq 0\} \subset c_{\max}$ be the closure of the positive Weyl chamber. Our next goal is to find the idempotents in $\overline{S \cap B^\sharp}$ which occur as limits of elements in $\exp \mathfrak{a}^+$. The results will be useful when we determine the isotropy group of a point in $\mathcal{M}_+^{cpt}$ under the action of G.

Suppose that $X \in \mathfrak{a}^+$. We write $E_X := (\mathbb{R}X - c_{\max}) \cap c_{\max}$ for the face of $c_{\max}$ generated by X and

$$F_X = X^\perp \cap -c^*_{\max} = E_X^\perp \cap -c^*_{\max} \tag{5.36}$$

for the (up to a minus sign) opposite face. For any subset Σ of Δ we consider $\mathrm{cone}(\Sigma) = \sum_{\alpha \in \Sigma} \mathbb{R}^+\alpha$ (cf. Remark 2.1.7). In particular, we have $\mathrm{cone}(\Delta_-) = -c^*_{\max}$. We set $\Delta_X := E_X^\perp \cap \Delta$, $\Delta_{X,\pm} := E_X^\perp \cap \Delta_\pm$, and $\Delta_{X,0} := E_X^\perp \cap \Delta_0$. Then Remark 2.1.7 shows that $F_X = \mathrm{cone}(\Delta_{X,-})$ because

$$\Delta_{X,+} = \{\alpha \in \Delta_+ \mid \alpha \in F_X\} = \{\alpha \in \Delta_+ : \alpha(X) = 0\}.$$

An element $X' \in E_X \cap \mathfrak{a}^+$ is said to be *relatively regular in* E_X if all roots in Δ_0 which do not vanish on E_X, are nonzero on X'. We note that if X is relatively regular in E_X, then $\Delta_X^+ = \Delta_X \cap \Delta^+ = X^\perp \cap \Delta^+$ because $\Delta_{X,+} = X^\perp \cap \Delta_+$, and $\Delta_{X,0}^+ = \Delta_0^+ \cap X^\perp$ follows from relative regularity (recall that $X \in \mathfrak{a}^+$).

Lemma 5.7.8 *Let $X \in \mathfrak{a}^+$ and $E_X \in \mathrm{Fa}(c_{\max})$ the face generated by X. Then the following assertions hold.*

1) $[\Delta^+ + (\Delta^+ \setminus E_X^\perp)] \cap \Delta^+ \subset \Delta^+ \setminus E_X^\perp = \Delta^+ \setminus \Delta_X^+$ *and*

$$\mathrm{cone}(\Delta_X^+) = E_X^\perp \cap \mathrm{cone}(\Delta^+) \in \mathrm{Fa}(\mathrm{cone}(\Delta^+)).$$

2) *There exists a relatively regular element $X' \in \mathfrak{a}^+ \cap \mathrm{algint}(E_X)$ with $E_X = E_{X'}$.*

Proof: 1) In view of Remark 2.1.7.5), we have that

$$E_X^\perp = F_X - F_X = \mathrm{cone}(\Delta_{X,-}) - \mathrm{cone}(\Delta_{X,-}). \tag{5.37}$$

If $\gamma \in \Delta_{X,0}^+$, then the reflection s_γ at $\ker\gamma$ leaves Δ_- invariant and $s_\gamma(X) = X$. Therefore $s_\gamma(\Delta_{X,-}) = \Delta_{X,-}$ and consequently $E_X^\perp$ is invariant under the subgroup W_X of the Weyl group $W_0 = W(\Delta_0)$, which is generated by the reflections leaving X fixed. If $X \in c_{\max}^o$, then $E_X = c_{\max}$, $E_X^\perp = \{0\}$, and $\Delta_X^+ = \emptyset$. So we may assume that $X \in \partial c_{\max}$. Let $\Sigma = \{\alpha_0, \alpha_1, ..., \alpha_l\}$ be a basis of Δ^+ with $\alpha_0 \in \Delta_+$ (cf. Lemma 5.5.10). Since $X \in \partial c_{\max}$, there exists a root $\gamma \in \Delta_+$ with $\gamma(X) = 0$. Therefore $X \in \mathfrak{a}^+$ and $\gamma = \alpha_0 + \sum_{i=1}^l n_i\alpha_i$ entail that

$$0 = \gamma(X) \geq \alpha_0(X) \geq 0,$$

consequently $\alpha_0 \in E_X^\perp$. Note that the coefficient of α_0 must be 1 because $\gamma(Y^0) = 1$. Hence we may assume that

$$\Sigma_X := X^\perp \cap \Sigma = \{\alpha_0, \alpha_1, ..., \alpha_k\}. \tag{5.38}$$

So $\alpha = \alpha_0 + \sum_{i=1}^l n_i\alpha_i \in \Delta_{X,+}$ is equivalent to $n_i = 0$ for $i > k$. Consequently, $\alpha_0 \in E_X^\perp$ and (5.37) imply that

$$E_X^\perp = \mathbb{R}\alpha_0 \oplus E_{X,0}$$

with

$$E_{X,0} := E_X^\perp \cap \mathrm{span}\{\alpha_1, ..., \alpha_k\} = E_X^\perp \cap \mathrm{span}\{\alpha_1, ..., \alpha_l\}. \tag{5.39}$$

Next we claim the existence of a set of simple roots which span $E_{X,0}$. To see this, we first note that $E_{X,0}$ is invariant under the finite group W_X because W_X also fixes $\mathrm{span}\{\alpha_1, ..., \alpha_l\} = \mathrm{span}\Delta_0$. We recall that W_X is generated by the reflections $s_{\alpha_1}, ..., s_{\alpha_k}$ at the hyperplanes $\ker\alpha_j$ (cf. [168], 1.1.2.8). Therefore

$$E_{X,0} = E_{X,0,\mathrm{eff}} \oplus E_{X,0,\mathrm{fix}},$$

where

$$E_{X,0,\mathrm{fix}} = \{Y \in E_{X,0} \mid (\forall w \in W_X)\ w \cdot Y = Y\}$$

and

$$E_{X,0,\mathrm{eff}} = \mathrm{span}\{w \cdot Y - Y \mid w \in W_X, Y \in E_{X,0}\}.$$

For $Y \in E_{X,0,\mathrm{fix}}$ the relations $s_{\alpha_i}(Y) = Y$ imply that $\alpha_i \perp Y$ for $i = 1, ..., k$. Hence

$$Y \in \mathrm{span}\{\alpha_1, ..., \alpha_k\}^\perp \cap E_{X,0} \subset E_{X,0}^\perp \cap E_{X,0} = \{0\}.$$

In addition, the fact that W_X is generated by the reflections s_{α_i}, $i = 1, ..., k$ implies that

$$\begin{aligned} E_{X,0} &= E_{X,0,\text{eff}} = \text{span}\{s_{\alpha_i}(Y) - Y \mid i = 1, ..., k, Y \in E_{X,0}\} \\ &= \text{span}\{\alpha_j \mid \alpha_j \notin E_{X,0}^{\perp}\}. \end{aligned}$$

This proves our claim, and from now on we may assume that $E_{X,0} = \text{span}\{\alpha_1, ..., \alpha_j\}$ with $j \leq k$. Now $E_X^{\perp} = \text{span}\{\alpha_0, ..., \alpha_j\}$ and therefore

$$\Delta^+ \setminus E_X^{\perp} = \left\{ \sum_{i=1}^{l} n_i \alpha_i \middle| \exists i > j \, : \; n_i > 0 \right\}.$$

This implies the first assertion of the lemma. The second assertion is trivial.

2) From $E_X^{\perp} = \text{span}\{\alpha_1, ..., \alpha_j\}$ it follows that

$$\{\beta|_{E_X} \mid \beta \in \Delta^+\} = \{\beta|_{E_X} \mid \beta \in \mathbb{R}^+\alpha_{j+1} + \ldots + \mathbb{R}^+\alpha_l\}$$

lies in a pointed cone. Whence

$$c := \{Y \in \text{span} E_X \mid \forall \beta \in \Delta^+ \, : \, \beta(Y) \geq 0\} = (\text{span} E_X) \cap \mathfrak{a}^+$$

has nonempty interior in $\text{span}E_X$. But $X \in \mathfrak{a}^+ \cap \text{algint}(E_X)$. Hence there exists

$$X' \in \text{algint} E_X \cap \text{algint}(c).$$

It follows that $E_X = E_{X'}$, $X' \in \mathfrak{a}^+$, and that X' is relatively regular. □

Lemma 5.7.9 *Let* $X \in \mathfrak{a}^+$ *and*

$$e_X = \lim_{t \to \infty} \exp(tX) = (1, \gamma) \in \overline{S \cap B^\sharp}.$$

Then the following assertions hold:

1) $\mathfrak{g}(\Delta^+ \setminus \Delta_X^+)^\sharp$ *is an ideal in* $\mathfrak{n}^\sharp = \mathfrak{g}(\Delta^+)^\sharp$ *and*

$$\mathfrak{n}^\sharp \cong \mathfrak{g}(\Delta^+ \setminus \Delta_X^+)^\sharp \rtimes \mathfrak{g}(\Delta_X^+)^\sharp.$$

2) $\text{Ad}(N_0^\sharp A) \ker \gamma \subset \ker \gamma \subset \mathfrak{n}_-$.

Proof: 1) is immediate from Lemma 5.7.8.

2) Note that (5.36) implies that

$$\begin{aligned}\ker\gamma &= \{Y\in\mathfrak{n}_- \mid \lim_{t\to\infty} e^{\operatorname{ad} tX}Y=0\}\\ &= \sum_{\alpha\in\Delta_-,\alpha(X)\neq 0}\mathfrak{g}_\alpha\\ &= \mathfrak{g}(\Delta_+\setminus(\Delta_+\cap X^\perp))^\sharp\\ &= \mathfrak{g}(\Delta_+\setminus(\Delta_+\cap E_X^\perp))^\sharp\\ &= \mathfrak{n}_-\cap\mathfrak{g}(\Delta^+\setminus\Delta_X^+)^\sharp.\end{aligned}$$

Thus $\ker\gamma$ is an ideal in $\mathfrak{n}^\sharp$ by 1) and therefore invariant under $N_0^\sharp$. Since it is a sum of root spaces, $\ker\gamma$ is also invariant under $\operatorname{Ad}(A)$. □

Lemma 5.7.10 *Let $X\in\mathfrak{a}^+$ and $e_X=\lim_{t\to\infty}\exp(tX)=(1,\gamma)\in\overline{S\cap B^\sharp}$. If $\lambda_{e_X},\rho_{e_X}\colon\overline{B^\sharp}\to\overline{B^\sharp}$ are the left and right multiplications with e_X in $\overline{B^\sharp}$, then we have:*

1) $\lambda_{e_X}^{-1}(e_X)\cap B^\sharp=\ker\gamma\rtimes\exp\left(\mathfrak{g}(\Delta_0^+\setminus\Delta_X^+)\right)^\sharp\exp(\Delta_{X,+}^\perp)$, where $\Delta_{X,-}^\perp$ is identified with the corresponding subset of $\mathfrak{a}$.

2) Let $\Sigma(X):=\left[(X^\perp\cap\Delta_0^+)\setminus\Delta_{X,0}^+\right]\cup\left[(\Delta_0^+\cap\Delta_{X,+}^\perp]\setminus X^\perp\right)$. Then

$$\rho_{e_X}^{-1}(e_X)\cap B^\sharp=\exp\left(\mathfrak{g}\left(\Sigma(X)\right)\right)^\sharp\exp(\Delta_{X,+}^\perp)$$

where $\Delta_0^+\cap\Delta_{X,+}^\perp=\{\alpha\in\Delta_0^+ \mid \forall\beta\in\Delta_{X,+}:\ (\alpha\mid\beta)=0\}$.

Proof: 1) The formula

$$e_X(g,\delta)=(1,\gamma)(g,\delta)=(\gamma(g),\gamma\delta)$$

shows that $\lambda_{e_X}(g,\delta)=e_X$ is equivalent to $g\in\ker\gamma$ and $\gamma\delta=\gamma$, i.e., $\delta(\ker\gamma)\subset\ker\gamma$ and $\delta(x)\in x\ker\gamma$ for $x\in\operatorname{Im}\gamma$. According to Lemma 5.7.9.2), the first condition on δ is satisfied if $\delta\in\operatorname{Ad}(B^\sharp)$. For $\delta=e^{\operatorname{ad}Y}$ with $Y\in\mathfrak{n}_0+\mathfrak{a}$ the second condition is satisfied by all elements of $e^{\operatorname{ad}\mathbb{R}Y}$ if and only if $[Y,\operatorname{Im}\gamma]\subset\ker\gamma$. The set $\{g\in\operatorname{Aut}(N_-)\,:\,\gamma g=\gamma\}$ is a pseudo-algebraic semigroup. Whence $\gamma e^{\operatorname{ad}Y}=\gamma$ implies that

$$\gamma e^{t\operatorname{ad}Y}=\gamma\qquad\forall t\in\mathbb{R}$$

whenever $\operatorname{Spec}(\operatorname{ad}Y)\subset\mathbb{R}$ ([94], Lemma 5.1). This implies in particular that $\lambda_{e_X}^{-1}(e_X)\cap N_0^\sharp A$ is a connected normal subgroup and therefore

$$\lambda_{e_X}^{-1}(e_X)\cap N_0^\sharp A=\exp\{Y\in\mathfrak{n}_0+\mathfrak{a}\mid[Y,\operatorname{Im}\gamma]\subset\ker\gamma\}.$$

For $Y \in \mathfrak{a}$ this condition means that $Y \in \Delta_{X,-}^{\perp}$ since

$$\operatorname{Im}\gamma = \mathfrak{g}(\Delta_{X,+})^{\sharp} = \gamma(\Delta_{X,-}). \tag{5.40}$$

If $Y \in \mathfrak{g}_\alpha$ with $\alpha \in \Delta_0^+ \setminus \Delta_X^+$, then clearly $[Y, \operatorname{Im}\gamma] \subset \ker\gamma$ (Lemma 5.7.9.1)). This shows the inclusion $\supseteq$. Conversely, suppose that $\alpha \in \Delta_{X,0}^+ \subset E_X^{\perp}$. By Remark 2.1.7 and (5.37) we have that

$$\Delta_{X,+}^{\perp} = \Delta_{X,-}^{\perp} = F_X^{\perp} = E_X - E_X.$$

Now $(E_X - E_X) \cap E_X^{\perp} = \{0\}$ implies the existence of $\beta \in \Delta_{X,-}$ with $(\alpha \mid \beta) \neq 0$. Hence Lemma 5.7.1.2) shows that $[Y, \operatorname{Im}\gamma] \neq \{0\}$ for $Y \in \mathfrak{g}_\alpha$ because $\bigoplus_{n\in\mathbb{Z}} \mathfrak{g}_{\beta+n\alpha} \subset \operatorname{Im}\gamma$. But then

$$(\Delta_{X,+} + \Delta_{X,0}^+) \cap \Delta \subset \Delta_{X,+}$$

shows that $[Y, \operatorname{Im}\gamma] \not\subset \ker\gamma$.

2) First we note that $ge_X = g'e_X = e_X$ implies that $(gg')e_X = e_X$ and $g^{-1}(ge_X) = g^{-1}e_X = e_X$ for $g, g' \in B^{\sharp}$. So $\rho_{e_X}^{-1}(e_X) \cap B^{\sharp}$ is a subgroup. Moreover, for $(n_-, \delta) \in B^{\sharp}$ the condition

$$(n_-, \delta)(1, \gamma) = (n_-, \delta\gamma) = (1, \gamma)$$

is equivalent to $n_- = 1$ and $\delta\gamma = \gamma$. Thus

$$\rho_{e_X}^{-1}(e_X) \cap B^{\sharp} = \{(1, \delta) : \delta|_{\operatorname{Im}\gamma} = \operatorname{id}_{\operatorname{Im}\gamma}\}. \tag{5.41}$$

This is a pseudo-algebraic subgroup of the group $\operatorname{Ad}(B^{\sharp})$ which consists of real upper triangular matrices. So it is connected by [94], Lemma 5.1, because the exponential function of $B^{\sharp}$ is surjective. Therefore it only remains to compute the Lie algebra of this analytic subgroup. First we note that $e^{\operatorname{ad}\mathbb{R}Y}|_{\operatorname{Im}\gamma} = \operatorname{id}_{\operatorname{Im}\gamma}$ is equivalent to $[Y, \operatorname{Im}\gamma] = \{0\}$.

For $Y \in \mathfrak{a}$ this means that $Y \in \Delta_{X,+}^{\perp}$. Let $Y \in \mathfrak{g}_\alpha$ with $\alpha \in \Delta_0^+$. We have to consider several cases:

a) $\alpha(X) = 0$ and $\alpha \notin E_X^{\perp}$. Let $\beta \in \Delta_{X,+} \subset E_X^{\perp}$. Then $\alpha + \beta \notin \Delta$ since otherwise (5.36) implies that

$$\beta + \alpha \in \Delta_+ \cap X^{\perp} = \Delta_{X,+} \subset E_X^{\perp}.$$

This shows that $[Y, \operatorname{Im}\gamma] = \{0\}$.

b) $\alpha \in E_X^{\perp}$. Then we have already seen in the proof of 1) that $[Y, \operatorname{Im}\gamma] \neq \{0\}$.

c) $\alpha(X) > 0$ and $\alpha \notin \Delta^{\perp}_{X,+}$. Let $\beta \in \Delta_{X,+}$ with $(\alpha \mid \beta) \neq 0$. Then $\beta(X) = 0$ and therefore $(\beta - \alpha)(X) < 0$. So $\beta - \alpha$ is no root because $X \in c_{\max}$. Hence $(\alpha \mid \beta) < 0$ and $\beta + \alpha \in \Delta_+$. Now Lemma 5.7.1.2) implies that $[Y, \mathfrak{g}_\beta] \neq \{0\}$.

d) $\alpha(X) > 0$ and $\alpha \in \Delta^{\perp}_{X,+}$. As in c) we see that $\beta - \alpha \notin \Delta$ for all $\beta \in \Delta_{X,+}$. The reflection s_α interchanges the two ends of the α-string through β and it fixes β because $(\alpha \mid \beta) = 0$. So β agrees also with the upper end of this root string and $\alpha + \beta \notin \Delta$. But this clearly implies that $[Y, \operatorname{Im}\gamma] = \{0\}$. □

Remark 5.7.11 The formula for $\rho_{e_X}^{-1}(e_X)$ is relatively complicated. This comes from the fact that in general X is not relatively regular in the face E_X. Since every face of $c_{\max}$ contains relatively regular elements (they form an open dense subset), every idempotent may be reached by such an element. An example for an element which is not relatively regular in general is Y^0. The face it generates is $c_{\max}$, all compact roots vanish on Y^0, but no compact root vanishes on $c_{\max}$.

Now suppose that X is relatively regular and in $\mathfrak{a}^+$. Then

$$X^\perp \cap \Delta^+ = \Delta^+_X = E^\perp_X \cap \Delta^+.$$

Therefore

$$(X^\perp \cap \Delta_0^+) \setminus \Delta^+_{X,0} = \emptyset \quad \text{and} \quad (\Delta_0 \cap \Delta^\perp_{X,+}) \cap X^\perp = \emptyset$$

because $\Delta_0 \cap X^\perp \subset E^\perp_X = \operatorname{span}\Delta_{X,+}$. So the formula for $\rho_{e_X}^{-1}(e_X)$ becomes easier:

$$\rho_{e_X}^{-1}(e_X) \cap B^\sharp = \exp\left(\mathfrak{g}(\Delta_0^+ \cap \Delta^\perp_{X,+})\right)^\sharp \exp(\Delta^\perp_{X,+}). \tag{5.42}$$

□

Lemma 5.7.12 *Let L be a connected Lie group and γ an idempotent endomorphism of L. Then*

$$L \cong \ker\gamma \rtimes \operatorname{Im}\gamma.$$

In particular, $\ker\gamma$ *is connected.*

Proof: It is clear that $\operatorname{Im}\gamma = \{g \in L \mid \gamma(g) = g\}$ and $\ker\gamma$ are closed subgroups. Obviously, $\ker\gamma$ is normal in L, so $\ker\gamma \cdot \operatorname{Im}\gamma$ is a subgroup and the intersection of $\operatorname{Im}\gamma$ and $\ker\gamma$ is trivial. If $g \in L$, then $g = \gamma(g)\,[\gamma(g)^{-1}g]$ and

$$\gamma\,[\gamma(g)^{-1}g] = \gamma^2(g)^{-1}\gamma(g) = \gamma(g)^{-1}\gamma(g) = 1.$$

Hence $L \cong \ker\gamma \cdot \operatorname{Im}\gamma$ is a semidirect product decomposition. □

Lemma 5.7.13 *Let L be a connected Lie group and $e = (1, \gamma)$ be an idempotent in $L \rtimes \mathrm{End}(L)$. Further, let $s = (a, \delta)$. Then L1)–L4) and R1)–R4) are equivalent:*

L1) $ese = es$.

L2) $e(es) = (es)e$.

L3) $es \in e\mathrm{Aff}(L)e$.

L4) $\delta(\ker \gamma) \subset \ker \gamma$.

R1) $ese = se$.

R2) $e(se) = (se)e$.

R3) $se \in e\mathrm{Aff}(L)e$.

R4) $a \in \mathrm{Im}\,\gamma$ *and* $\delta(\mathrm{Im}\,\gamma) \subset \mathrm{Im}\,\gamma$.

Proof: The equivalence of L1)–L3) and R1)–R3) is trivial. To see that L1) is equivalent to L4), we compute

$$es = (1, \gamma)(a, \delta) = (\gamma(a), \gamma \circ \delta)$$

and

$$ese = (\gamma(a), \gamma \circ \delta)\,(1, \gamma) = (\gamma(a), \gamma \circ \delta \circ \gamma).$$

So $es = ese$ is equivalent to $\gamma\delta = \gamma\delta\gamma$. On the image of γ this relation is trivial. According to Lemma 5.7.12, it holds if and only if it holds on the kernel of γ, i.e., if and only if $\delta(\ker \gamma) \subset \ker \gamma$.

For the equivalence of R1) and R4), we compute $se = (a, \delta\gamma)$. So $se = ese$ is equivalent to $\gamma(a) = a$ and $\delta\gamma = \gamma\delta\gamma$. The first condition means that $a \in \mathrm{Im}\,\gamma$. On the kernel of γ, the second relation is trivial. In view of Lemma 5.7.12, it holds if and only if it holds on the image of γ, i.e., if and only if $\delta(\mathrm{Im}\,\gamma) \subset \mathrm{Im}\,\gamma$. □

A *face* F of a topological monoid (i.e., semigroup with identity) T is a closed subsemigroup whose complement $T \setminus F$ is a semigroup ideal. The lattice of faces of T is denoted $\mathrm{Fa}(T)$. The group of units in T will be denoted $U(T)$. In particular, if $e \in T$ is idempotent, then eTe is a monoid with unit e and $U(eTe)$ is the unit group of this monoid.

Lemma 5.7.14 *Let T be a topological semigroup. For $e \in E(T)$ we set*

$$T_e := \{t \in T : et \in eTe\}$$

and

$$F_e := \{t \in T : et \in U(eTe)\}.$$

Suppose that $U(eTe)$ is closed in eTe. Then T_e is a closed subsemigroup of T with identity e, the mapping

$$\lambda_e : T_e \to eTe, \quad t \mapsto et$$

is a semigroup homomorphism, and F_e is a face of T_e.

Proof: Let $t, t' \in T_e$. Then

$$e(tt')e = (et)(t'e) = (ete)(t'e) = (et)(et'e) = (et)(et') = ett'.$$

Hence T_e is a closed subsemigroup of T and the mapping λ_e is a homomorphism. Now it is clear that $F_e := \lambda_e^{-1}(U(eTe))$, as the inverse image of a face, is a face of T_e. □

Lemma 5.7.15 *Let $X \in \mathfrak{a}^+$ and $e_X = \lim_{t\to\infty} \exp(tX) = (1, \gamma) \in \overline{B^\sharp}$. Then the following assertions hold:*

1) $\overline{B^\sharp}_{e_X} := \{s \in \overline{B^\sharp} : e_X s = e_X s e_X\} = \overline{B^\sharp}$.

2) *e_X is an identity element in the semigroup $e_X\overline{B^\sharp}$.*

3) $\gamma(\Omega) = \Omega \cap \operatorname{Im}\gamma$.

4) $e_X\overline{S \cap B^\sharp} = \{e_X s \in e_X\overline{B^\sharp} \mid e_X s(\gamma(\overline{\Omega}_-)) \subset \gamma(\overline{\Omega}_-)\}$.

5) *$B^\sharp \cong (\ker \lambda_{e_X} \cap B^\sharp) \rtimes B_X^\sharp$, where*

$$B_X^\sharp := \operatorname{Im}\gamma \rtimes \exp\left(\mathfrak{g}(\Delta_{X,0}^+)\right)^\sharp \exp(E_X^\perp).$$

6) $e_X\overline{B^\sharp} = e_X\overline{B_X^\sharp}$.

7) *$U(e_X\overline{B^\sharp}) = e_X B^\sharp$ and $U(e_X\overline{S \cap B^\sharp}) = \{e_X\}$.*

Proof: 1) In view of Lemma 5.7.13, we only have to recall from Lemma 5.7.9 that

$$\delta(\ker\gamma) \subset \ker\gamma \qquad \forall \delta \in \operatorname{Ad}(B^\sharp).$$

2) Lemma 5.7.14 shows that $e_X\overline{B^\sharp}$ is a semigroup. The other assertion is a consequence of 1).

3) Since $e_X \in \overline{S \cap B^\sharp}$, we have that

$$e_X \cdot \overline{\Omega}_- = \gamma(\overline{\Omega}_-) \subset \overline{\Omega}_- \cap \operatorname{Im}\gamma.$$

But $\gamma|_{\operatorname{Im}\gamma} = \operatorname{id}_{\operatorname{Im}\gamma}$. Hence

$$\overline{\Omega}_- \cap \operatorname{Im}\gamma \subset \gamma(\overline{\Omega}_-).$$

4) Let $s \in \overline{S \cap B^\sharp}$. Then

$$e_X s \cdot \gamma(\overline{\Omega}_-) = e_X s e_X \cdot \overline{\Omega}_- \subset \overline{\Omega}_-$$

because $e_X s e_X \in \overline{S \cap B^\sharp}$. Thus $\subset$ holds. If, conversely, s is contained in the right-hand side, then, in view of 2),

$$s(\overline{\Omega}_-) = s e_X(\overline{\Omega}_-) = s \cdot \gamma(\overline{\Omega}_-) \subset \gamma(\overline{\Omega}_-) \subset \overline{\Omega}_-.$$

5) It follows from (5.40) that $B^\sharp_X$ is a subgroup of $B^\sharp$. Since $B^\sharp_X \cap \lambda_{e_X}^{-1}(e_X) = \{1\}$ by Lemma 5.7.10, we conclude that

$$B^\sharp \cong (\ker \lambda_{e_X} \cap B^\sharp) \rtimes B^\sharp_X.$$

6) The relation $e_X \overline{B^\sharp_X} \subset e_X \overline{B^\sharp}$ is trivial. But $e_X \overline{B^\sharp_X}$ is closed. Therefore 5) implies that

$$e_X \overline{B^\sharp_X} = \overline{e_X B^\sharp_X} = \overline{e_X B^\sharp} \supset e_X \overline{B}^\sharp.$$

7) First we prove that $U(e_X \overline{B^\sharp_X}) = e_X B^\sharp_X$. The inclusion $e_X B^\sharp_X \subset U(e_X \overline{B^\sharp_X})$ is trivial. For the converse, we consider the homomorphism $\phi : e_X \overline{B^\sharp_X} \to \operatorname{Aff}(\operatorname{Im}\gamma)$ defined by $\phi(s) := s|_{\operatorname{Im}\gamma}$. Since $s(\ker\gamma) = s e_X(\ker\gamma) = s(1) = \{1\}$ for all $s \in e_X \overline{B^\sharp_X}$, it follows that ϕ is a homeomorphism onto a closed subsemigroup of $\operatorname{Aff}(\operatorname{Im}\gamma)$ which satisfies $\phi(e_X) = \operatorname{id}$. Therefore

$$U(e_X \overline{B^\sharp_X}) = \phi^{-1}\left(H\left(\phi(e_X \overline{B^\sharp_X})\right)\right).$$

On the other hand, $\phi(e_X B^\sharp_X) = \operatorname{Im}\gamma \rtimes Q$, where Q is a subgroup of $\operatorname{Aut}(\operatorname{Im}\gamma)$ consisting of upper triangular matrices with respect to the root decomposition of the Lie algebra of $\operatorname{Im}\gamma$. Hence Q is closed in $\operatorname{Aut}(\operatorname{Im}\gamma)$ and it suffices to show that $U(\overline{Q}) = Q$, where $\overline{Q}$ is the closure of Q in $\operatorname{End}(\operatorname{Im}\gamma)$. But if

$$\lim u_n = u \in U(\operatorname{End}(\operatorname{Im}\gamma)) = \operatorname{Aut}(\operatorname{Im}\gamma),$$

then $u \in Q$. This proves that $U(e_X \overline{B^\sharp_X}) = e_X B^\sharp_X$. Therefore

$$U\left((e_X(\overline{S \cap B^\sharp})\right) \subset U\left(e_X \overline{B^\sharp_X}\right) \subset e_X B^\sharp_X.$$

The semigroup $e_X(\overline{S \cap B^\sharp})$ is compact since $\overline{S \cap B^\sharp}$ is. Thus 4) shows that

$$\phi\left(U\left((e_X(\overline{S \cap B^\sharp})\right)\right)$$

is a compact subgroup of the simply connected solvable group $\operatorname{Im}\gamma \rtimes Q$ and therefore trivial. □

Notes for Chapter 5

The material in Section 5.1 was first proved in [129]. Part of it can also be found in the work of Ol'shanskii, e.g., [138]. There is extensive literature on the exponential map for symmetric spaces, e.g., in [29, 44, 104]. The semigroup $H \exp C$ was first introduced by Ol'shanskii [137, 138] for the group case. In recent years they have become increasingly important in geometry and analysis, as we will see in the next chapters. References to applications will be given in the notes to those chapters. Further sources are [50, 52, 64, 63, 93, 129, 130] and the work of Ol'shanskii and Paneitz. That $\mathcal{M}$ is globally hyperbolic was first proved by J. Faraut in [25] for the case $G_{\mathbb{C}}/G$. This was generalized in [129] to arbitrary noncompactly causal symmetric spaces using the causal embedding from Lemma 5.2.8. The proof presented here is an adaptation of that in [25]. A different approach can be found in [114]. The characterization of $G^r \exp(C_{\max})$ as a compression semigroup was noted first by Ol'shanskii. The proof presented here appeared in [58]. The nonlinear convexity theorem was proven by Neeb in [116]. The proof given is taken from [124]. The results on $B^\sharp$ have been proved in [55].

Chapter 6

The Order Compactification of Noncompactly Causal Symmetric Spaces

The order compactification of ordered homogeneous spaces defined in Section 2.4 is a fairly abstract construction. In this chapter we show that for the special case of noncompactly causal symmetric spaces, many features of the order compactification can be made quite explicit. In particular, the orbit structure can be determined completely and described in terms of the restricted root system. The basic idea is to identify $gH \in \mathcal{M}$ with the compact set $g \cdot \overline{\mathcal{O}} \subset \mathcal{X}$, where $\mathcal{O}$ is the open domain in the flag manifold $\mathcal{X}$ defined in Section 5.1. Similarly as for the order compactification, this yields a compactification of $\mathcal{M}$ via the suitable Vietoris topology. The point is that this compactification is essentially the same as the order compactification but easier to treat, since it deals with bounded convex sets in a finite-dimensional linear space rather than translates of a "nonlinear cone."

6.1 Causal Galois Connections

In this section we suppose that G is a connected Lie group and S an extended Lie subsemigroup of G with unit group H. Let $\mathcal{M} = G/H$ and consider the order $\leq$ on $\mathcal{M}$ induced by $\leq_S$ (cf. Section 2.4). In the following, $\mathcal{X}$ denotes a metrizable compact G-space and $\mathcal{O} \subset \mathcal{X}$ an open subset with the property

$$S = \{g \in G \mid g \cdot \mathcal{O}\} \quad \text{and} \quad S^o = \{g \in G \mid g \cdot \overline{\mathcal{O}} \subset \mathcal{O}\}. \tag{6.1}$$

(cf. Corollary 5.4.21 for an example of this situation.)

We endow the set $\mathcal{F}(\mathcal{X})$ of closed (hence compact) subsets of $\mathcal{X}$ with the Vietoris topology (cf. Appendix B). We write 2^G for the set of all subsets of G, and define the mappings

$$\Gamma : \mathcal{F}(\mathcal{X}) \to \mathcal{F}(G), \quad F \mapsto \{g \in G \mid g^{-1} \cdot F \subset \overline{\mathcal{O}}\} \tag{6.2}$$

and

$$\hat{\Gamma} : 2^G \to \mathcal{F}(\mathcal{X}), \quad A \mapsto \bigcap_{a \in A} a \cdot \overline{\mathcal{O}}. \tag{6.3}$$

We call Γ a *causal Galois connection*. That this is no misnomer is a consequence of the following lemma.

Lemma 6.1.1 *The mappings*

$$\hat{\Gamma} : (\mathcal{F}(G), \subset) \to (\mathcal{F}(\mathcal{X}), \subset) \quad \textit{and} \quad \Gamma : (\mathcal{F}(\mathcal{X}), \subset) \to (\mathcal{F}(G), \subset)$$

are antitone and define a Galois connection between the above partially ordered sets. Moreover, the following assertions hold:

1) $\downarrow \Gamma(F) = \Gamma(F)$ *for every* $F \in \mathcal{F}(\mathcal{X})$.

2) $\Gamma(F)^o = \{g \in G \mid g^{-1} \cdot F \subset \mathcal{O}\}$.

3) *For every subset* $A \subset G$ *we have that* $\hat{\Gamma}(A) = \hat{\Gamma}(\overline{A})$ *and* $\hat{\Gamma}(\downarrow A) = \hat{\Gamma}(A)$.

4) $\hat{\Gamma}(\bigcup_{i \in I} A_i) = \bigcap_{i \in I} \hat{\Gamma}(A_i)$.

5) $\Gamma(\bigcup_{i \in I} F_i) = \bigcap_{i \in I} \Gamma(F_i)$.

6) $\Gamma(\bigcap_{n \in \mathbb{N}} F_n) = \overline{\bigcup_{n \in \mathbb{N}} \Gamma(F_n)}$ *for every decreasing sequence* F_n *in* $\mathcal{F}(\mathcal{X})$.

7) $\hat{\Gamma}(g \cdot A) = g \cdot \hat{\Gamma}(A)$ *and* $\Gamma(g \cdot F) = g \cdot \Gamma(F)$ *for all* $g \in G, A \in 2^G, F \in \mathcal{F}(\mathcal{X})$.

8) $\Gamma(g \cdot \overline{\mathcal{O}}) = \downarrow g$ *for all* $g \in G$.

9) $\hat{\Gamma}(\downarrow g) = g \cdot \overline{\mathcal{O}}$ *for all* $g \in G$.

Proof: The fact that Γ and $\hat{\Gamma}$ define a Galois connection follows from the fact that

$$\begin{aligned} A \subset \Gamma(F) \quad &\Leftrightarrow A \subset \{g \in G \mid F \subset g \cdot \overline{\mathcal{O}}\} \\ &\Leftrightarrow F \subset \textstyle\bigcap_{g \in A} g \cdot \overline{\mathcal{O}} = \hat{\Gamma}(A). \end{aligned}$$

1) Let $g \in \hat{\Gamma}(F)$ and $s \in S$. Then $g^{-1} \cdot F \subset \overline{\mathcal{O}}$ and therefore

$$(gs^{-1})^{-1} \cdot F = sg^{-1} \cdot F \subset s \cdot \overline{\mathcal{O}} \subset \overline{\mathcal{O}},$$

hence $\downarrow g = gS^{-1} \subset \Gamma(F)$.

2) Let $g \in \Gamma(F)^o$. Then, since S has dense interior, there exists an $s \in S^o$ with $gs \in \Gamma(F)$. Therefore

$$g^{-1} \cdot F = s(gs)^{-1} \cdot F \subset s \cdot \overline{\mathcal{O}} \subset \mathcal{O}.$$

Conversely, suppose that $g^{-1} \cdot F \subset \mathcal{O}$. Then we find a neighborhood U of g in G such that $U^{-1} \cdot F \subset \mathcal{O}$ because F is compact and $\mathcal{O}$ is open. Thus $g \in U \subset \Gamma(F)$.

3) It is clear that $\hat{\Gamma}(\overline{A}) \subset \hat{\Gamma}(A) = \hat{\Gamma}(\downarrow A)$ because

$$as^{-1} \cdot \overline{\mathcal{O}} = a \cdot (s^{-1} \cdot \overline{\mathcal{O}}) \supset a \cdot \overline{\mathcal{O}} \quad \text{for every } a \in G, s \in S.$$

Let $x \in \hat{\Gamma}(A)$ and $a \in \overline{A}$ with $a = \lim_{n\to\infty} a_n$ and $a_n \in A$. For every $n \in \mathbb{N}$ we find an element $f_n \in \overline{\mathcal{O}}$ with $x = a_n \cdot f_n$. We may assume that $f := \lim_{n\to\infty} f_n$ exists in the compact set $\overline{\mathcal{O}}$. Then

$$x = \lim_{n\to\infty} a_n \cdot f_n = a \cdot f \in a \cdot \overline{\mathcal{O}}.$$

4), 5) These assertions are trivial.

6) Set $F := \bigcap_{n\in\mathbb{N}} F_n$. First we note that $\Gamma(F_n) \subset \Gamma(F)$ and therefore that the right-hand side of 6) is contained in the left. Let $g \in \Gamma(F)$ and $s_n \in S^o$ be a sequence with $\lim_{n\to\infty} s_n = 1$. For every $n \in \mathbb{N}$ we have that

$$s_n g^{-1} \cdot F \subset s_n \cdot \overline{\mathcal{O}} \subset \mathcal{O},$$

i.e., $F \subset gs_n^{-1} \cdot \mathcal{O}$. Note that $F_n \to F$ in the Vietoris topology. Hence we find $n_0 \in \mathbb{N}$ such that $F_{n_0} \subset gs_n^{-1} \cdot \mathcal{O}$, which implies that $gs_n^{-1} \in \Gamma(F_{n_0})$. This shows that $gs_n^{-1} \in \bigcup_{k\in\mathbb{N}} \Gamma(F_k)$ for every $n \in \mathbb{N}$. Now

$$g = \lim_{n\to\infty} gs_n^{-1} \in \overline{\bigcup_{k\in\mathbb{N}} \Gamma(F_k)}.$$

7) This is immediate from

$$\Gamma(g\cdot F) = \{x \in G \mid x^{-1}g\cdot F \subset \overline{\mathcal{O}}\} = \{x \in G \mid (g^{-1}x)^{-1}\cdot F \subset \overline{\mathcal{O}}\} = g\cdot\Gamma(F)$$

and from

$$\hat{\Gamma}(g\cdot A) = \bigcap_{a\in g\cdot A} a\cdot\overline{\mathcal{O}} = \bigcap_{a\in A} ga\cdot\overline{\mathcal{O}} = g\cdot\hat{\Gamma}(A).$$

8) We deduce from 7) that $\Gamma(g\overline{\mathcal{O}}) = g\cdot\Gamma(\overline{\mathcal{O}})$, and therefore it remains to show that $S^{-1} = \{g \in G \mid g^{-1}\cdot\overline{\mathcal{O}} \subset \overline{\mathcal{O}}\}$. First, according to our assumptions about $\mathcal{O}$ and $\mathcal{X}$, we have that $S^{-1} \subset \Gamma(\overline{\mathcal{O}})$ because $g^{-1}\cdot\mathcal{O} \subset \mathcal{O}$ implies that $g^{-1}\cdot\overline{\mathcal{O}} \subset \overline{\mathcal{O}}$. We claim that $g \in S^{-1}$ for every $g \in \Gamma(\overline{\mathcal{O}})$. Let $s_n \in S^o$ be a sequence with $\lim_{n\to\infty} s_n = 1$. Then

$$s_n g^{-1}\cdot\overline{\mathcal{O}} = s_n\cdot(g^{-1}\cdot\overline{\mathcal{O}}) \subset s_n\cdot\overline{\mathcal{O}} \subset \mathcal{O}$$

and therefore $s_n g^{-1} \in S^o$. Thus $g^{-1} = \lim_{n\to\infty} s_n g^{-1} \in \overline{S^o} = S$.

9) In view of 3), we have $\hat{\Gamma}(\downarrow g) = \hat{\Gamma}(\{g\}) = g\cdot\overline{\mathcal{O}}$. □

Corollary 6.1.2 *For two elements $g, g' \in G$ we have*

$$g \leq_S g' \iff g'\cdot\overline{\mathcal{O}} \subset g\cdot\overline{\mathcal{O}}$$

and

$$g \in (\downarrow g')^o \iff g'\cdot\overline{\mathcal{O}} \subset g\cdot\mathcal{O}.$$

Proof: These are direct consequences of the fact that

$$S^o = \{g \in G \mid g\cdot\overline{\mathcal{O}} \subset \mathcal{O}\} \quad \text{and} \quad S^{-1} = \Gamma(\overline{\mathcal{O}}).$$ □

Lemma 6.1.3 *The mapping $\Gamma: \mathcal{F}(\mathcal{X}) \to \mathcal{F}(G)$ is continuous with respect to the Vietoris topologies on $\mathcal{F}(\mathcal{X})$ and $\mathcal{F}(G)$.*

Proof: Suppose that $F_n \to F$ in $\mathcal{F}(\mathcal{X})$. We split up the proof into two steps.

1) $\Gamma(F)^o \subset \liminf_{n\to\infty}\Gamma(F_n)$: Let $g \in \Gamma(F)^o$. According to Lemma 6.1.1.2), we have $g^{-1}\cdot F \subset \mathcal{O}$, hence $F \subset g\cdot\mathcal{O}$. Consequently, we find $n_0 \in \mathbb{N}$ such that $F_n \subset g\cdot\mathcal{O}$ for $n \geq n_0$. Thus $g^{-1}\cdot F_n \subset \mathcal{O} \subset \overline{\mathcal{O}}$ and $g \in \Gamma(F_n)$ for $n \geq n_0$.

2) $\limsup_{n\to\infty}\Gamma(F_n) \subset \Gamma(F)$: Let $g \in \limsup_{n\to\infty}\Gamma(F_n)$ and choose a subsequence F_{n_k} and $g_k \in \Gamma(F_{n_k})$ with $g_k \to g$. Then $F = \lim_{k\to\infty} F_{n_k}$. Pick $f \in F$. If $f_k \in F_{n_k}$ with $f_k \to f$, then $f_k \in g_k\cdot\overline{\mathcal{O}}$ so $f \in \lim_{k\to\infty} g_k\cdot\overline{\mathcal{O}} = g\cdot\overline{\mathcal{O}}$. Thus $g \in \Gamma(F)$, since f was arbitrary. As $\liminf_{n\to\infty}\hat{F}(f_n)$ is closed, we now have

$$\overline{\Gamma(F)^o} \subset \liminf\Gamma(F_n) \subset \limsup\Gamma(F_n) \subset \Gamma(F).$$

But $\Gamma(F)$ is closed and satisfies $\Gamma(F) = \downarrow\Gamma(F)$, so Lemma 2.4.7 implies $\Gamma(F) = \overline{\Gamma(F)^o}$ and hence Lemma B.1.1 shows that $\Gamma(F_n) \to \Gamma(F)$. □

Definition 6.1.4 We set $\mathcal{M}^{\mathcal{O}} := \overline{\{g\cdot\overline{\mathcal{O}} \mid g \in G\}} \subset \mathcal{F}(\mathcal{X})$. The map

$$\iota: G \to \mathcal{M}^{\mathcal{O}}, g \mapsto g\cdot\overline{\mathcal{O}}$$

is called a *causal orbit map.* □

Note that Lemma 6.1.1.8) implies that

$$\Gamma \circ \iota = \overline{\eta} \circ \pi, \tag{6.4}$$

where $\pi: G \to G/H$ is the quotient map and $\overline{\eta}$ the order compactification from Lemma 2.4.2.

Lemma 6.1.5 (Fixed Points of the Galois Connection) *The following assertions hold:*

1) $\Gamma \circ \hat{\Gamma}(A) = A$ *for every* $A \in \Gamma(\mathcal{M}^{\mathcal{O}})$.

2) $\hat{\Gamma} \circ \Gamma(F) = F$ *for every* $F \subset \mathcal{X}$ *for which there exists a decreasing sequence* $F_n = g_n \cdot \overline{\mathcal{O}}$ *with*

$$F = \lim_{n\to\infty} F_n = \bigcap_{n\in\mathbb{N}} g_n \cdot \overline{\mathcal{O}}.$$

Proof. 1) There exists a compact subset $F \in \mathcal{M}^{\mathcal{O}}$ with $A = \Gamma(F)$. The fact that $\hat{\Gamma}$ and Γ define a Galois connection (Lemma 6.1.1) implies that

$$\Gamma\hat{\Gamma}(A) = \Gamma\hat{\Gamma}\Gamma(F) = \Gamma(F) = A.$$

2) We use Lemma 6.1.1.6) to see that $\Gamma(F) = \overline{\bigcup_{n\in\mathbb{N}} \Gamma(F_n)}$. According to Lemma 6.1.1.3), this leads to

$$\hat{\Gamma}\Gamma(F) = \hat{\Gamma}\left(\overline{\bigcup_{n\in\mathbb{N}} \Gamma(F_n)}\right) = \hat{\Gamma}(\bigcup_{n\in\mathbb{N}} \downarrow g_n) = \bigcap_{n\in\mathbb{N}} \hat{\Gamma}(\downarrow g_n) = \bigcap_{n\in\mathbb{N}} g_n \cdot \overline{\mathcal{O}} = F. \quad \square$$

Proposition 6.1.6 *The mapping* $\Gamma : \mathcal{M}^{\mathcal{O}} \to \mathcal{M}^{cpt}$ *is a quotient morphism of compact G-spaces, where G acts on* $\mathcal{M}^{\mathcal{O}}$ *by* $(g, F) \mapsto g \cdot F$. *Moreover, the action of G on* $\mathcal{M}^{\mathcal{O}}$ *is continuous.*

Proof. We show first that $\Gamma(\mathcal{M}^{\mathcal{O}}) = \mathcal{M}^{cpt}$. In fact, consider $g \cdot \overline{\mathcal{O}} \in \mathcal{M}^{\mathcal{O}}$. Then, according to Lemma 6.1.1.8),

$$\Gamma(g\cdot\overline{\mathcal{O}}) = \downarrow g = gS^{-1} = \overline{\eta}(gH).$$

This proves the claim, since $\Gamma\colon \mathcal{M}^{\mathcal{O}} \to \Phi(G)$ and $\overline{\eta}\colon \mathcal{M} \to \Phi(G)$ are continuous and $\{g \cdot \overline{\mathcal{O}} \mid g \in G\}$ is dense in $\mathcal{M}^{\mathcal{O}}$ just as $\mathcal{M}$ is dense in $\mathcal{M}^{cpt}$.

It follows from Lemma 6.1.1.7) that Γ is equivariant. Lemma 6.1.3 together with the above implies that Γ is a quotient mapping of compact spaces. The action of G on $\Phi(\mathcal{X})$ is continuous by Lemma B.1.2. Thus $\mathcal{M}^{\mathcal{O}}$ is G-invariant and the restriction is obviously a continuous action on $\mathcal{M}^{\mathcal{O}}$. □

Theorem 6.1.7 *Let*

$$\mathcal{M}_+^{\mathcal{O}} := (\Gamma|_{\mathcal{M}^{\mathcal{O}}})^{-1}(\mathcal{M}_+^{cpt}). \tag{6.5}$$

Then the following holds:

1) $\overline{\iota(S)} = \mathcal{M}_+^{\mathcal{O}}$.

2) $S = \{g \in G \mid g \cdot (\mathcal{M}_+^{\mathcal{O}})^o \subset (\mathcal{M}_+^{\mathcal{O}})^o\}$.

3) $S^o = \{g \in G \mid g \cdot \mathcal{M}_+^{\mathcal{O}} \subset (\mathcal{M}_+^{\mathcal{O}})^o\}$.

Proof: 1) This follows from $\Gamma\left(\iota(S)\right) = \overline{\eta}(\mathcal{M}_+)$.

2) and 3): First we note that $\Gamma(\mathcal{M}_+^{\mathcal{O}}) = \mathcal{M}_+^{cpt}$ by Proposition 6.1.6. We claim that $[\mathcal{M}_+^{cpt}]^o = \Gamma([\mathcal{M}_+^{\mathcal{O}}]^o)$ and $[\mathcal{M}_+^{\mathcal{O}}]^o = \Gamma^{-1}([\mathcal{M}_+^{cpt}]^o)$. In fact, we have

$$F \in \mathcal{M}_+^{\mathcal{O}} \Leftrightarrow \Gamma(F) \in \mathcal{M}_+^{cpt} \Leftrightarrow 1 \in \Gamma(F) \Leftrightarrow F \subset \overline{\mathcal{O}}$$

because of Lemma 2.4.4, and, using Lemma B.1.2, we see that $F \in [\mathcal{M}_+^{\mathcal{O}}]^o$ implies that there exists a neighborhood U of 1 in G such that $U \cdot F \subset \overline{\mathcal{O}}$. On the other hand, $\Gamma(F) \in [\mathcal{M}_+^{cpt}]^o$ holds if and only if there exists a neighborhood U of 1 in G such that $U \cdot F \subset \overline{\mathcal{O}}$ because of Proposition 2.4.4. This shows that

$$\Gamma([\mathcal{M}_+^{\mathcal{O}}]^o) \subset [\mathcal{M}_+^{cpt}]^o.$$

The reverse inclusion follows from the continuity of Γ. Now we have $\Gamma([\mathcal{M}_+^{\mathcal{O}}]^o) = [\mathcal{M}_+^{cpt}]^o$, which upon taking the preimage under Γ also shows that $[\mathcal{M}_+^{\mathcal{O}}]^o = \Gamma^{-1}([\mathcal{M}_+^{cpt}]^o)$. Now we see that $g \in S$ is equivalent to

$$\begin{aligned} g \cdot [\mathcal{M}_+^{cpt}]^o \subset [\mathcal{M}_+^{cpt}]^o &\iff g \cdot \Gamma([\mathcal{M}_+^{\mathcal{O}}]^o) \subset \Gamma([\mathcal{M}_+^{\mathcal{O}}]^o) \\ &\iff \Gamma(g \cdot [\mathcal{M}_+^{\mathcal{O}}]^o) \subset \Gamma([\mathcal{M}_+^{\mathcal{O}}]^o) \\ &\iff g \cdot [\mathcal{M}_+^{\mathcal{O}}]^o \subset [\mathcal{M}_+^{\mathcal{O}}]^o. \end{aligned}$$

Similarly, $g \cdot \mathcal{M}_+^{cpt} \subset [\mathcal{M}_+^{cpt}]^o \iff g \cdot \mathcal{M}_+^{\mathcal{O}} \subset [\mathcal{M}_+^{\mathcal{O}}]^o$. An application of Proposition 2.4.4 completes the proof of the last two claims. □

6.2 An Alternative Realization of $\mathcal{M}_+^{cpt}$

From now on we assume that $\mathcal{M} = G/H$ is a noncompactly causal symmetric space such that G is contained in a simply connected complexification $G_{\mathbb{C}}$. Moreover, the semigroup $S = S(C_{\max})$, the flag manifolds $\mathcal{X}G/P_{\max}$ and $\mathcal{X}_{\mathbb{C}} = G_{\mathbb{C}}/(P_{\max})_{\mathbb{C}}$, and the open domain $\mathcal{O} \subset \mathcal{X}$, are the ones from Section 6.1.

Recall the causal orbit map $\iota: G \to \mathcal{M}^{\mathcal{O}}$ from Definition 6.1.4.

Theorem 6.2.1 $\overline{\iota(S)} = \left[KA \cdot \overline{\iota(S \cap A)}\right] \cap \mathcal{F}(\overline{\mathcal{O}})$.

Proof: $\subset$: Let $E = \lim_{n\to\infty} s_n \cdot \overline{\mathcal{O}} \in \overline{\iota(S)} \subset \mathcal{F}(\overline{\mathcal{O}})$. From [115], 2.9, we know that $G = KAH$. Therefore we find elements $k_n \in K, a_n \in A$, and $h_n \in H$ such that $s_n = k_n a_n h_n$. Then $s_n \cdot \overline{\mathcal{O}} = k_n a_n \cdot \overline{\mathcal{O}}$ because $H \cdot \overline{\mathcal{O}} = \overline{\mathcal{O}}$. According to [66], p. 198, the group K is compact since $\mathfrak{k} + i\mathfrak{p}$ is a compact real form of the complex semisimple Lie algebra $\mathfrak{g}_{\mathbb{C}}$ and $G \subset G_{\mathbb{C}}$. Therefore we may assume that $k_0 = \lim_{n\to\infty} k_n$ exists in K and we find that

$$E = \lim_{n\to\infty} k_n a_n \cdot \overline{\mathcal{O}} = k_0 \cdot \lim_{n\to\infty} a_n \cdot \overline{\mathcal{O}}$$

because $k_0^{-1} \cdot E = \lim_{n\to\infty} k_0^{-1} k_n a_n \cdot \overline{\mathcal{O}} = \lim_{n\to\infty} a_n \cdot \overline{\mathcal{O}}$. Thus

$$\emptyset \neq \Gamma(k_o^{-1} \cdot E) = k_o^{-1}\Gamma(E) = \lim_{n\to\infty} \Gamma(a_n \cdot \overline{\mathcal{O}}) = \lim_{n\to\infty} \downarrow a_n = \lim_{n\to\infty} \eta(a_n)$$

since $\Gamma(E) \in \overline{\eta(S)}$ contains S^{-1}. Now we use Lemma 5.6.5 to find an $a \in A$ such that $a_n a^{-1} \in S \cap A$ for all $n \geq n_0$. Then

$$E = k_o a \lim_{n\to\infty} (a^{-1} a_n) \cdot \overline{\mathcal{O}} \in KA \cdot \overline{\iota(S \cap A)} \cap \mathcal{F}(\overline{\mathcal{O}}).$$

$\supset$: Let $E = ka \cdot \lim_{n\to\infty} a_n \cdot \overline{\mathcal{O}} \subset \overline{\mathcal{O}}$. Taking $s \in S^o$, we find that

$$\lim_{n\to\infty} skaa_n \cdot \overline{\mathcal{O}} \subset s \cdot \overline{\mathcal{O}} \subset \mathcal{O}.$$

Therefore we find an $n_0 \in \mathbb{N}$ such that $skaa_n \cdot \mathcal{O} \subset \mathcal{O}$ for $n \geq n_0$. Thus $skaa_n \in S$ and consequently

$$s \cdot E = \lim_{n\to\infty} skaa_n \cdot \overline{\mathcal{O}} \in \overline{\iota(S)}.$$

The fact that $1 \in \overline{S^o}$ now implies that $E \in \overline{\iota(S)}$. □

Remark 6.2.2 Consider the commutative diagram

$$\begin{array}{ccccc} \mathcal{O} & \longrightarrow & N_- \cdot o_{\mathcal{X}} & \longrightarrow & \mathcal{X} \\ \uparrow & & \uparrow & & \\ \Omega_- & \longrightarrow & \mathfrak{n}_- & \simeq & N_- \end{array}$$

The vertical maps are G-equivariant homeomorphisms and $\mathcal{O}$, respectively Ω_-, is relatively compact in the open set $N_- \cdot o_{\mathcal{X}}$, respectively $\mathfrak{n}_-$. Therefore $\mathcal{F}(\overline{\Omega}_-)$ can be identified with the compact set

$$\mathcal{F}(\overline{\mathcal{O}}) = \{F \in \mathcal{F}(\mathcal{X}) \mid F \subset \overline{\mathcal{O}}\}$$

(cf. Lemma B.1.1). In particular, we have an action of $\mathrm{Aff}_{com}(N_-)$ on $\mathcal{F}(\overline{\mathcal{O}})$ and can consider $\mathcal{M}^{\mathcal{O}}$ as a subset of $\mathcal{F}(\overline{\Omega}_-)$. □

Next we consider the set $\overline{\iota(S \cap A)} \subset \mathcal{F}(\overline{\mathcal{O}})$.

Lemma 6.2.3 *Recall the closure $\overline{B^\sharp}$ of $B^\sharp$ in* Aff(N_-) *and the closure S_A^{cpt} of $S \cap A$ in $\overline{B^\sharp}$. Then the following assertions hold:*

1) $\overline{S \cap B^\sharp} = \{\gamma \in \overline{B^\sharp} \mid \gamma \cdot \overline{\mathcal{O}} \subset \overline{\mathcal{O}}\}$ *is a compact semigroup.*

2) $\mathcal{M}_+^{\mathcal{O}} = \overline{\iota(S)} = \overline{S \cap B^\sharp} \cdot \overline{\mathcal{O}}$.

3) $\mathcal{M}_+^{\mathcal{O}} =\subset KA \cdot (S_A^{cpt} \cdot \overline{\mathcal{O}})$.

Proof. 1) In view of Theorem 5.4.8 and Remark 5.7.4, we only have to show that $\overline{S \cap B^\sharp} \supset \{\gamma \in \overline{B^\sharp} \mid \gamma \cdot \overline{\mathcal{O}} \subset \overline{\mathcal{O}}\}$, because the other inclusion is clear. Let $\gamma \in \{\alpha \in \overline{B^\sharp} \mid \alpha \cdot \overline{\mathcal{O}} \subset \overline{\mathcal{O}}\}$ and $\gamma_n \in B^\sharp$ with $\gamma_n \to \gamma$. We choose $X \in c_{\max}^o$ and set $a(t) = \exp(tX)$. Then $e^{\mathrm{ad}\, tX}(\overline{\Omega}_-) \subset \Omega_-$ for every $t > 0$. Therefore $\overline{\Omega}_- \subset \left[e^{-\mathrm{ad}\, tX}(\overline{\Omega}_-)\right]^o$ entails the existence of $n_0 \in \mathbb{N}$ such that $\gamma_n(\overline{\Omega}_-) \subset e^{-\mathrm{ad}\, tX}(\overline{\Omega}_-)$ for all $n \geq n_0$. Then $a(t)\gamma_n \in S \cap B^\sharp$. We conclude that

$$\gamma = \lim_{n\to\infty} a(t)^{-1}[a(t)\gamma_n] \in a(t)^{-1}\overline{S \cap B^\sharp}.$$

Letting $t \to 0$, we find that $\gamma \in \overline{S \cap B^\sharp}$ because $\overline{S \cap B^\sharp}$ is compact.

2) Note that compactness shows

$$\overline{(S \cap B^\sharp) \cdot \overline{\Omega}_-} \subset \overline{\overline{S \cap B^\sharp} \cdot \overline{\Omega}_-} = \overline{S \cap B^\sharp} \cdot \overline{\Omega}_- \subset \overline{(S \cap B^\sharp) \cdot \overline{\Omega}_-}$$

so that

$$\overline{\iota(S)} = \overline{(S \cap B^\sharp) \cdot \overline{\Omega}_-} = \overline{S \cap B^\sharp} \cdot \overline{\Omega}_-.$$

The first equality follows from Theorem 6.1.7.

3) Since S_A^{cpt} is a compact subsemigroup of $\overline{S \cap B^\sharp}$ which contains $S \cap A$, Theorem 6.2.1 and Proposition 5.7.5 entail that

$$\overline{\iota(S)} \subset KA \cdot \overline{\iota(S \cap A)} = KA \cdot (S_A^{cpt} \cdot \overline{\mathcal{O}}). \qquad \square$$

Lemma 6.2.3 shows that information on the orbit structure of $\overline{\iota(S)} = \mathcal{M}_+^{\mathcal{O}}$ may be obtained from $S_A^{cpt} \cdot \overline{\mathcal{O}}$, which is an orbit of a compact abelian semigroup. The topological structure of $\overline{\iota(S)}$ is encoded in the compact semigroup $\overline{S \cap B^\sharp}$.

For a face $F \in \mathrm{Fa}(\mathrm{cone}(\Delta_-))$, we set

$$\overline{\Omega}_F := e_F \cdot \overline{\Omega}_- \tag{6.6}$$

and note that we may view $\overline{\Omega}_F$ as a subset,

$$\overline{\mathcal{O}}_F \subset \mathcal{X} \subset \mathcal{X}_\mathbb{C}. \tag{6.7}$$

Theorem 6.2.4 *The causal Galois connection* $\Gamma\colon \mathcal{F}(\mathcal{X}) \to \mathcal{F}(G)$ *defined in* (6.2) *induces a homeomorphism* $\mathcal{M}_+^{\mathcal{O}} \to \mathcal{M}_+^{cpt}$.

Proof: We note first that

$$\overline{\iota(S)} \subset KA \cdot \{\overline{\Omega}_F \mid F \in \mathrm{Fa}(\mathrm{cone}(\Delta_-))\} \tag{6.8}$$

which is a consequence of Lemma 6.2.3.3) and recall that

$$S_A^{cpt} = \exp(c_{\max}) E(S_A^{cpt}) = \exp(c_{\max})\{e_F \mid F \in \mathrm{Fa}(\mathrm{cone}(\Delta_-))\}$$

from Theorem 5.7.6

Further, we know from Theorem 6.1.7 that $\mathcal{M}_+^{cpt} = \Gamma(\mathcal{M}_+^{\mathcal{O}})$. Since $\Gamma\colon \mathcal{M}^{\mathcal{O}} \to \mathcal{M}^{cpt}$ is a quotient map by Theorem 6.1.6, it only remains to prove that $\Gamma|_{\overline{\iota(S)}}$ is injective. Let $E \in \overline{\iota(S)}$. Then there exists $g \in G$ and $F \in \mathrm{Fa}(\mathrm{cone}\,\Delta_-)$ such that $E = g \cdot \overline{\Omega}_F \subset \overline{\mathcal{O}}$. Thus

$$\hat{\Gamma}\Gamma(E) = \hat{\Gamma}\Gamma(g \cdot \overline{\Omega}_F) = g \cdot \hat{\Gamma}\Gamma(\overline{\Omega}_F) = g \cdot \overline{\Omega}_F = E$$

because $\overline{\Omega}_F = \lim_{t\to\infty} \exp(tX) \cdot \overline{\mathcal{O}}$ for $X \in \mathrm{Int}_{F^\perp}(c_{\max} \cap F^\perp)$ (cf. Theorem 5.7.6, Lemma 6.1.5). This proves the claim. $\square$

6.3 The Stabilizers for $\mathcal{M}_+^{cpt}$

We remain in the situation of Section 6.2. This section is devoted to the study of the stabilizer groups in G of points in $\mathcal{M}_+^{cpt}$.

Let X be an element of $\mathfrak{a}^+$. We determine the connected component of the isotropy group of the element $\lim_{t\to\infty}(\exp tX)\cdot\overline{\mathcal{O}}$ in $\mathcal{M}_+^{\mathcal{O}}$. We endow the space of vector subspaces of $\mathfrak{g}$ with the Vietoris topology which coincides with the usual topology coming from the differentiable structure on the Graßmann manifold.

Proposition 6.3.1 *The limit* $\mathfrak{h}_X := \lim_{t\to\infty} e^{\operatorname{ad} tX}\mathfrak{h}$ *exists and*

$$\mathfrak{h}_X = \mathfrak{z}_\mathfrak{h}(X) + \mathfrak{g}(\Delta^+ \setminus X^\perp).$$

Proof: We write $\mathfrak{g} = \mathfrak{m} + \mathfrak{a} + \bigoplus_{\alpha\in\Delta}\mathfrak{g}_\alpha$ and

$$\operatorname{pr}_\mathfrak{h} : \mathfrak{g} \to \mathfrak{h}, \quad Y \mapsto \frac{1}{2}[Y + \tau(Y)]$$

for the projection of $\mathfrak{g}$ onto $\mathfrak{h}$. Then

$$\mathfrak{h} = \mathfrak{m} + \bigoplus_{\alpha\in\Delta^+} \operatorname{pr}_\mathfrak{h}(\mathfrak{g}_\alpha)$$

because $\operatorname{pr}_\mathfrak{h}(\mathfrak{g}_\alpha) = \operatorname{pr}_\mathfrak{h}(\mathfrak{g}_{-\alpha})$. The subspace $\mathfrak{m} = \mathfrak{z}_\mathfrak{h}(\mathfrak{a})$ is fixed under A and for $Y \in \mathfrak{g}_\alpha, \alpha \in \Delta^+$ we find that

$$\begin{aligned}
\lim_{t\to\infty} e^{\operatorname{ad} tX}\operatorname{pr}_\mathfrak{h}(\mathbb{R}Y) &= \lim_{t\to\infty} e^{\operatorname{ad} tX}\mathbb{R}\,[Y + \tau(Y)] \\
&= \lim_{t\to\infty} \mathbb{R}\left[e^{t\alpha(X)}Y + e^{-t\alpha(X)}\tau(Y)\right] \\
&= \begin{cases} \operatorname{pr}_\mathfrak{h}(\mathbb{R}Y), & \text{if } \alpha(X) = 0 \\ \mathbb{R}Y, & \text{if } \alpha(X) > 0. \end{cases}
\end{aligned}$$

We conclude that $\lim_{t\to\infty} e^{t\operatorname{ad} X}\mathfrak{h}$ exists and equals

$$\mathfrak{h}_X = \mathfrak{m} \oplus \bigoplus_{\alpha\in\Delta^+\cap X^\perp} \operatorname{pr}_\mathfrak{h}(\mathfrak{g}_\alpha) \oplus \bigoplus_{\alpha\in\Delta^+\setminus X^\perp} \mathfrak{g}_\alpha = \mathfrak{z}_\mathfrak{h}(X) \oplus \mathfrak{g}(\Delta^+ \setminus X^\perp). \qquad \square$$

Lemma 6.3.2 $\mathfrak{h}_X \cap (\mathfrak{a} + \mathfrak{n}^\sharp) = \{0\}$ *and* $\mathfrak{g} = \mathfrak{h}_X + \mathfrak{a} + \mathfrak{n}^\sharp$.

Proof: Let $Y = Y_1 + Y_2 = Z_1 + Z_2 \in \mathfrak{h}_X \cap (\mathfrak{a} + \mathfrak{n}^\sharp)$ with $Y_1 \in \mathfrak{z}_\mathfrak{h}(X)$, $Y_2 \in \mathfrak{g}(\Delta^+ \setminus X^\perp)$, $Z_1 \in \mathfrak{a}$, and $Z_2 \in \mathfrak{n}^\sharp = \mathfrak{g}(\Delta^+)^\sharp$. Then

$$Y_1 = Z_1 + Z_2 - Y_2 \in (\mathfrak{a} + \mathfrak{n}^\sharp + \mathfrak{n}) \cap \mathfrak{z}_\mathfrak{g}(X) = \mathfrak{a} \oplus \bigoplus_{\alpha\in\Delta\cap X^\perp} \mathfrak{g}_\alpha.$$

We conclude that $Y_2 = 0$ because the sum $\mathfrak{a} + \mathfrak{n}^\sharp + \mathfrak{n}$ is direct. Then $Y_1 \in (\mathfrak{a} + \mathfrak{n}^\sharp) \cap \mathfrak{h} = \{0\}$ and therefore $Y = 0$. This proves that

$\mathfrak{h}_X \cap (\mathfrak{n}^\sharp + \mathfrak{a}) = \{0\}$. The second assertion follows from $\dim \mathfrak{h} = \dim \mathfrak{h}_X = \dim \mathfrak{g} - (\dim \mathfrak{a} + \dim \mathfrak{n}^\sharp)$. □

Note that Lemma B.1.5 implies

$$H_X \subset H_F := \{g \in G \mid g \cdot \overline{\Omega}_F = \overline{\Omega}_F\}, \tag{6.9}$$

where $H_X = \langle \exp \mathfrak{h}_X \rangle$ because $\overline{\Omega}_F = \lim_{t\to\infty} \exp(tX) \cdot \overline{\mathcal{O}}$.

Proposition 6.3.3 *Let $\mathfrak{h}_F$ be the Lie algebra of H_F. Then*

$$\mathfrak{h}_F = \mathfrak{h}_X + [\mathfrak{h}_F \cap (\mathfrak{a} + \mathfrak{n}^\sharp)]$$

and

$$H_F \cap B^\sharp = \rho_{e_X}^{-1}(e_X) \cap B^\sharp.$$

Proof: The first assertion follows from Lemmas 6.3.2 and B.1.5. Write $e_X = (1, \gamma)$. We know already that $\overline{\Omega}_F$ corresponds to $\gamma(\overline{\Omega}_-) \subset \mathfrak{n}_-$ (cf. Lemma 5.7.2 and Theorem 5.4.8) and that

$$H_F \cap B^\sharp = \{(n_-, \delta) \in B^\sharp \mid (n_-, \delta) \cdot \gamma(\overline{\Omega}_-) = \gamma(\overline{\Omega}_-)\}.$$

Let $(n_-, \delta) \in H_F \cap B^\sharp$. Then

$$e_X(n_-, \delta) \cdot \gamma(\overline{\Omega}_-) = e_X \cdot \gamma(\overline{\Omega}_-) = \gamma(\overline{\Omega}_-).$$

Therefore $e_X(n_-, \delta) \in U\left(e_X(\overline{S \cap B^\sharp})\right) = \{e_X\}$ (Lemma 5.7.15.7)). We conclude that $(n_-, \delta) \in \lambda_{e_X}^{-1}(e_X) \subset \ker \gamma \rtimes N_0^\sharp A$ (Lemma 5.7.10.1)). Now $(a, \delta) \cdot 1 = a \in \gamma(\overline{\Omega}_-) \subset \operatorname{Im} \gamma$ implies that

$$a \in \operatorname{Im} \gamma \cap \ker \gamma = \{1\}.$$

Next we have $\delta\left(\gamma(\overline{\Omega}_-)\right) \subset \gamma(\overline{\Omega}_-)$, which implies that $\delta(\operatorname{Im}\gamma) \subset \operatorname{Im}\gamma$ and $\delta \in \lambda_{e_X}^{-1}(e_X)$ entails that $\gamma\delta|_{\operatorname{Im}\gamma} = \gamma$. Thus $\delta|_{\operatorname{Im}\gamma} = \operatorname{id}_{\operatorname{Im}\gamma}$, so (5.41) implies that

$$H_F \cap B^\sharp \subset \rho_{e_X}^{-1}(e_X).$$

Conversely, $(n_-, \delta)e_X = e_X$ entails that

$$(n_-, \delta) \cdot \gamma(\overline{\Omega}_-) = (n_-, \delta)e_X \cdot \overline{\Omega}_- = e_X \cdot \overline{\Omega}_- = \gamma(\overline{\Omega}_-).$$ □

Corollary 6.3.4 *Suppose that $X \in \mathfrak{a}^+$ is relatively regular. Then*

$$\mathfrak{h}_F = \mathfrak{m} \oplus \Delta_{X,+}^\perp \oplus \mathfrak{g}(\Delta^+ \setminus X^\perp) \oplus \mathfrak{g}(\Delta_0^+ \cap \Delta_{X,+}^\perp)^\sharp \oplus \bigoplus_{\alpha \in \Delta_X^+} p_{\mathfrak{h}}(\mathfrak{g}_\alpha)^\sharp.$$

Proof: Remark 5.7.11, Proposition 6.3.1, and Proposition 6.3.3. □

6.4 The Orbit Structure of $\mathcal{M}^{cpt}$

We retain the hypotheses of Section 6.2. In this section we finally determine the orbit structure of the order compactification of $\mathcal{M}$.

Proposition 6.4.1 *Let $w \in N_{K\cap H}(\mathfrak{a})$ and $X \in c_{\max}$ with*

$$\lim_{t\to\infty} \exp(tX)\cdot\overline{\mathcal{O}} = \overline{\Omega}_F .$$

Then

$$\lim_{t\to\infty} \exp\left(t\,\mathrm{Ad}(w)X\right)\cdot\overline{\mathcal{O}} = w\cdot\overline{\Omega}_F .$$

Proof: We have that

$$\begin{aligned} \lim_{t\to\infty} \exp\left(t\,\mathrm{Ad}(w)\cdot X\right)\cdot\overline{\mathcal{O}} &= \lim_{t\to\infty} w\exp\left(tX\right)w^{-1}\cdot\overline{\mathcal{O}} \\ &= \lim_{t\to\infty} w\exp\left(tX\right)\cdot\overline{\mathcal{O}} = w\cdot\overline{\Omega}_F . \end{aligned}$$

which proves the assertion. □

Corollary 6.4.2 *For every orbit $G\cdot\overline{\Omega}_F$ there exists $X \in \mathfrak{a}^+$ such that*

$$\lim_{t\to\infty} \exp(tX)\cdot\overline{\mathcal{O}} \in G\cdot\overline{\Omega}_F .$$

Proof: Since $c_{\max} = W_0\cdot\mathfrak{a}^+$ this follows from Proposition 6.4.1. □

Let $X \in \mathfrak{a}^+$, $e_X = (1,\gamma)$ the corresponding idempotent of $\overline{S\cap B^\natural}$, and $F = F_X$. We set

$$\mathfrak{n}_{X,-} := \mathfrak{g}(\Delta_{X,-}) = \mathrm{Im}\,\gamma \quad\text{and}\quad N_{X,-} := \exp(\mathfrak{n}_{X,-}). \tag{6.10}$$

Remark 6.4.3 Let L be a group acting on a space $\mathcal{X}$ and $\mathcal{Y}\subset\mathcal{X}$ be a subset. We define

$$\begin{aligned} N_{\mathcal{Y}}(L) &:= \{g\in L \mid g\cdot\mathcal{Y} = \mathcal{Y}\}, \\ Z_{\mathcal{Y}}(L) &:= \{g\in L \mid \forall y\in\mathcal{Y} : g\cdot y = y\}. \end{aligned}$$

Then the subgroup $Z_{\mathcal{Y}}(L)$ of $N_{\mathcal{Y}}(L)$ is normal. In fact, let $y \in \mathcal{Y}$, $h \in N_{\mathcal{Y}}(L)$ and $g \in Z_{\mathcal{Y}}(L)$. But then

$$hgh^{-1}\cdot y = h\cdot\left[g\cdot(h^{-1}\cdot y)\right] = h\cdot(h^{-1}\cdot y) = y$$

because $h^{-1}\cdot y \in \mathcal{Y}$ and this implies the claim. □

Recall from (6.7) that the $\overline{\Omega}_F$ correspond to compact subsets $\overline{\mathcal{O}}_F$ of $\overline{\mathcal{O}}\subset\mathcal{X}$.

Lemma 6.4.4 *The following assertions hold:*

1) $\overline{\Omega}_F$ *is a compact subset of* $N_{X,-}\cdot \mathbf{o}_{\mathcal{X}}$ *with dense interior.*

2) Set $Z_F := \{g \in G \mid \forall x \in \overline{\Omega}_F : g\cdot x = x\}$. *Then*

$$\begin{aligned} Z_F &= \{g\in G \mid \forall x \in N_{X,-}\cdot \mathbf{o} : g\cdot x = x\} \\ &= \bigcap_{g\in N_{X,+}} g\overline{P}_{\max} g^{-1}. \end{aligned}$$

3) The normalizer of Z_F *contains* A, M, *and* H_F.

Proof 1) This follows from the fact that $\overline{\Omega}_F = [e_F\cdot \exp(\overline{\Omega}_-)]\cdot \mathbf{o}_{\mathcal{X}}$ and that $e_F\cdot\exp(\Omega) = \exp(\Omega)\cap N_{X,-}$ is open in $N_{X,-}$ (cf. Lemma 5.7.15).

2) The equality

$$\{g\in G \mid \forall y\in N_{X,-}\cdot \mathbf{o}_{\mathcal{X}} : g\cdot y = y\} = \bigcap_{g\in N_{X,+}} gP_{\max}g^{-1}$$

is clear, since $gP_{\max}g^{-1}$ is the stabilizer of $g\cdot \mathbf{o}_{\mathcal{X}}$. It is also clear that Z_F contains this subgroup. To see that the converse inclusion holds, let $g\in Z_F$. Then the mapping

$$\Phi : N_{X,-} \to \mathcal{X}, \quad n_- \mapsto n_-^{-1} g n_- \cdot \mathbf{o}_{\mathcal{X}}$$

is analytic and constant on the open subset $\exp(\Omega)\cap N_{X,-}$. Therefore it is constant because $N_{X,-}$ is connected. We conclude that $g\cdot(n_-\cdot\mathbf{o}_{\mathcal{X}}) = n_-\cdot\mathbf{o}_{\mathcal{X}}$ for all $n_-\in N_{X,-}$.

3) Lemma 6.4.3 shows that Z_F is a normal subgroup of H_F. That MA normalizes Z_F follows from 2) and the invariance of $N_{X,-}\cdot\mathbf{o}_{\mathcal{X}}$ under MA. To see this invariance, let $g\in MA$. Then $g\cdot\mathbf{o}_{\mathcal{X}} = \mathbf{o}_{\mathcal{X}}$ because $MA\subset P_{\max}$ and therefore

$$gN_{X,-}\cdot\mathbf{o}_{\mathcal{X}} = gN_{X,-}g^{-1}\cdot\mathbf{o}_{\mathcal{X}} = N_{X,-}\cdot\mathbf{o}_{\mathcal{X}}. \qquad \square$$

Lemma 6.4.5 *Let* P_F *be the normalizer of* $\mathfrak{z}_F$. *Then* P_F *is a parabolic subgroup of* G *containing* $P_{\min} = MAN$.

Proof: Let $\mathfrak{p}_F$ be the Lie algebra of P_F. Then, by Lemma 6.4.4.3), $\mathfrak{h}_F + \mathfrak{a} \subset \mathfrak{p}_F$ and $MA\subset P_F$. The subalgebra generated by $\mathrm{pr}_{\mathfrak{h}}(\mathfrak{g}_\alpha)$ and $\mathfrak{a}$ contains $\mathfrak{g}_\alpha + \mathfrak{g}_{-\alpha}$ for every $\alpha\in\Delta_X^+$ with X relatively regular (cf. Corollary 6.3.4). So

$$\mathfrak{m}+\mathfrak{a}+\mathfrak{g}(\Delta^+) = \mathfrak{m}+\mathfrak{a}+\mathfrak{n}\subset\mathfrak{p}_F.$$

Therefore $P_{\min} = MAN \subset P_F$ and consequently P_F is a parabolic subgroup. $\square$

Lemma 6.4.6 *Let $\overline{\Omega}_F$, $\overline{\Omega}_{F'}$ be such that $F \neq F'$ and $F^\perp \cap c_{\max}$, $F'^\perp \cap c_{\max}$ are faces of $c_{\max}$ generated by relatively regular elements X, X' in $\mathfrak{a}^+$. Then $G \cdot \overline{\Omega}_F \neq G \cdot \overline{\Omega}_{F'}$.*

Proof: Suppose that $G \cdot \overline{\Omega}_F = G \cdot \overline{\Omega}_{F'}$. Then there exists $g \in G$ such that $g \cdot \overline{\Omega}_F = \overline{\Omega}_{F'}$. Therefore $H_{F'} = gH_Fg^{-1}$ and $gZ_Fg^{-1} = Z_{F'}$. This implies also that $gP_Fg^{-1} = P_{F'}$. But P_F and $P_{F'}$ are parabolic subgroups of G containing $P_{\min}$. Hence they are equal ([168], p. 46). Moreover, P_F is its own normalizer. So $g \in P_F$ and $Z_{F'} = Z_F$. Thus

$$\Delta^\perp_{X,+} = \mathfrak{z}_F \cap \mathfrak{a} = \mathfrak{z}_{F'} \cap \mathfrak{a} = \Delta^\perp_{X',+}.$$

Then the faces $\Delta^\perp_{X,+} \cap c_{\max}$ and $\Delta^\perp_{X',+} \cap c_{\max}$ are equal and $F = F'$ follows from Remark 2.1.7. □

Proposition 6.4.1 and Lemma 6.4.6 now yield the following.

Theorem 6.4.7 *The G-orbits of $\overline{\Omega}_F$ and $\overline{\Omega}_{F'}$ for $F, F' \in \mathrm{Fa}(\mathrm{cone}(\Delta_-))$ agree if and only if F and F' are conjugate under the Weyl group W_0.*

Lemma 6.4.8 *1) $\overline{G \cdot \overline{\Omega}_-} \subset KA \cdot \{\overline{\Omega}_F \mid F \in \mathrm{Fa}(\mathrm{cone}(\Delta_-))\}$.*

2) For all $F \in \mathrm{Fa}(\mathrm{cone}(\Delta_-))$ we have $G \cdot \overline{\Omega}_F = KA \cdot \overline{\Omega}_F$.

Proof: 1) In view of Theorem 5.7.6.2) and Theorem 5.7.6.4) this is exactly what was shown in the first part of the proof of Theorem 6.2.1.

2) Using Corollary 6.4.2 and 1), we find

$$G \cdot \overline{\Omega}_F \subset \overline{G \cdot \overline{\Omega}_-} = KA \cdot \{\overline{\Omega}_{F'} \mid F' \in \mathrm{Fa}(\mathrm{cone}(\Delta_-))\}.$$

If $g \cdot \overline{\Omega}_F \in KA \cdot \overline{\Omega}_{F'}$, then we have $G \cdot \overline{\Omega}_F = G \cdot \overline{\Omega}_{F'}$ so Lemma 6.4.6 implies that $F = w \cdot F'$ for some element $w \in W_0$. But $W_0 \cdot A = A$ and the action of W_0 on A is induced by conjugation with elements from K, so

$$KA \cdot \overline{\Omega}_F = KA \cdot \overline{\Omega}_{w \cdot F'} = KAw \cdot \overline{\Omega}_{F'} = Kw(w^{-1}Aw) \cdot \overline{\Omega}_{F'} = KA \cdot \overline{\Omega}_{F'}.$$

It follows in particular that $g \cdot \overline{\Omega}_F \in KA \cdot \overline{\Omega}_F$. □

Lemma 6.4.9 *Suppose that $G \cdot \Gamma(\overline{\Omega}_{F'}) \subset \overline{G \cdot \Gamma(\overline{\Omega}_F)}$ for some $F', F \in \mathrm{Fa}(\mathrm{cone}(\Delta_-))$. Then there exists a Weyl group element $w \in W_0$ with $w \cdot F' \subset F$, i.e., $w \cdot F'$ is a face of F.*

Proof: The hypothesis means that there exists a sequence of elements $g_n \in G$ such that $\overline{\Omega}_{F'} = \lim_{n\to\infty} g_n \cdot \overline{\Omega}_F$. Lemma 6.4.8 shows that we can find $k_n \in K$ and $a_n \in A$ satisfying $\overline{\Omega}_{F'} = \lim_{n\to\infty} k_n a_n \cdot \overline{\Omega}_F$. Without loss of generality we may assume that the k_n converge to some $k_o \in K$, so that $k_o^{-1} \cdot \overline{\Omega}_{F'} = \lim_{n\to\infty} a_n \cdot \overline{\Omega}_F \in \overline{A \cdot \overline{\Omega}_F}$. This implies that $e_F(k_o^{-1} \cdot \overline{\Omega}_{F'}) = k_o^{-1} \cdot \overline{\Omega}_{F'}$. On the other hand, we know that $\Gamma(k_o^{-1} \cdot \overline{\Omega}_F) = k_o^{-1} \cdot \Gamma(\overline{\Omega}_F) \neq \emptyset$. Therefore Lemma 5.6.5 tells us that there exists an $a_o \in A$ with $k_o^{-1}\overline{\Omega}_{F'} \in a_o\overline{\Omega}_{F'} \subset a_o(S \cap A)\overline{\mathcal{O}}$. Thus

$$k_o^{-1} \cdot \overline{\Omega}_{F'} \in e_F A S_A^{cpt} \cdot \overline{\mathcal{O}} = A S_A^{cpt} \cdot \overline{\Omega}_F = \bigcup_{F'' \subset F} A \cdot \overline{\Omega}_{F''}.$$

Pick $F'' \subset F$ with $k_o^{-1} \cdot \overline{\Omega}_{F'} \in A \cdot \overline{\Omega}_{F''}$. Then $G \cdot \overline{\Omega}_{F'} = G \cdot \overline{\Omega}_{F''}$ entails that $F'' \in W_0 \cdot F'$ (cf. Theorem 6.4.7). □

Lemma 6.4.10 *Let $F \in \mathrm{Fa}(\mathrm{cone}(\Delta_+))$. Then the following assertions are equivalent.*

1) $F = \tilde{F} \cap \mathrm{cone}(\Delta_-)$ *for an* $\tilde{F} \in \mathrm{Fa}(\mathrm{cone}(-\Delta^+))$.

2) *There exists an $X \in \mathfrak{a}^+$ such that* $F = X^\perp \cap \mathrm{cone}(\Delta_-)$.

3) *There exists a relatively regular $X \in \mathfrak{a}^+$ such that*

$$F = X^\perp \cap \mathrm{cone}(\Delta_-).$$

4) $F^\perp \cap c_{\max}$ *is generated by a relatively regular element of* $\mathfrak{a}^+$.

Proof. 1) $\Rightarrow$ 2): $\tilde{F} \in \mathrm{Fa}(\mathrm{cone}(-\Delta^+))$ means that $\tilde{F} = \mathrm{op}(E)$ for some $E \in \mathrm{Fa}(\mathrm{cone}(-\Delta^+)^*) = \mathrm{Fa}(-\mathfrak{a}^+)$. If $X \in \mathrm{algint}(E)$, then $\tilde{F} = X^\perp \cap \mathrm{cone}(-\Delta^+)$ and hence $F = \tilde{F} \cap \mathrm{cone}(\Delta_-) = X^\perp \cap \mathrm{cone}(\Delta_-)$.

2) $\Rightarrow$ 3) follows from Lemma 5.7.8.2).

3) $\Rightarrow$ 4): If $F = X^\perp \cap \mathrm{cone}(\Delta_-)$, then

$$\begin{aligned} c_{\max} \cap F^\perp &= c_{\max} \cap \mathfrak{a}^{[\mathbb{R}X + \mathrm{cone}(\Delta_-]^*)} \\ &= c_{\max} \cap \mathfrak{a}^{[\mathbb{R}X + c_{\max}]} = c_{\max} \cap [\mathbb{R}^+ X - c_{\max}], \end{aligned}$$

where $\mathfrak{a}^C$ for cone C in $\mathfrak{a}$ is the edge of the cone, is generated by X (cf. Section 2.1 for the notation).

4) $\Rightarrow$ 1): If $F^\perp \cap c_{\max}$ is generated by $X \in \mathfrak{a}^+$, then $F = X^\perp \cap \mathrm{cone}(\Delta_-)$. Let $\tilde{F} := X^\perp \cap \mathrm{cone}(-\Delta^+) \in \mathrm{Fa}(\mathrm{cone}(-\Delta^+))$. Then $\tilde{F} \cap \mathrm{cone}(\Delta_-) = F$. □

Remark 6.4.11 For all $F \in \mathrm{Fa}(\mathrm{cone}(\Delta_-))$ one can find a $\gamma \in \mathcal{W}$ such that $\gamma(F)$ satisfies the conditions from Lemma 6.4.10. If $\gamma(F)$ and $\gamma'(F)$ both satisfy these conditions, then Lemma 6.4.6 shows that $\gamma(F) = \gamma'(F)$. □

The space Fa(cone(Δ_-)) carries a natural partial order $\preceq$. It is given by inclusion. The corresponding strict order will be denoted by $\prec$. On the other hand, we introduce an ordering on the set of G-orbits in $\mathcal{M}^{cpt}$ via

$$G\cdot m \prec G\cdot m' \quad :\Leftrightarrow \quad \overline{G\cdot m}\subset\overline{G\cdot m'},\quad \overline{G\cdot m}\neq\overline{G\cdot m'};$$

$$G\cdot m \preceq G\cdot m' \quad :\Leftrightarrow \quad \begin{cases} G\cdot m\prec G\cdot m' \\ \text{or} \\ G\cdot m = G\cdot m'. \end{cases}$$

It is clear that the relation $\preceq$ on $G\backslash\mathcal{M}^{cpt}$ is reflexive and transitive. For the antisymmetry, note that $G\cdot m\prec G\cdot m'\prec G\cdot m$ implies that $\overline{G\cdot m}$ is strictly contained in $\overline{G\cdot m'}$, which in turn is contained in $\overline{G\cdot m}$. This shows that $\overline{G\cdot m}=\overline{G\cdot m'}$, in contradiction to the hypothesis. Thus $\preceq$ is a partial order. Note that $G\cdot m\prec G\cdot m'$ means that $m\in\overline{G\cdot m'}$ and the G-orbit of m is not dense in $\overline{G\cdot m'}$.

Lemma 6.4.12 *Let $F'\subset F$ be faces of $c^*_{\max}$. Then $\Gamma(\overline{\Omega}_{F'})\in\overline{G\cdot\Gamma(\overline{\Omega}_F)}$.*

Proof: $F\subset F'$ implies that

$$e_Fe_{F'}=e_{F'}e_F=e_{F'}=\lim_{t\to\infty}\exp tX'$$

for $X'\in\operatorname{Int}_{(F')^\perp}(c_{\max}\cap(F')^\perp)$ (cf. Theorem 5.7.6). Using Lemma 5.7.2, we can calculate

$$\begin{aligned}\overline{\Omega}_{F'}=e_{F'}\overline{\Omega}_- &= \lim_{t\to\infty}\exp tX'\cdot e_F\overline{\Omega}_- \\ &= \lim_{t\to\infty}\exp tX'\cdot\overline{\Omega}_F\in\overline{G\cdot\overline{\Omega}_F}.\end{aligned}$$

Thus $\Gamma(\overline{\Omega}_{F'})\in\overline{G\cdot\Gamma(\overline{\Omega}_F)}$. □

For an element $E\in\mathcal{M}_+^{\mathcal{O}}$ we define the *degree* $d(E):=\dim\ker e_F$, where $E\in G\cdot\overline{\Omega}_F$. We note that $e_F:\mathfrak{n}_-\to\mathfrak{n}_-$ is a projection and $m:=\dim\mathfrak{n}_- - \dim\ker e_F$ agrees with the topological dimension of the set $E\cong\overline{\Omega}_F\cong e_F\cdot\overline{\Omega}_-$, which is an m-dimensional compact convex set. Upon identification of $\mathcal{M}_+^{\mathcal{O}}$ with $\mathcal{M}_+^{cpt}$ via Γ, we have the function d also on $\mathcal{M}_+^{cpt}$.

Proposition 6.4.13 *Suppose that $F,F'\in$ Fa(cone(Δ_-)) and F' is strictly contained in F. Then*

$$G\cdot\Gamma(\overline{\Omega}_{F'})\prec G\cdot\Gamma(\overline{\Omega}_F).$$

Proof: Recall that $-c^*_{\max}$ is a polyhedral cone spanned by the elements of Δ_-. Thus the hypothesis shows that $F' \cap \Delta_-$ is strictly contained in $F \cap \Delta_-$. But then we have

$$d(\overline{\Omega}_{F'}) = \dim \ker e_{F'} < \dim \ker e_F = d(\overline{\Omega}_F).$$

So $\overline{\Omega}_{F'} \in \{y \in \mathcal{M}^O_+ \mid d(y) \geq d(\overline{\Omega}_F) + 1\}$ and hence $G \cdot \overline{\Omega}_{F'} \cap \mathcal{M}^O_+ \subset \{y \in \mathcal{M}^O_+ \mid d(y) \geq d(\overline{\Omega}_F) + 1\}$ or

$$G \cdot \Gamma(\overline{\Omega}_{F'}) \cap \mathcal{M}^{cpt}_+ \subset \{x \in \mathcal{M}^{cpt}_+ \mid d(x) \geq d\left(\Gamma(\overline{\Omega}_F)\right) + 1\}. \tag{6.11}$$

But since $\Gamma(\overline{\Omega}_F)$ is not contained in the right-hand side of this inclusion, it is also not contained in $G \cdot \Gamma(\overline{\Omega}_{F'}) \cap \mathcal{M}^{cpt}_+$, whence $G \cdot \Gamma(\overline{\Omega}_{F'}) \neq G \cdot \Gamma(\overline{\Omega}_F)$ and the assertion follows from Lemma 6.4.12. □

Theorem 6.4.14 *Let $\mathcal{F} := \{F \in \mathrm{Fa}(\mathrm{cone}(\Delta_-)) \mid \exists \tilde{F} \in \mathrm{Fa}(\mathrm{cone}(-\Delta^+)) : \tilde{F} \cap \mathrm{cone}(\Delta_-) = F\}$. Then the mapping*

$$\Upsilon : \mathcal{F} \to (\mathcal{M}^{cpt} \setminus \{\emptyset\})/G, \quad F \mapsto G \cdot \overline{\Omega}_F$$

is an order isomorphism.

Proof: The surjectivity of Υ follows from Lemma 6.4.8 and the injectivity from Lemma 6.4.6 and Remark 6.4.11. According to to Proposition 6.4.13 it is order-preserving, and that the inverse is also order-preserving is a consequence of Lemma 6.4.9. □

We describe the structure of the lattice $\mathcal{F}$ in more detail. Note first that $\mathcal{F}$ is isomorphic to $-\mathcal{F}$. Since $\mathrm{cone}(\Delta^+)$ is polyhedral and Δ^+ is generated by a set $\Upsilon = \{\alpha_0, \ldots, \alpha_l\}$ of linearly independent elements (cf. Lemma 5.5.10), it is clear that the mapping

$$\mathrm{Fa}(\mathrm{cone}(\Delta^+)) \to 2^\Upsilon, \quad F \mapsto F \cap \Upsilon$$

is an order isomorphism. So we have to determine the image of the mapping

$$2^\Upsilon \to \mathcal{F}, \quad D \mapsto \mathrm{cone}(D) \cap \mathrm{cone}(\Delta_+).$$

To each subset $D \subset \Upsilon$ there corresponds a subgraph of the Dynkin graph of Δ. Recall that Υ contains only one noncompact root α_0. Let $D_0 \subset D$ correspond to the connected component of α_0 in the Dynkin graph. Then

$$(\mathrm{span} D) \cap \Delta = [(\mathrm{span} D_0) \cap \Delta] \dot{\cup} [\mathrm{span}(D \setminus D_0) \cap \Delta]$$

and therefore

$$\mathrm{cone}(D) \cap \mathrm{cone}(\Delta_+) = \mathrm{cone}(D_0) \cap \mathrm{cone}(\Delta_+).$$

Hence it suffices to consider subsets $D \subset \Upsilon$ such that $\alpha_0 \in D$ and the corresponding subgraph is connected. The resulting lattices have been listed in [55].

Example 6.4.15 In the situation of the SL(2, $\mathbb{R}$) Example 5.7.7 we have

$$\overline{\Omega}_{F_1} = \{0\} \quad \text{and} \quad \overline{\Omega}_{F_2} = \overline{\Omega}_-$$

and $\mathcal{F} = \{F_1, F_2\}$. In particular, $\mathcal{M}^{cpt} \setminus \{\emptyset\}$ has two SL(2, $\mathbb{R}$) orbits. □

Corollary 6.4.16 $\overline{G \cdot \Gamma(\overline{\Omega}_F)} = \bigcup_{F' \subset F} G \cdot \Gamma(\overline{\Omega}_{F'})$.

Proof. $\overline{G \cdot \Gamma(\overline{\Omega}_F)}$ is the disjoint union of G-orbits. For each of these G-orbits $G \cdot x$ we have $G \cdot x \subset \overline{G \cdot \Gamma(\overline{\Omega}_F)}$, so Lemma 6.4.9 implies that $G \cdot x = G \cdot \Gamma(\overline{\Omega}_{F'})$ for some face F' of F. This proves the inclusion $\subset$, whereas the reverse inclusion is clear from Corollary 6.4.2. □

Theorem 6.4.17 *For $F \in \mathrm{Fa}(\mathrm{cone}(\Delta_-))$ we have*

$$g \cdot \overline{\Omega}_F \in \mathcal{M}_+^{O} \quad \Leftrightarrow \quad g \cdot \overline{\Omega}_F \subset \overline{\Omega}_-.$$

Proof. Recall that Theorem 5.7.6.3) shows that $\overline{\Omega}_F = \lim_{t \to \infty} \exp(tX) \cdot \overline{\Omega}_-$ for $X \in \mathrm{Int}_{F^\perp}(c_{\max} \cap F^\perp)$.

$\Leftarrow$: If $g \cdot \overline{\Omega}_F \in \mathcal{M}_+^{O}$ and $s \in S^o$, then we have $sg \cdot \overline{\Omega}_F \in (\mathcal{M}_+^{O})^o$ by Theorem 6.1.7 and hence $sg \exp(tX) \cdot \overline{\Omega}_- \in \mathcal{M}_+^{O}$ for large t. We choose $s_n \in S^o$ with $s_n \to 1$ and t_n such that $t_n \to \infty$ and

$$s_n g \exp(t_n X) \cdot \overline{\Omega}_- \in \mathcal{M}_+^{O}. \tag{6.12}$$

Now the calculation in the proof of Lemma 6.1.5 shows that (6.12) holds precisely when $\overline{\eta}(s_n g \exp(t_n X) H) \in \mathcal{M}_+^{cpt}$ because of Theorem 6.1.7. But this in turn is equivalent to $s_n g \exp t_n X \in S$ and hence to $s_n g \exp(t_n X) \cdot \overline{\Omega}_- \subset \overline{\Omega}_-$. If we let n tend to ∞ we obtain $g \cdot \overline{\Omega}_F \subset \overline{\Omega}_-$.

$\Rightarrow$: This time we choose $s_n \in S^o$ and $t_n \in \mathbb{R}$ with $s_n \to 1$, $t_n \to \infty$ and $s_n g \exp(t_n X) \cdot \overline{\Omega}_- \subset \overline{\Omega}_-$. Reading the argument in the first part of the proof backwards, we find $s_n g \exp(t_n X) \cdot \overline{\Omega}_- \in \mathcal{M}_+^{O}$ and upon taking the limit, $g \cdot \overline{\Omega}_F \in \mathcal{M}_+^{O}$. □

Corollary 6.4.18 *Let $F \in \mathrm{Fa}(\mathrm{cone}(\Delta_-))$. Then*

$$\overline{G \cdot \Gamma(\overline{\Omega}_F) \cap \mathcal{M}_+^{cpt}} = \overline{G \cdot \Gamma(\overline{\Omega}_F)} \cap \mathcal{M}_+^{cpt}.$$

Proof: The inclusion $\subset$ is obvious. To show the reverse inclusion, let F' be a face of F and note that $\Gamma(\overline{\Omega}_{F'}) = \lim_{t\to\infty} \exp tX' \cdot \Gamma(\overline{\Omega}_F)$ for $X' \in \mathrm{Int}_{F'^{\perp}}(c_{\max} \cap F'^{\perp})$. If $x \in G\cdot\Gamma(\overline{\Omega}_{F'}) \cap \mathcal{M}_+^{cpt}$, then there exists a $g \in G$ such that $x = g \cdot \Gamma(\overline{\Omega}_{F'}) \in \mathcal{M}_+^{cpt}$ and Theorem 6.4.17 implies that $g \cdot \overline{\Omega}_{F'} \subset \overline{\Omega}_-$. Choose $s_n \in S^o$ and $t_n > 0$ with $s_n \to 1$, $t_n \to \infty$ and $s_n g \exp t_n X' \cdot \overline{\Omega}_F \subset \overline{\Omega}_-$. Then Theorem 6.4.17 shows that $s_n g \exp t_n X' \cdot \Gamma(\overline{\Omega}_F) \in \mathcal{M}_+^{cpt}$ and in the limit we find $g \cdot \Gamma(\overline{\Omega}_{F'}) \in \mathcal{M}_+^{cpt}$. In other words,

$$G \cdot \Gamma(\overline{\Omega}_{F'}) \cap \mathcal{M}_+^{cpt} \subset \overline{G \cdot \Gamma(\overline{\Omega}_F) \cap \mathcal{M}_+^{cpt}}.$$

Finally, Corollary 6.4.16 shows that

$$\mathcal{M}_+^{cpt} \cap \overline{G \cdot \Gamma(\overline{\Omega}_F)} = \bigcup_{F' \subset F} [G \cdot \Gamma(\overline{\Omega}_{F'}) \cap \mathcal{M}_+^{cpt}] \subset \overline{G \cdot \Gamma(\overline{\Omega}_F) \cap \mathcal{M}_+^{cpt}} \qquad \square$$

For $k \in \mathbb{N}_o$ we set

$$(\mathcal{M}_+^{cpt})_k := \{x \in \mathcal{M}_+^{cpt} \mid d(x) = k\}. \tag{6.13}$$

Corollary 6.4.19 *1)* $(\mathcal{M}_+^{cpt})_k = \bigcup_{d(\overline{\Omega}_F)=k} (G \cdot \Gamma(\overline{\Omega}_F) \cap \mathcal{M}_+^{cpt})$.

2) Each $G \cdot \Gamma(\overline{\Omega}_F)$ *with* $d(\overline{\Omega}_F) = k$ *is open in* $(\mathcal{M}_+^{cpt})_k$ *with respect to the induced topology.*

Proof: 1) is obvious.

2) It suffices to show that each $G \cdot \Gamma(\overline{\Omega}_F) \cap \mathcal{M}_+^{cpt}$ is closed in $\mathcal{M}_+^{cpt}$, since the union is finite. Using Corollary 6.4.16 and Corollary 6.4.18, we calculate

$$\begin{aligned} \overline{G \cdot \Gamma(\overline{\Omega}_F) \cap \mathcal{M}_+^{cpt}} \cap (\mathcal{M}_+^{cpt})_k &= \overline{G \cdot \Gamma(\overline{\Omega}_F)} \cap (\mathcal{M}_+^{cpt})_k \\ &= \bigcup_{F' \subset F} [G \cdot \Gamma(\overline{\Omega}_{F'}] \cap (\mathcal{M}_+^{cpt})_k) \\ &= G \cdot \Gamma(\overline{\Omega}_F) \cap \mathcal{M}_+^{cpt} \end{aligned}$$

since $d(\overline{\Omega}_{F'}) < d(\overline{\Omega}_F)$ for F' strictly contained in F. $\square$

6.5 The Space SL(3, R)/ SO(2, 1)

We conclude this chapter with a detailed discussion of the space $\mathcal{M} = \mathrm{SL}(3,\mathbb{R})/\mathrm{SO}(2,1)$. Thus we let $G = \mathrm{SL}(3,\mathbb{R})$, $\mathfrak{g} = \mathfrak{sl}(3,\mathbb{R})$. As a Cartan involution $\mathfrak{g}$ we use $\theta : \mathfrak{g} \to \mathfrak{g}, X \mapsto -{}^tX$, which yields $K = \mathrm{SO}(3)$, $\mathfrak{k} = \mathfrak{so}(3,\mathbb{R})$ and

$$\mathfrak{p} = \left\{ \begin{pmatrix} A & c \\ {}^tc & -\operatorname{Tr} A \end{pmatrix} \middle| \, {}^tA = A, c \in \mathbb{R}^2 \right\}.$$

As a maximal abelian subspace $\mathfrak{a}$ of $\mathfrak{p}$ we choose

$$\mathfrak{a} = \left\{ \begin{pmatrix} r+s & 0 & 0 \\ 0 & -r+s & 0 \\ 0 & 0 & -2s \end{pmatrix} \middle| r, s \in \mathbb{R} \right\},$$

so that

$$A = \exp \mathfrak{a} = \left\{ \begin{pmatrix} a_1 & 0 & 0 \\ 0 & a_2 & 0 \\ 0 & 0 & a_3 \end{pmatrix} \middle| a_i > 0, a_1 a_2 a_3 = 1 \right\}.$$

Moreover, we have

$$\begin{aligned} M &= Z_K(\mathfrak{a}) \\ &= \left\{ I_3, \begin{pmatrix} -I_2 & 0 \\ 0 & 1 \end{pmatrix}, \begin{pmatrix} I_{1,1} & 0 \\ 0 & -1 \end{pmatrix}, \begin{pmatrix} 1 & 0 \\ 0 & -I_2 \end{pmatrix} \right\}. \end{aligned}$$

Now consider the involution $\tau : \mathfrak{g} \to \mathfrak{g}$ given by

$$\begin{aligned} X = \begin{pmatrix} A & b \\ {}^t c & d \end{pmatrix} &\mapsto \begin{pmatrix} -{}^t A & c \\ {}^t b & -d \end{pmatrix} \\ &= \begin{pmatrix} 1 & 0 \\ 0 & -1 \end{pmatrix} \theta(X) \begin{pmatrix} 1 & 0 \\ 0 & -1 \end{pmatrix} \end{aligned}$$

and its global counterpart, which is given by the same formula. Then

$$\begin{aligned} H &= \mathrm{SO}(2,1) \\ &= \left\{ \begin{pmatrix} A & b \\ {}^t c & d \end{pmatrix} \in \mathrm{SL}(3,\mathbb{R}) \middle| \begin{matrix} b, c \in \mathbb{R}^2, \\ d \in \mathbb{R}, {}^t AA - c{}^t c = 1, \\ {}^t Ab = dc, \|b\|^2 - |d|^2 = -1 \end{matrix} \right\}, \\ \mathfrak{h} &= \mathfrak{so}(2,1) = \left\{ \begin{pmatrix} A & b \\ {}^t b & 0 \end{pmatrix} \middle| {}^t A = -A, b \in \mathbb{R}^2 \right\}, \\ \mathfrak{q} &= \left\{ \begin{pmatrix} A & c \\ -{}^t c & -\operatorname{Tr} A \end{pmatrix} \middle| {}^t A = A, c \in \mathbb{R}^2 \right\}. \end{aligned}$$

The c-dual objects are

$$\begin{aligned} G^c &= \mathrm{SU}(2,1) \\ &= \left\{ \begin{pmatrix} A & b \\ {}^t c & d \end{pmatrix} \in \mathrm{SL}(3,\mathbb{C}) \middle| \begin{matrix} A \in \mathrm{M}(2,\mathbb{C}), \\ b, c \in \mathbb{C}^2, d \in \mathbb{C}, \\ A^* A - \bar{c}{}^t c = 1 \\ A^* b = d\bar{c}, \|b\|^2 - |d|^2 = -1 \end{matrix} \right\}, \end{aligned}$$

$$\begin{aligned}
\mathfrak{g}^c = \mathfrak{h} + i\mathfrak{q} &= \mathfrak{su}(2,1) \\
&= \left\{ \begin{pmatrix} A + iB & b + ic \\ {}^t b - i {}^t c & -i \operatorname{Tr} B \end{pmatrix} \middle| {}^t A = -A, {}^t B = B \right\}, \\
K^c &= \left\{ \begin{pmatrix} A & 0 \\ 0 & \det A^{-1} \end{pmatrix} \middle| A \in \mathrm{U}(2) \right\}, \\
\mathfrak{k}^c &= \left\{ \begin{pmatrix} A & 0 \\ 0 & -\operatorname{Tr} A \end{pmatrix} \middle| A^* = -A \right\}.
\end{aligned}$$

We write $\mathfrak{a} = \mathbb{R}H_1 + \mathbb{R}H_2$ with

$$H_1 := \begin{pmatrix} 1 & 0 & 0 \\ 0 & -1 & 0 \\ 0 & 0 & 0 \end{pmatrix} \quad \text{and} \quad H_2 := \begin{pmatrix} 1 & 0 & 0 \\ 0 & 1 & 0 \\ 0 & 0 & -2 \end{pmatrix}$$

and note that $X^0 = \frac{1}{3}H_2$ is a cone-generating element in $\mathfrak{q}_p$. As a system of positive roots we choose

$$\Delta^+ = \{\alpha_{13}, \alpha_{23}, \alpha_{12}\}$$

and note that

$$\begin{aligned}
\alpha_{12}(rH_1 + sH_2) &= 2r \\
\alpha_{23}(rH_1 + sH_2) &= -r + 3s \\
\alpha_{13}(rH_1 + sH_2) &= r + 3s,
\end{aligned}$$

so that

$$\Delta_+ = \{\alpha_{13}, \alpha_{23}\}, \quad \Delta_0^+ = \{\alpha_{12}\}.$$

The corresponding root vectors are

$$\begin{aligned}
E_{12} &= \begin{pmatrix} 0 & 1 & 0 \\ 0 & 0 & 0 \\ 0 & 0 & 0 \end{pmatrix}, \\
E_{23} &= \begin{pmatrix} 0 & 0 & 0 \\ 0 & 0 & 1 \\ 0 & 0 & 0 \end{pmatrix}, \\
E_{13} &= \begin{pmatrix} 0 & 0 & 1 \\ 0 & 0 & 0 \\ 0 & 0 & 0 \end{pmatrix}.
\end{aligned}$$

Therefore we have

$$\mathfrak{n}_+ = \left\{ \begin{pmatrix} 0 & 0 & x \\ 0 & 0 & y \\ 0 & 0 & 0 \end{pmatrix} \middle| x, y \in \mathbb{R} \right\},$$

$$\begin{aligned}
N_+ &= \left\{ \begin{pmatrix} 1 & 0 & x \\ 0 & 1 & y \\ 0 & 0 & 1 \end{pmatrix} \middle| x, y \in \mathbb{R} \right\}, \\
\mathfrak{n}_0 &= \left\{ \begin{pmatrix} 0 & x & 0 \\ 0 & 0 & 0 \\ 0 & 0 & 0 \end{pmatrix} \middle| x \in \mathbb{R} \right\}, \\
N_0 &= \left\{ \begin{pmatrix} 1 & x & 0 \\ 0 & 1 & 0 \\ 0 & 0 & 1 \end{pmatrix} \middle| x \in \mathbb{R} \right\}, \\
N &= \left\{ \begin{pmatrix} 1 & * & * \\ 0 & 1 & * \\ 0 & 0 & 1 \end{pmatrix} \right\}
\end{aligned}$$

and

$$N^\sharp = \left\{ \begin{pmatrix} 1 & 0 & 0 \\ * & 1 & 0 \\ * & * & 1 \end{pmatrix} \right\}.$$

Moreover,

$$\mathfrak{a}_{\mathbb{C}} = \left\{ \begin{pmatrix} a & 0 & 0 \\ 0 & b & 0 \\ 0 & 0 & -a-b \end{pmatrix} \middle| a, b \in \mathbb{C} \right\}$$

is a Cartan algebra of $\mathfrak{g}_{\mathbb{C}} = \mathfrak{sl}(3, \mathbb{C})$. This leads to

$$\begin{aligned}
(P^c)^+ &= \left\{ \begin{pmatrix} 1 & b \\ 0 & 1 \end{pmatrix} \middle| b \in \mathbb{C}^2 \right\}, \qquad (\mathfrak{p}^c)^+ = \left\{ \begin{pmatrix} \mathbf{0} & b \\ 0 & 0 \end{pmatrix} \middle| b \in \mathbb{C}^2 \right\}, \\
(P^c)^- &= \left\{ \begin{pmatrix} 1 & 0 \\ {}^tc & 1 \end{pmatrix} \middle| c \in \mathbb{C}^2 \right\}, \qquad (\mathfrak{p}^c)^- = \left\{ \begin{pmatrix} \mathbf{0} & 0 \\ {}^tc & 0 \end{pmatrix} \middle| c \in \mathbb{C}^2 \right\}, \\
K^c_{\mathbb{C}} &= \left\{ \begin{pmatrix} A & 0 \\ 0 & \det A^{-1} \end{pmatrix} \middle| A \in \mathrm{GL}(2, \mathbb{C}) \right\},
\end{aligned}$$

and

$$K^c_{\mathbb{C}}(P^c)^+ = \left\{ \begin{pmatrix} A & c \\ 0 & \det A^{-1} \end{pmatrix} | A \in \mathrm{GL}(2, \mathbb{C}), c \in \mathbb{C}^2 \right\}.$$

Now we can write down the maximal parabolic,

$$P_{\max} = G \cap K^c_{\mathbb{C}}(P^c)^+ = \left\{ \begin{pmatrix} A & b \\ 0 & \det A^{-1} \end{pmatrix} \middle| A \in \mathrm{GL}(2, \mathbb{R}), b \in \mathbb{R}^2 \right\},$$

and identify the corresponding flag manifold as

$$G/P_{\max} = K/K \cap P_{\max} = \mathrm{SO}(3)/\mathrm{O}(2) = \mathbb{RP}^2.$$

The domain $(\Omega_-)_{\mathbb{C}}$ is given by

$$(\Omega_-)_{\mathbb{C}} = \left\{ \begin{pmatrix} 1 & 0 \\ {}^t z & 1 \end{pmatrix} \middle| z \in \mathbb{C}^2 \right\}$$

and the action of $G_{\mathbb{C}} = \mathrm{SL}(3, \mathbb{C})$ on $(\Omega_-)_{\mathbb{C}}$ can be described by

$$\begin{pmatrix} A & b \\ {}^t c & d \end{pmatrix} \cdot z = ({}^t c + d^t z)(A + b^t z)^{-1}$$

because

$$\begin{pmatrix} A & b \\ {}^t c & d \end{pmatrix} \cdot \begin{pmatrix} 1 & 0 \\ {}^t z & 1 \end{pmatrix} = \begin{pmatrix} 1 & 0 \\ ({}^t c + d^t z)(A + b^t z)^{-1} & 1 \end{pmatrix} \begin{pmatrix} A' & b' \\ 0 & d' \end{pmatrix}.$$

Next we compute the H-orbit of $0 \in \mathfrak{n}_-$. Since $(H \cap K) \cdot 0 = \{0\}$, we have to consider only symmetric elements $g = \begin{pmatrix} A & b \\ {}^t b & d \end{pmatrix} \in H$. Since $H = -H$, we may also assume that $d > 0$. Then $d = \sqrt{1 + \|b\|^2}$ and ${}^t A = A$ implies that $A = \sqrt{1 + b^t b}$, where $b \in \mathbb{R}^2$ is arbitrary. In particular, we find that

$$g \cdot 0 = {}^t b(\sqrt{1 + \|b\|^2})^{-1}.$$

Since the spectrum of $b^t b$ is $\{0, \|b\|^2\}$ and Ω_- is rotations-invariant, we see that

$$\Omega_- = H \cdot 0 = \{z \in \mathbb{R}^2 : \|z\| < 1\}.$$

In order to compute the affine semigroup we consider the group

$$B^\sharp = N^\sharp A \cong N_- \times_{sdir} (N_0^\sharp A) = \left\{ \begin{pmatrix} a_1 & 0 & 0 \\ 0 & a_2 & * \\ * & * & (a_1 a_2)^{-1} \end{pmatrix} \middle| a_i > 0 \right\},$$

which acts on $N_- \cong \mathbb{R}^2$ via

$$\begin{pmatrix} a_1 & 0 & 0 \\ z & a_2 & 0 \\ x & y & (a_1 a_2)^{-1} \end{pmatrix} \cdot (x', y') = \frac{1}{a_1 a_2} \left(\frac{1}{a_1} x' - z y' + (a_2 - a_1) x, \frac{1}{a_2} y' \right).$$

The linear contractions A of Ω_- are those with $\|A\| \leq 1$. Using the fact that $\|A\|^2 = \|AA^*\|$, one obtains for example that

$$\left\| \begin{pmatrix} a & b \\ 0 & c \end{pmatrix} \right\|^2 = \frac{1}{2} \left((a^2 + b^2 + c^2) + \sqrt{|(a^2 + b^2 + c^2)^2 - 4a^2 c^2|} \right).$$

For $g = \begin{pmatrix} a_1 & 0 & 0 \\ z & a_2 & 0 \\ 0 & 0 & (a_1a_2)^{-1} \end{pmatrix} \in N_0^\sharp A$ we have that

$$g \cdot (x, y) = (a_1a_2)^{-1} \begin{pmatrix} a_1^{-1} & -z \\ 0 & a_2^{-1} \end{pmatrix} \begin{pmatrix} x \\ y \end{pmatrix}.$$

So $g \cdot \Omega_- \subset \Omega_-$ if and only if

$$\frac{1}{2a_1^2a_2^2}\left((a_1^{-2} + a_2^{-2} + z^2) + \sqrt{|(a_1^{-2} + a_2^{-2} + z^2)^2 - 4(a_1a_2)^{-2}|}\right) \leq 1.$$

For $z = 0$ this is equivalent to $\max\{a_1^{-2}, a_2^{-2}\}/(a_1^2a_2^2) \leq 1$. So we find for

$$g = \exp(rH_1 + sH_2) = \begin{pmatrix} e^{r+s} & 0 & 0 \\ 0 & e^{-r+s} & 0 \\ 0 & 0 & e^{-2s} \end{pmatrix}$$

the condition

$$e^{2|r|-6s} \leq 1 \quad \Longleftrightarrow \quad |r| - 3s \leq 0.$$

Because of

$$\begin{aligned} c_{\max} &= \{rH_1 + sH_2 \mid |r| \leq 3s\} \\ &= \mathbb{R}^+(3H_1 + H_2) + \mathbb{R}^+(-3H_1 + H_2), \\ c_{\min} &= c_{\max}^* = \{rH_1 + sH_2 \mid |r| \leq s\} \\ &= \mathbb{R}^+(H_1 + H_2) + \mathbb{R}^+(-H_1 + H_2), \end{aligned}$$

this condition is also equivalent to $rH_1 + sH_2 \in c_{\max}$.

The cone $\text{cone}(\Delta_-) = -c_{\max}^* = -c_{\min}$ has four faces:

$$\begin{aligned} F_0 &= \{0\}, \\ F_1 &= -\mathbb{R}^+(H_1 + H_2), \\ F_1' &= -\mathbb{R}^+(H_1 - H_2), \end{aligned}$$

and

$$F_2 = -c_{\min}.$$

The faces F_1 and F_1' are conjugate under the Weyl group which acts by reflections on the line $\mathbb{R}^+H_1$. Corresponding to these faces we have four idempotents in S_A^{cpt}. Clearly e_{F_0} is the identity, whereas $e_{F_2} = 0$. We have $e_{F_1} = e_X$ with

$$X = -3H_1 - H_2 = \begin{pmatrix} -4 & 0 & 0 \\ 0 & 2 & 0 \\ 0 & 0 & 2 \end{pmatrix} \in \mathfrak{a}^-.$$

An interesting feature in this example is that the causal Galois connection $\Gamma: \mathcal{F}(\mathcal{X}) \to \mathcal{F}(G)$ is not injective. To see this, let

$$a_t := \begin{pmatrix} e^{-4t} & 0 & 0 \\ 0 & e^{2t} & 0 \\ 0 & 0 & e^{2t} \end{pmatrix} \quad \text{and} \quad a_t' := \begin{pmatrix} e^{2t} & 0 & 0 \\ 0 & e^{-4t} & 0 \\ 0 & 0 & e^{2t} \end{pmatrix}.$$

Then $a_t \cdot (x, y) = (e^{6t}x, y)$,

$$\lim_{t\to\infty} a_t \cdot \Omega_- = \mathbb{R} \times [-1, 1],$$

and $a_t' \cdot (x, y) = (x, e^{6t}y)$, so

$$\lim_{t\to\infty} a_t' \cdot \Omega_- = [-1, 1] \times \mathbb{R}.$$

The sequences a_n and a_n' are not bounded in the order induced on A, hence

$$\Gamma(a_t \cdot \overline{\Omega_-}) = \eta(a_t) \to \emptyset$$

and

$$\Gamma(a_t' \cdot \overline{\Omega_-}) = \eta(a_t') \to \emptyset$$

in $\mathcal{F}(G)$. since

$$\lim_{t\to\infty} a_t \cdot \overline{\Omega_-} \neq \lim_{t\to\infty} a_t' \cdot \overline{\Omega_-},$$

we see that Γ is not injective.

We now describe the groups H_X and H_F. Let $X = -3H_1 - H_2$ be as before. Then $\Delta_{X,+} = \{\alpha_{23}\}$ and

$$\begin{aligned} \mathfrak{h}_X &= \lim_{t\to\infty} e^{\operatorname{ad} tX}\eta = \operatorname{pr}_{\mathfrak{h}}(\mathfrak{g}_{\alpha_{23}}) \oplus \mathfrak{g}_{\alpha_{12}} \oplus \mathfrak{g}_{\alpha_{13}} \\ &= \mathfrak{z}_{\mathfrak{h}}(X) + \mathfrak{g}_{\alpha_{12}} \oplus \mathfrak{g}_{\alpha_{13}} \\ &= \left\{ \begin{pmatrix} 0 & x & y \\ 0 & 0 & z \\ 0 & z & 0 \end{pmatrix} \middle| x, y, z \in \mathbb{R} \right\}. \end{aligned}$$

This is a solvable Lie algebra of the type $\mathbb{R}^2 \times_{sdir} \mathbb{R}$, where $\mathbb{R}^2$ is the sum of two real root spaces. The stabilizer algebra is given by

$$\mathfrak{h}_F = \mathbb{R}X + \mathfrak{h}_X = \mathbb{R}X \oplus \operatorname{pr}_{\mathfrak{h}}(\mathfrak{g}_{\alpha_{23}}) \oplus \mathfrak{g}_{\alpha_{12}} \oplus \mathfrak{g}_{\alpha_{13}}.$$

Let us write $e_X = e_{F_1} = (1, \gamma)$. Then

$$\begin{aligned} \operatorname{Im}\gamma &= \mathfrak{g}_{-\alpha_{23}}, \\ \ker\gamma &= \mathfrak{g}_{-\alpha_{13}}, \\ \rho_{e_X}^{-1}(e_X) &= \exp(\mathbb{R}X), \end{aligned}$$

and

$$\mathbf{L}(\ker \lambda_{e_X}) = \mathfrak{g}_{-\alpha_{13}} \times_{sdir} (\mathfrak{g}_{-\alpha_{12}} \oplus \mathbb{R}X).$$

The maximal parabolic Lie algebra $\overline{\mathfrak{p}}_{\max}$ of dimension 6 is

$$\begin{aligned}\overline{\mathfrak{p}}_{\max} &= \mathfrak{m} \oplus \mathfrak{a} \oplus \mathfrak{g}(\Delta_0) \oplus \mathfrak{g}(\Delta_+)\\ &= \left\{ \begin{pmatrix} a_{11} & a_{12} & a_{13} \\ a_{21} & a_{22} & a_{23} \\ 0 & 0 & -a_{11} - a_{22} \end{pmatrix} \middle| a_{ij} \in \mathbb{R} \right\}.\end{aligned}$$

The set Ω_{F_1} is $\gamma(\Omega_-) =]-1, 1[\cdot E_{32}$, and its pointwise stabilizer has the Lie algebra

$$\bigcap_{y \in \operatorname{Im} \gamma} \operatorname{Ad}(y)\overline{\mathfrak{p}}_{\max} = \mathbb{R}X \oplus \mathfrak{g}_{\alpha_{12}} \oplus \mathfrak{g}_{\alpha_{13}}.$$

The normalizer of its nilradical is the maximal parabolic algebra

$$\mathfrak{p}_F = \mathfrak{m} \oplus \mathfrak{a} \oplus \mathfrak{g}(\Delta^+) \oplus \mathfrak{g}_{-\alpha_{23}}.$$

Notes for Chapter 6

The material of this chapter has been developed in [55] in order to get a hold of the ideal structure of the groupoid C^*-algebra naturally associated to any ordered homogeneous space (cf. [54], [108], and the notes for Chapter 9).

Chapter 7

Holomorphic Representations of Semigroups, and Hardy Spaces

In the next three chapters we will give an overview of some of the applications of the theory of semigroups and causal symmetric spaces to harmonic analysis and representation theory. We present here only a broad outline of the theory, i.e., the main definitions and results, but mostly without proofs. We refer to the original literature for more detailed information. In the notes following each chapter the reader will find comments on the history of the subject and detailed references to the original works.

In this chapter we deal with highest-weight modules, holomorphic representations of semigroups, the holomorphic discrete series, and Hardy spaces on compactly causal symmetric spaces. The original idea of the theory goes back to the seminal article by Gelfand and Gindikin in 1977 [34], in which they proposed a new approach for studying the Plancherel formula for semisimple Lie group G. Their idea was to consider functions in $\mathbf{L}^2(G)$ as the sum of boundary values of holomorphic functions defined on domains in $G_{\mathbb{C}}$. The first deep results in this direction are due to Ol'shanskii [139] and Stanton [159], who realized the holomorphic discrete series of the group G in a Hardy space of a local tube domain containing G in the boundary. The generalization to noncompactly causal symmetric spaces was carried out in [63, 133, 135]. This program was carried out for solvable groups in

[64] and for general groups in [91].

7.1 Holomorphic Representations of Semigroups

Let $G_{\mathbb{C}}$ be a complex Lie group with Lie algebra $\mathfrak{g}_{\mathbb{C}}$ and let $\mathfrak{g}$ be a real form of $\mathfrak{g}_{\mathbb{C}}$. We assume that G, the analytic subgroup of $G_{\mathbb{C}}$ with Lie algebra $\mathfrak{g}$, is closed in $G_{\mathbb{C}}$. Let C be a regular G-invariant cone in $\mathfrak{g}$ such that the set $S(C) = G \exp iC$ is a closed semigroup in $G_{\mathbb{C}}$. Moreover, we assume that the map

$$G \times C \ni (a, X) \mapsto a \exp iX \in S(C)$$

is a homeomorphism and even a diffeomorphism when restricted to $G \times C^o$. Finally, we assume that there exists a real automorphism σ of $G_{\mathbb{C}}$ whose differential is the complex conjugation of $\mathfrak{g}_{\mathbb{C}}$ with respect to $\mathfrak{g}$, i.e., $\sigma(X + iY) = X - iY$ for $X, Y \in \mathfrak{g}$. All of those hypotheses are satisfied for Hermitian Lie groups and also for some solvable Lie groups; cf. [64].

Let $\pi : G \to U(\mathbf{V})$ be a unitary, strongly continuous representation of G in a Hilbert space $\mathbf{V}$. A vector $v \in \mathbf{V}$ is called a *smooth* or C^∞*-vector* if the map

$$\mathbb{R} \ni t \mapsto \hat{v}(t) := \pi(\exp tX)v \in \mathbf{V}$$

is smooth for all $X \in \mathfrak{g}$. Here a map $f : U \to \mathbf{V}$, $U \subset \mathbb{R}^n$ open, is called *differentiable* at the point $x_o \in \mathbf{V}$ if there exists a linear map $Df(x_o) := T_{x_o} f : \mathbf{V} \to \mathbf{V}$ such that

$$f(x) = f(x_o) + Df(x_o)(x - x_o) + o(\|x - x_o\|)\,.$$

The function f is of class C^1 if f is differentiable at every point in V and $x \mapsto Df(x) \in \mathrm{Hom}(\mathbf{V}, \mathbf{V})$ is continuous. Moreover, f is of class C^2 if $Df(x)$ is of class C^1. We denote the differential of $Df(x)$ by $D^2 f(x)$. We say that f is of class C^{k+1} if $D^k f$ is of class C^1. In that case we define $D^{k+1} f := D(D^k f)$. Finally, we say that f is *smooth* if f is of class C^k for every $k \in \mathbb{N}$.

Let $\mathbf{V}^\infty$ be the set of smooth vectors. $\mathbf{V}^\infty$ is a G-invariant dense subspace of $\mathbf{V}$. We define a representation of $\mathfrak{g}$ on $\mathbf{V}^\infty$ by

$$\pi^\infty(X)v = \lim_{t \to 0} \frac{\pi(\exp tX)v - v}{t}\,.$$

We denote this representation simply by π or use the module notation, $\pi(X)v = X \cdot v$. We extend this representation to $\mathfrak{g}_{\mathbb{C}}$ by complex linearity, $\pi(X + iY) = \pi(X) + i\pi(Y)$, $X, Y \in \mathfrak{g}$. Let $U(\mathfrak{g})$ denote the universal enveloping algebra of $\mathfrak{g}_{\mathbb{C}}$. Then π extends to a representation on $U(\mathfrak{g})$, again denoted by π. The set $\mathbf{V}^\infty$ is a topological vector space in a natural way, cf. [168]. Furthermore $\mathbf{V}^\infty$ is G-invariant and $U(\mathfrak{g})$-invariant. As $\pi(g)\exp(tX)v = \exp(t\,\mathrm{Ad}(g)X)\pi(g)v$, we get

$$\pi(g)\pi(X)v = \pi(\mathrm{Ad}(g)X)\pi(g)v$$

for all $g \in G$ and all $X \in \mathfrak{g}$. Define $Z^* = -\sigma(Z)$, $Z \in \mathfrak{g}_{\mathbb{C}}$. Then a simple calculation shows that for the densely defined operator $\pi(Z)$, $Z \in \mathfrak{g}_{\mathbb{C}}$, we have $\pi(Z)^* = \pi(Z^*)$. Define

$$C(\pi) := \{X \in \mathfrak{g} \mid \forall u \in \mathbf{V}^\infty \ : \ (i\pi(X)u|u) \leq 0\},$$

where $(\cdot|\cdot)$ is the inner product on $\mathbf{V}$. Thus $C(\pi)$ is the set of elements of $\mathfrak{g}$ for which $\pi(iX)$ is *negative*. The elements of $C(\pi)$ are called *negative elements* for the representation π.

Lemma 7.1.1 *$C(\pi)$ is a closed G-invariant convex cone in $\mathfrak{g}$.* □

Let C be an invariant cone in $\mathfrak{g}$. We denote the set of all unitary representations $\pi : G \to U(\mathbf{V})$ with $C(\pi) \subset C$ by $\mathcal{A}(C)$. A unitary representation π is called *C-admissible* if $\pi \in \mathcal{A}(C)$.

Let S be a semigroup with unit and let $^\sharp : S \to S$ be a bijective involutive antihomomorphism, i.e.,

$$(ab)^\sharp = b^\sharp a^\sharp \quad \text{and} \quad a^{\sharp\sharp} = a$$

We call $^\sharp$ an *involution* on the semigroup S and we call the pair $(S, ^\sharp)$ for a *semigroup with involution* or *involutive semigroup.*

Example 7.1.2 The most important example will be the semigroup $S(C)$ with the involution

$$s^\sharp = \sigma(s)^{-1}$$

(this is a special case of the involution $^\sharp$ on p. 121 for general symmetric pair). In this case $(a \exp iX)^\sharp = a^{-1} \exp i\,\mathrm{Ad}(a)X \in S(C)$. □

Example 7.1.3 (Contraction Semigroups on a Hilbert Space) Another example is the semigroup $C(\mathbf{V})$ of *contractions* on a complex Hilbert space $\mathbf{V}$:

$$C(\mathbf{V}) = \{T \in \mathrm{Hom}(\mathbf{V}) \mid \|T\| \leq 1\}$$

Denote by T^* the adjoint of T with respect to the inner product on $\mathbf{V}$. Then $(C(\mathbf{V}), *)$ is a semigroup with involution. □

Definition 7.1.4 Let $(S, \sharp)$ be a topological semigroup with involution, then a semigroup homomorphism $\rho : S \to C(\mathbf{V})$ is called a contractive representation of $(S, \sharp)$ if $\rho(g^\sharp) = \rho(g)^*$ and ρ is continuous w.r.t. the weak operator topology of $C(\mathbf{V})$. A contractive representation is called irreducible if there is no closed nontrivial subspace of $\mathbf{V}$ invariant under $\rho(S)$. □

Definition 7.1.5 Let ρ be a contractive representation of the semigroup $S(C) \subset G_{\mathbf{C}}$. Then ρ is holomorphic if the function $\rho: S(C)^o \to \mathrm{Hom}(\mathbf{V})$ is holomorphic. □

The following lemma shows that, if a unitary representation of the group G extends to a holomorphic representation of $S(C)$, then this extension is unique.

Lemma 7.1.6 *If $f: S(C) \to \mathbf{C}$ is continuous and $f|_{S(C)^o}$ is holomorphic such that $f|_G = 0$, then $f = 0$.* □

To construct a holomorphic extension ρ of a representation π we have to assume that $\pi \in \mathcal{A}(C)$. Then for any $X \in C$, the operator $i\pi(X)$ generates a self adjoint contraction semigroup which we denote by

$$T_X(t) = e^{ti\pi(X)}\,.$$

For $s = g \exp iX \in S(C)$ we define

$$\rho(s) := \rho(g) T_X(1)$$

Theorem 7.1.7 *ρ is a contractive and holomorphic representation of the semigroup $S(C)$. In particular, every representation $\pi \in \mathcal{A}(C)$ extends uniquely to a holomorphic representation of $S(C)$ which is uniquely determined by π.* □

For the converse of Theorem 7.1.7, we remark the following simple fact. Let $(S, \sharp)$ be a semigroup with involution and let ρ be a contractive representation of S. Let

$$G(S) := \{s \in S \mid s^\sharp s = s s^\sharp = 1\}$$

Then $G(S)$ is a closed subgroup of S and $\pi := \rho|_{G(S)}$ is a unitary representation of $G(S)$. Obviously,

$$G \subset G(S(C))\,.$$

Thus every holomorphic representation of $S(C)$ defines a unique unitary representation of G by restriction.

Theorem 7.1.8 *Let ρ be a holomorphic representation of $S(C)$. Then $\rho|_G \in \mathcal{A}(C)$ and the ρ agrees with the extension of $\rho|_G$ to $S(C)$.* □

Two representations ρ and π of the Ol'shanskii semigroup $S(C)$ are (unitarily) equivalent if there exists a unitary isomorphism $U: \mathbf{V}_\rho \to \mathbf{V}_\pi$ such that

$$U\rho(s) = \pi(s)U \quad \forall s \in S(C)$$

In particular, two contractive representations ρ and π of $S(C)$ are equivalent if and only if $\rho|_G$ and $\pi|_G$ are unitarily equivalent. We call a holomorphic contractive representation ρ of $S(C)$ *admissible* if $\rho|_G \in \mathcal{A}(C)$ and write $\rho \in \mathcal{A}(C)$. We denote by $\widehat{S(C)}$ the set of equivalence classes of irreducible holomorphic representations of $S(C)$.

Lemma 7.1.9 *Let π be an irreducible holomorphic representation of $S(C)$. Then the function*

$$\Theta_\pi(s) := \operatorname{Tr} \pi(s)$$

is well defined for every $s \in S(C)^o$. Furthermore $\Theta_\pi : S(C)^o \to \mathbb{C}$ is holomorphic and positive definite. □

Theorem 7.1.10 (Neeb) *Let π and ρ be irreducible holomorphic representations of $S(C)$. Then π and ρ are equivalent if and only if $\Theta_\pi = \Theta_\rho$.*
□

A nonzero function $\alpha: S(C) \to \mathbb{R}^+$ is called an *absolute value* if for all $s, t \in S(C)$ we have $\alpha(st) \leq \alpha(s)\alpha(t)$ and $\alpha(s^\sharp) = \alpha(s)$. Let α be an absolute value. A representation ρ of $S(C)$ is *α-bounded* if

$$\|\rho(s)\| \leq \alpha(s)$$

for all $s \in S(C)$. Note that this depends only on the unitary equivalence class of ρ. We denote by $\widehat{S(C)}(\alpha)$ the subset in $\widehat{S(C)}$ of α-bounded representations. If $\pi \in \widehat{S(C)}$, then, by abuse of notation, $\alpha(s) := \|\pi(s)\|$ is an absolute value of $S(C)$. Let $(\rho, \mathbf{V})$ and $(\pi, \mathbf{W})$ be holomorphic representations of $S(C)$. Let $\mathbf{V}\hat{\otimes}\mathbf{W}$ be the Hilbert space tensor product of $\mathbf{V}$ and $\mathbf{W}$. Define a representation of $S(C)$ in $\mathbf{V}\hat{\otimes}\mathbf{W}$ by

$$[\rho \otimes \pi](s) := \rho(s) \otimes \pi(s)$$

Then $\rho \otimes \pi \in \mathcal{A}(C)$. We denote the representation $s \mapsto \mathrm{id}$ by ι.

Theorem 7.1.11 (Neeb, Ol'shanskii) *Let $(\rho, \mathbf{V})$ be a holomorphic representation of the Ol'shanskii semigroup $S(C)$ and let $\alpha(s) = \|\rho(s)\|$. Then* *there exists a Borel measure μ on $\widehat{S(C)}$ supported on $\widehat{S(C)}(\alpha)$ and a direct integral of representations*

$$\left(\int_{\widehat{S(C)}(\alpha)}^{\oplus} \rho_\omega d\mu(\omega), \int_{\widehat{S(C)}(\alpha)}^{\oplus} \mathbf{V}_\omega d\mu(\omega)\right)$$

such that:

1) *$(\rho, \mathbf{V})$ is equivalent to $\left(\int_{\widehat{S(C)}(\alpha)}^{\oplus} \rho_\omega d\mu(\omega), \int_{\widehat{S(C)}(\alpha)}^{\oplus} \mathbf{V}_\omega d\mu(\omega)\right)$.*

2) *There exists a subset $N \subset \widehat{S(C)}(\alpha)$ such that $\mu(N) = 0$ and if $\omega \in \widehat{S(C)}(\alpha) \setminus N$, then $(\rho_\omega, \mathbf{V}_\omega)$ is equivalent to $(\pi_\omega \otimes \iota, \mathbf{H}_\omega \hat{\otimes} \mathbf{W}_\omega)$, where $\pi_\omega \in \omega$ and $\mathbf{W}_\omega$ is a Hilbert space.*

3) *If $\omega \in \widehat{S(C)}(\alpha)$ then set $n(\omega) := \dim \mathbf{W}_\omega$. Then n is a μ-measurable function from $\widehat{S(C)}(\alpha)$ to the extended positive axis $[0, \infty]$ which is called the multiplicity function.* □

7.2 Highest-Weight Modules

Representations with negative elements can exist only if there exists a nontrivial G-invariant cone in the Lie algebra $\mathfrak{g}$. If $\mathfrak{g}$ is simple, this implies in particular that $\mathfrak{g}$ is Hermitian. We will thus assume from now on that $\mathfrak{g}$ is a semisimple Hermitian Lie algebra. Thus G is a semisimple Hermitian Lie group. For simplicity we will assume that G is contained in a simply connected complexification $G_{\mathbb{C}}$. Then $G = G_{\mathbb{C}}^{\sigma}$ and $(G_{\mathbb{C}}, G)$ is a noncompactly causal symmetric pair. Let Z^0 be a central element in $\mathfrak{k}$ defining a complex structure on $\mathfrak{p}$. Then $X^0 = -iZ^0$ is a cone-generating element for $(G_{\mathbb{C}}, G)$ and the corresponding eigenspace decomposition is

$$\mathfrak{g}_{\mathbb{C}} = \mathfrak{n}_+ \oplus \mathfrak{h}^a \oplus \mathfrak{n}_-$$

as before. A comparision with the standard notation for Hermitian Lie algebras (cf. Example 5.1.10, p. 124) yields

$$\mathfrak{n}_+ = \mathfrak{p}^+, \quad \mathfrak{n}_- = \mathfrak{p}^-, \quad \text{and} \quad \mathfrak{h}^a = \mathfrak{k}_{\mathbb{C}}.$$

Thus N_+ corresponds to $P^+ = \exp \mathfrak{p}^+$, N_- corresponds to $P^- := \exp \mathfrak{p}^-$, and G_0 corresponds to $K_{\mathbb{C}} = G_{\mathbb{C}}^{\theta}$. Thus

$$G \subset P^+ K_{\mathbb{C}} P^-$$

(cf. Lemma 5.1.4, Remark 5.1.9, and Example 5.1.10). Let us recall some of the notation introduced in Section 2.6.1. For $x = pkq \in P^+K_{\mathbb{C}}P^-$ we write $p^+(x) = p$, $p^-(x) = q$ and $k_{\mathbb{C}}(x) = k$. By Lemma 5.1.2, the set $GK_{\mathbb{C}}P^-$ is an open submanifold of the complex flag manifold $\mathcal{X}_{\mathbb{C}} = G_{\mathbb{C}}/K_{\mathbb{C}}P^-$ and $G \cap K_{\mathbb{C}}P^- = K$. This implies that G/K is holomorphically equivalent to an open submanifold D of $\mathcal{X}_{\mathbb{C}}$. We also have the map $x \mapsto \zeta(x) = \log(x)$, which maps G/K biholomorphically into an open symmetric domain $\Omega_+ \subset \Omega_+$ (cf. Theorem 5.1.8). If $Z \in \mathfrak{p}^+$ and $g \in G_{\mathbb{C}}$ is such that $g \exp Z \in P^+K_{\mathbb{C}}P^-$, then $g \cdot Z = z(g \exp Z)$. Moreover, we have the universal automorphic factor $j(g, Z) := k_{\mathbb{C}}(g \exp Z)$. For j we find

$$\begin{aligned} j(k, Z) &= k, \\ j(p, Z) &= 1, \\ j(ab, Z) &= j(a, b \cdot Z) j(b, Z), \end{aligned}$$

if $k \in K_{\mathbb{C}}$, $Z \in \mathfrak{p}^+$, $p \in P^+$, and $a, b \in G$ are such that the expressions above are defined.

A $(\mathfrak{g}, K)$-module is a complex vectorspace $\mathbf{V}$ such that

1) $\mathbf{V}$ is a $\mathfrak{g}$-module.

2) $\mathbf{V}$ carries a representation of K, and the span of $K \cdot v$ is finite-dimensional for every $v \in \mathbf{V}$.

3) For $v \in \mathbf{V}$ and $X \in \mathfrak{k}$ we have

$$X \cdot v = \lim_{t \to 0} \frac{\exp(tX)v - v}{t}.$$

4) For $Y \in \mathfrak{g}$ and $k \in K$ the following holds for every $v \in \mathbf{V}$:

$$k \cdot (X \cdot v) = (\mathrm{Ad}(k)X) \cdot [k \cdot v].$$

Note that (3) makes sense, as $K \cdot v$ is contained in a finite dimensional vector space and this space contains a unique Hausdorff topology as a topological vector space.

The $(\mathfrak{g}, K)$-module is called *admissible* if the multiplicity of every irreducible representation of K in $\mathbf{V}$ is finite. If $(\pi, \mathbf{V})$ is an irreducible unitary representation of G, then the space of K-finite elements in $\mathbf{V}$, denoted by $\mathbf{V}_K$, is an admissible $(\mathfrak{g}, K)$-module.

Let $\mathfrak{t}$ be a Cartan subalgebra of $\mathfrak{k}$ and $\mathfrak{g}$. In this section, let Δ denote the root system $\Delta(\mathfrak{g}_{\mathbb{C}}, \mathfrak{t}_{\mathbb{C}})$.

Definition 7.2.1 Let $\mathbf{V}$ be a $(\mathfrak{g}, K)$-module. Then $\mathbf{V}$ is a ***highest-weight module*** if there exists a nonzero element $v \in \mathbf{V}$ and a $\lambda \in \mathfrak{t}_{\mathbb{C}}^*$ such that

1) $X \cdot v = \lambda(X)v$ for all $X \in \mathfrak{t}$.

2) There exists a positive system Δ^+ in Δ such that $\mathfrak{g}_{\mathbb{C}}(\Delta^+) \cdot v = 0$.

3) $\mathbf{V} = U(\mathfrak{g}) \cdot v$.

The element v is called a *primitive element* of weight λ. □

Let now $C \in \mathrm{Cone}_G(\mathfrak{g})$ and let $(\rho, \mathbf{V}) \in \mathcal{A}(C)$. We assume that $-Z^o \in C^o$. We assume furthermore that ρ is irreducible. Then $\mathbf{V}_K$ is an irreducible admissible $(\mathfrak{g}, K)$-module, and

$$\mathbf{V}_K = \bigoplus_{\lambda \in \mathfrak{t}_{\mathbb{C}}^*} \mathbf{V}_K(\lambda)$$

where $\mathbf{V}_K(\lambda) = \mathbf{V}_K(\lambda, \mathfrak{t}_{\mathbb{C}})$. Let $v \in \mathbf{V}_K(\lambda)$ be nonzero. Let $\alpha \in \Delta(\mathfrak{p}^+, \mathfrak{t}_{\mathbb{C}})$ and let $X \in \mathfrak{p}_\alpha^+ \setminus \{0\}$. Then

$$X^k \cdot v \in \mathbf{V}_K(\lambda + k\alpha).$$

In particular,

$$-iZ^0 \cdot (X^k \cdot v) = [-i\lambda(Z^0) + k]v.$$

This yields the following lemma.

Lemma 7.2.2 *Let the notation be as above. Then the following holds:*

1) $-i\lambda(Z^0) \leq 0$.

2) There exists a λ *such that* $\mathfrak{p}^+ \cdot \mathbf{V}_K(\lambda) = \{0\}$. □

Furthermore, the following holds.

Lemma 7.2.3 *Let* $\mathbf{W}^\lambda$ *be the* K*-module generated by* $\mathbf{V}_K(\lambda)$*. Then* $\mathbf{W}^\lambda$ *is irreducible and* $\mathbf{V}_K = U(\mathfrak{p}^-)\mathbf{W}^\lambda$. □

Let $\Delta_k = \Delta(\mathfrak{k}_{\mathbb{C}}, \mathfrak{t}_{\mathbb{C}})$ and let Δ_k^+ be a positive system in Δ_K. Let μ be the highest weight of $\mathbf{W}^\lambda$ with respect to Δ_k^+ and let v^λ be a nonzero highest weight vector. Then v^λ is a primitive element with respect to the positive system $\Delta_k^+ \cup \Delta_n^+$, where $\Delta_n = \Delta(\mathfrak{p}_{\mathbb{C}}, \mathfrak{t}_{\mathbb{C}})$ and $\Delta_n^+ = \Delta(\mathfrak{p}^+, \mathfrak{t}_{\mathbb{C}})$. As $-iZ^0$ is a positive linear combination of the vectors H_α, $\alpha \in \Delta_n^+$, we get the following theorem.

Theorem 7.2.4 *Let $(\rho, \mathbf{V}) \in \mathcal{A}(C)$ be irreducible. Then the corresponding $(\mathfrak{g}, K)$-module is a highest-weight module and equals $U(\mathfrak{p}^-)\mathbf{W}^\lambda$. In particular, every weight of $\mathbf{V}_K$ is of the form*

$$\nu - \sum_{\alpha \in \Delta(\mathfrak{p}^+, \mathfrak{t}_\mathbb{C})} n_\alpha \alpha .$$

Furthermore, $(\nu|\alpha) \leq 0$ for all $\alpha \in \Delta_n^+$. □

We will now show how to realize highest-weight modules in a space of holomorphic functions on G/K. We follow here the geometric construction by M. Davidson and R. Fabec [18]. For a more general approach, see [119]. To explain the method we start with $G = \mathrm{SU}(1,1)$. We set according to Example 2.6.16: $E = E_1 = \begin{pmatrix} 0 & 1 \\ 0 & 0 \end{pmatrix}$, $F = E_{-1} = \begin{pmatrix} 0 & 0 \\ 1 & 0 \end{pmatrix}$ and $H := H_1 = \begin{pmatrix} 1 & 0 \\ 0 & -1 \end{pmatrix}$. Thus $Z^0 = \frac{i}{2}H$ and $\mathfrak{p}^+ = \mathbb{C}E$, $\mathfrak{t}_\mathbb{C} = \mathbb{C}H$ and $\mathfrak{p}^- = \mathbb{C}F$. Let

$$g = \begin{pmatrix} \bar{\alpha} & \beta \\ \bar{\beta} & \alpha \end{pmatrix} \in \mathrm{SU}(1,1)$$

and let

$$\gamma = \begin{pmatrix} a & b \\ c & d \end{pmatrix} \in \mathrm{SL}(2, \mathbb{C}) .$$

We identify $\mathfrak{p}^+$ with $\mathbb{C}$ by $zE \mapsto z$ and similarily $\mathfrak{t}_\mathbb{C} \simeq \mathbb{C}$ by $zH \mapsto z$. Then ζ induces an isomorphism $\zeta(g) = \frac{\beta}{\alpha}$ of G/K onto the unit disc $D = \{z \in \mathbb{C} \mid |z| < 1\}$. Furthermore,

$$\gamma \cdot z = \frac{az + b}{cz + d}$$

and

$$j(g, z) = (cz + d)^{-1} .$$

The finite-dimensional holomorphic representations of $K_\mathbb{C}$ are the characters

$$\chi_n(\exp ziH) = e^{inz} .$$

In particular,

$$\chi_n(Z^0) = \frac{in}{2}$$

or

$$-i\chi_n(Z^0) = \frac{n}{2} .$$

Let $(\pi, \mathbf{V})$ be a unitary highest-weight representation of SU(1,1) and assume that $(\pi, \mathbf{V}) \in \mathcal{A}(C)$. Then $n \leq 0$. Let $\mathbf{V}(n)$ be the one-dimensional space of χ_n-isotropic vectors. Then

$$\mathbf{V}_K = \bigoplus_{k \in \mathbf{N}} \mathbf{V}(n-2k),$$

and the spaces $\mathbf{V}(m)$ and $\mathbf{V}(k)$ are orthogonal if $m \neq k$.

Let σ be the conjugation of $\mathfrak{sl}(2,\mathbb{C})$ with respect to $\mathfrak{su}(1,1)$. Then σ is given by

$$\sigma\left(\begin{pmatrix} a & b \\ c & -a \end{pmatrix}\right) = \begin{pmatrix} -\bar{a} & \bar{c} \\ \bar{b} & -\bar{a} \end{pmatrix}$$

so that $\sigma(E) = F$. By $\pi(T)^* = -\pi(\sigma(T))$ for all $T \in \mathfrak{sl}(2,\mathbb{C})$ we get

$$\pi(F)^* = \pi(-E).$$

Finally, it follows from $[F, E] = -H$ that for $v \in V(n)$:

$$\begin{aligned} \|\pi(F)^k v\|^2 &= (\pi(F)^k v \mid \pi(F)^k v) \\ &= (\pi((-X)^k F) v \mid v) \end{aligned}$$

Lemma 7.2.5 *Let the notation be as above. Then*

$$\begin{aligned} \pi(-E)^k \pi(F)^k v &= (-1)^k k! \frac{\Gamma(n+1)}{\Gamma(n-k+1)} v \\ &= (-n)_k v \end{aligned}$$

where $(a)_k = a(a+1)\cdots(a+k-1)$. □

As

$$\sum_{k=0}^{\infty} (-n)_k \frac{|z^2|^k}{k!} = (1-|z|^2)^n,$$

(cf. [36]) converges if and only if $|z| < 1$, it follows that

$$q_{(zE)} v := \sum_{k=0}^{\infty} \bar{z}^k \frac{F^k v}{n!}$$

converges if and only if $zX \in \Omega_+$.

Let now G be arbitrary. Let $\sigma : \mathfrak{g}_{\mathbb{C}} \to \mathfrak{g}_{\mathbb{C}}$ be the conjugation with respect to $\mathfrak{g}$. We use the notation from earlier in this section. Using the usual $\mathfrak{sl}(2,\mathbf{C})$ reduction, we get the following theorem.

Theorem 7.2.6 (Davidson–Fabec) *Let $T \in \mathfrak{p}^+$. Define $q_T : \mathbf{W}^\lambda \to \mathbf{V}$ by the formula*

$$q_T v := \sum_{k=0}^{\infty} \frac{\sigma(T)^k v}{n!} v\,.$$

1) *If $v \neq 0$, then the series that defines q_T converges in the Hilbert space $\mathbf{V}$ if and only if $T \in \Omega_+$.*

2) *Let π_λ be the representation of K on $\mathbf{W}^\lambda$. Let*

$$J_\lambda(g, Z) := \pi_\lambda(j(g, Z))\,.$$

Then

$$\pi(g)v = q_{g\cdot 0} J_\lambda(g, 0)^{*-1} v$$

for $g \in G$ and $v \in \mathbf{W}^\lambda$. □

It follows that the span of the $q_Z \mathbf{W}^\lambda$ with $Z \in \Omega_+$ is dense in $\mathbf{V}$, since $\mathbf{V}$ is assumed to be irreducible. Define $Q : \Omega_+ \times \Omega_+ \to \mathrm{GL}(\mathbf{W}^\lambda)$ by

$$Q(W, Z) = q_W^* q_Z\,.$$

Then the following theorem holds.

Theorem 7.2.7 (Davidson–Fabec) *Let the notation be as above. Then the following hold:*

1) $Q(W, Z) = J_\lambda(\exp(-\sigma(W)), T)^{*-1}$.

2) *$Q(W, Z)$ is holomorphic in the first variable and antiholomorphic in the second variable.*

3) $(Q(W, Z)u|v) = (q_Z u | q_W v)$ *for all $u, v \in \mathbf{W}_\lambda$.*

4) *Q is a positive-definite reproducing kernel.*

5) $Q(g \cdot W, g \cdot Z) = J_\lambda(g, W) Q(W, Z) J_\lambda(g, Z)^*$. □

For $Z \in \Omega_+$ and $u \in \mathbf{W}^\lambda$, let $F_{Z,u} : \Omega_+ \to \mathbf{W}^\lambda$ be the holomorphic function

$$F_{Z,u}(W) := Q(W, Z)u$$

and define

$$(F_{Z,u}|F_{T,w})_Q := (Q(T, Z)u, w)\,.$$

Let $\mathbf{H}(\Omega_+, \mathbf{W}^\lambda)$ be the completion of the span of $\{F_{Z,u} \mid Z \in \Omega_+, u \in \mathbf{W}^\lambda\}$ with respect to this inner product. Then $\mathbf{H}(\Omega_+, \mathbf{W}^\lambda)$ is an Hilbert space consisting of $\mathbf{W}^\lambda$-valued holomorphic functions. Define a representation of G in $\mathbf{H}(\Omega_+, \mathbf{W}^\lambda)$ by

$$(\rho(g)F)(W) := J_\lambda(g^{-1}, W)^{-1}F(g^{-1} \cdot W).$$

Then ρ is a unitary representation of G in $\mathbf{H}(\Omega_+, \mathbf{W}^\lambda)$ called the *geometric realization* of $(\pi, \mathbf{V})$.

Theorem 7.2.8 (Davidson–Fabec) *The map $q_Z v \mapsto F_{Z,v}$ extends to a unitary intertwining operator U between $(\pi, \mathbf{V})$ and the geometric realization $(\rho, \mathbf{H}(\Omega_+, \mathbf{W}^\lambda))$. It can be defined globally by*

$$[Uw](Z) = q_Z^* w, \quad w \in \mathbf{V},\ Z \in \Omega_+. \qquad \square$$

As the theorem stands, it gives a geometric realization for every unitary highest-weight module. What is missing is a natural analytic construction of the inner product on $\mathbf{H}(\Omega_+, \mathbf{W}^\lambda)$. This is known only for some special representations, e.g., the *holomorphic discrete series* of the group G. For that, let $\rho = \frac{1}{2}\sum_{\alpha \in \Delta^+} \alpha$ and let μ denote the highest weight of the representation of K on $\mathbf{W}^\lambda$. Furthermore, let dZ denote the usual Euclidean measure on $\mathfrak{p}^+$. For $f, g \in \mathbf{H}(\Omega_+, \mathbf{W}^\lambda)$, let

$$(f|g)_\lambda := \int_{G/K} (Q(Z,Z)f(Z)|g(Z))_{\mathbf{W}^\lambda} dZ$$

Theorem 7.2.9 (Harish-Chandra) *Assume that $(\mu + \rho|\alpha) < 0$ for all $\alpha \in \Delta_n^+$. Then $(f|g)_\lambda$ is finite for $f, g \in \mathbf{H}(\Omega_+, \mathbf{W}^\lambda)$ and there exists a positive constant c_λ such that*

$$(f|g)_Q = c_\lambda (f|g)_\lambda .$$

Moreover, $(\rho, \mathbf{H}(\Omega_+, \mathbf{W}_\lambda))$ is unitarily equivalent to a discrete sumand in $\mathbf{L}^2(G)$. $\square$

7.3 The Holomorphic Discrete Series

In this section we explain the construction of the holomorphic discrete series of a compactly causal symmetric space $\mathcal{M} = G/H$. In the next section we will see that those are the admissible representations of the Ol'shanskii semigroup that can be realized as discrete summands in $\mathbf{L}^2(\mathcal{M})$. We start with a simple structural fact about SU(1,1).

Define an involution on $\mathrm{SU}(1,1)$ by $\tau(a) = \bar{a}$, cf. Section 2.6.16:

$$\begin{aligned} H_{\mathbb{C}} := \mathrm{SL}(2,\mathbb{C})^{\tau} &= \left\{ h_z = \begin{pmatrix} \cosh z & \sinh z \\ \sinh z & \cosh z \end{pmatrix} \middle| \, z \in \mathbb{C} \right\} \\ &= \left\{ \begin{pmatrix} a & b \\ b & a \end{pmatrix} \middle| \, a, b \in \mathbb{C}, \, a^2 - b^2 = 1 \right\} \end{aligned}$$

and $H = \mathrm{SU}(1,1)^{\tau} = \pm\{h_t \mid t \in \mathbb{R}\}$. Let $\mathfrak{a}_p$ be the maximal abelian subalgebra of $\mathfrak{q}_p$ given by $\mathfrak{a}_p = \mathbb{R}X_o$. Let

$$a_t = \exp tX_o = \begin{pmatrix} \cosh\left(\frac{t}{2}\right) & -i\sinh\left(\frac{t}{2}\right) \\ i\sinh\left(\frac{t}{2}\right) & \cosh\left(\frac{t}{2}\right) \end{pmatrix}, \qquad t \in \mathbb{R}.$$

Define

$$\begin{aligned} a &= \frac{\cosh(t/2)}{\sqrt{\cosh t}}, \\ b &= \frac{-i\sinh(t/2)}{\sqrt{\cosh t}}, \\ \gamma &= \frac{1}{\sqrt{\cosh t}}, \end{aligned}$$

and

$$z = \frac{\gamma\cosh(t/2) - a\gamma^2}{b} = \frac{i\gamma\sinh(t/2) - b\gamma^2}{a}.$$

Then

$$\begin{aligned} h &:= \begin{pmatrix} a & b \\ b & a \end{pmatrix} \in H_{\mathbb{C}}, \\ k &:= \begin{pmatrix} \gamma & 0 \\ 0 & \gamma^{-1} \end{pmatrix} \in K_{\mathbb{C}}, \\ p &:= \begin{pmatrix} 1 & 0 \\ z & 1 \end{pmatrix} \in P^-, \end{aligned}$$

and $a_t = hkp$.

Now we go back to the general case. Let $\mathfrak{a}$ be a maximal abelian subalgebra of $\mathfrak{q}_k$. Then $i\mathfrak{a}$ is a maximal abelian subspace of $\mathfrak{q}_p^c$. Since $G^c/H \subset G_{\mathbb{C}}/H_{\mathbb{C}}$ is a noncompactly causal symmetric space as in Section 4.1 and 4.2, we find homomorphisms $\varphi_j : \mathrm{SL}(2,\mathbb{C}) \to G_{\mathbb{C}}$ intertwining the above involution on $\mathrm{SL}(2,\mathbb{C})$ and the given involution τ on $G_{\mathbb{C}}$. We may assume that our algebra $\mathfrak{a}_p$ is spanned by the elements

$$X_j := -i(E_j - E_{-j}) = \varphi_j\left(\frac{-i}{2}\begin{pmatrix} 0 & i \\ -i & 0 \end{pmatrix}\right),$$

where $\gamma_1, \ldots, \gamma_r$ are the strongly orthogonal roots. Let $A_p := \exp \mathfrak{a}_p$. Using $G = HA_pK$ (cf. [97]) and the fact that $K_{\mathbb{C}}$ normalizes P^-, $\sigma(P^+) = P^-$ and $G = G^{-1} = \sigma(G)$, we get the following theorem.

Theorem 7.3.1 *Let the notation be as above. Then*

1) *Let* $a(t_1, \ldots, t_r) = \exp(\sum_{j=1}^{r} t_j) \in A_p = \exp \mathfrak{a}_p$. *Then*

$$a(t_1, \ldots, t_r) \in H_{\mathbb{C}} \left[\exp \frac{1}{2} \sum_{j=1}^{r} - \log(\cosh t_j)\, H_j \right] P^-$$

2) $G \subset H_{\mathbb{C}}K_{\mathbb{C}}P^- \cap H_{\mathbb{C}}K_{\mathbb{C}}P^+ \cap P^-K_{\mathbb{C}}H_{\mathbb{C}} \cap P^+K_{\mathbb{C}}H_{\mathbb{C}}$. □

If $x = hkp \in H_{\mathbb{C}}K_{\mathbb{C}}P^-$, we write

$$h(x) = h, \quad k_H(x) = k, \quad \text{and} \quad p_H^-(x) = p.$$

Note that $h(x)$ and $k_H(x)$ are only well defined modulo $H_{\mathbb{C}} \cap K_{\mathbb{C}}$.

Let π be a holomorphic representation of $K_{\mathbb{C}}$ with nonzero $(H_{\mathbb{C}} \cap K_{\mathbb{C}})$-fixed vector. Denote the highest weight of π by μ_π. A simple generalization of Theorem A.2.2 and Lemma A.2.5 to the reductive Lie group $(K \cap H) \exp i\mathfrak{q}_k$ gives the following lemma.

Lemma 7.3.2 $\mu_\pi \in i\mathfrak{a}$ *and* $\mathbf{V}_\pi^{H_{\mathbb{C}} \cap K_{\mathbb{C}}}$ *is one-dimensional.* □

Let v_o be a nonzero $(H_{\mathbb{C}} \cap K_{\mathbb{C}})$-fixed vector. Define $\Phi_\pi : P^-K_{\mathbb{C}}H_{\mathbb{C}} \to \mathbf{V}_\pi$ by

$$\Phi_\pi(x) := \pi(k_H(x^{-1})^{-1})v_o.$$

We define a map $\mathbf{V}_\pi \to \mathcal{O}(P^+K_{\mathbb{C}}H_{\mathbb{C}})$ $v \mapsto \varphi(\pi, v)$ by

$$\varphi(\pi, v)(x) := (v|\Phi_\pi(\bar{x})), \tag{7.1}$$

where $\mathcal{O}(P^+K_{\mathbb{C}}H_{\mathbb{C}})$ denotes the holomorphic functions on $P^+K_{\mathbb{C}}H_{\mathbb{C}}$. By construction we have the following lemma.

Lemma 7.3.3 *Let the notation be as above. Then the following hold:*

1) *Let* $p \in P^+$, $h \in H_{\mathbb{C}}$, *and* $x \in P^+K_{\mathbb{C}}H_{\mathbb{C}}$. *Then* $\varphi(\pi, v)(pxh) = \varphi(\pi, v)(x)$.

2) *For* $k \in K_{\mathbb{C}}$ *we have* $\varphi(\pi, v)(k^{-1}x) = \varphi(\pi, \pi(k)v)(x)$; *i.e., the map* $v \mapsto \varphi(\pi, v)$ *is a* $K_{\mathbb{C}}$*-intertwiner.* □

As $\varphi(\pi, v)$ is right $H_{\mathbb{C}}$-invariant, we view it as a function on

$$P^+K_{\mathbb{C}}H_{\mathbb{C}}/H_{\mathbb{C}} = P^+K_{\mathbb{C}} \cdot \mathbf{o} \subset \mathcal{M}_{\mathbb{C}}.$$

As $G \subset P^+K_{\mathbb{C}}H_{\mathbb{C}}$, we can by restriction view $\varphi(\pi, v)$ as a function on $\mathcal{M}$. To decide when $\varphi(\pi, v)$ is in $\mathbf{L}^2(\mathcal{M})$, we write the G-invariant measure on $\mathcal{M}$ in polar coordinates using $G = KA_pH$. Let $\alpha \in \Delta(\mathfrak{g}, \mathfrak{a}_p)$. Let

$$\begin{aligned} p_\alpha &= \dim\{X \in \mathfrak{g}_\alpha \mid \theta\tau(X) = X\}, \\ q_\alpha &= \dim\{X \in \mathfrak{g}_\alpha \mid \theta\tau(X) = -X\}. \end{aligned}$$

Then there is a positive constant c such that

$$dx = c\prod\nolimits_{\alpha\in\Delta^+(\mathfrak{g},\mathfrak{a}_p)} \left|\sinh\left(\sum\nolimits_j s_j\alpha(L_j)\right)\right|^{p_\alpha}\left[\cosh\left(\sum\nolimits_j s_j\alpha(L_j)\right)\right]^{q_\alpha} ds_1\ldots ds_r\,dk$$

cf. [32], where $\Delta^+(\mathfrak{g}, \mathfrak{a}_p)$ is a positive system in $\Delta(\mathfrak{g}, \mathfrak{a}_p)$. Recall that $\Delta = \Delta(\mathfrak{g}_{\mathbb{C}}, \mathfrak{a}_{\mathbb{C}}) = \Delta(\mathfrak{g}_{\mathbb{C}}, \mathfrak{t}_{\mathbb{C}})$. Then

$$\Delta = \Delta(\mathfrak{k}_{\mathbb{C}}, \mathfrak{a}_{\mathbb{C}})\dot{\cup}\Delta(\mathfrak{p}^+, \mathfrak{a}_{\mathbb{C}})\dot{\cup}\Delta(\mathfrak{p}^-, \mathfrak{a}_{\mathbb{C}}).$$

Let Δ_k^+ be a positive system in $\Delta(\mathfrak{k}_{\mathbb{C}}, \mathfrak{a}_{\mathbb{C}})$, let $\Delta_n^+ := \Delta(\mathfrak{p}^+, \mathfrak{a}_{\mathbb{C}})$, and finally, let $\Delta^+ = \Delta_k^+ \cup \Delta_n^+$. Let

$$\rho = \frac{1}{2}\sum_{\alpha\in\Delta^+} [\dim_{\mathbb{C}}(\mathfrak{g}_{\mathbb{C}})_\alpha]\,\alpha.$$

By Lemma 7.3.1.1), we get the following theorem.

Theorem 7.3.4 *Let $v \in \mathbf{V}_\pi$, $v \neq 0$. Then $\varphi(\pi, v)|_{\mathcal{M}} \in \mathbf{L}^2(\mathcal{M})$ if and only if $(\mu_\pi + \rho|\alpha) < 0$ for all $\alpha \in \Delta_n^+$.* □

Theorem 7.3.5 *Assume that $(\mu_\pi + \rho|\alpha) < 0$ for all $\alpha \in \Delta_n^+$. Let $\mathbf{E}_\pi$ be the closed, G-invariant module in $\mathbf{L}^2(\mathcal{M})$ generated by $\{\varphi(\pi, v) \mid v \in \mathbf{V}_\pi\}$. Then the following hold.*

1) *$\mathbf{E}_\pi$ is irreducible.*

2) *$\mathbf{E}_\pi$ is a highest-weight module with a primitive element $\varphi(\pi, w)$, where w is a nonzero highest-weight vector for π.*

3) *The multiplicity of $\mathbf{E}_\pi$ in $\mathbf{L}^2(\mathcal{M})$ is 1.* □

Denote the representation of G in $\mathbf{E}_\pi$ by ρ_π. The representations $(\rho_\pi, \mathbf{E}_\pi)$ are called *holomorphic discrete series of $\mathcal{M}$.*

As $\mathbf{E}_\pi$ is a highest-weight module, there exists an G-intertwining operator

$$T_\pi : \mathbf{E}_\pi \to \mathbf{H}(\Omega_+, \mathbf{V}_\pi).$$

To construct such an operator, define

$$\Psi_\pi(g,x) := \pi(k_{\mathbb{C}}(g))\Phi_\pi(g^{-1}x),\ g \in G,\ x \in P^+K_{\mathbb{C}}H_{\mathbb{C}}.$$

Then $g \mapsto \Psi_\pi(g,x)$ is right K-invariant and defines a holomorphic function on G/K.

Theorem 7.3.6 *The map* $T : \mathbf{E}_\pi \to \mathbf{H}(\Omega_+, \mathbf{V}_\pi)$, *defined by*

$$Tf(z) := \int_{\mathcal{M}} f(x)\Psi_\pi(z,x)\,dx,$$

is a nonzero intertwining operator and

$$\mathrm{Hom}_G(\mathbf{E}_\pi, \mathbf{H}(\Omega_+, \mathbf{V}_\pi)) = \mathbb{C}T. \qquad \square$$

Example 7.3.7 Let G be a connected semisimple Lie group and let $G_1 = G \times G$, $H = \mathrm{diag}(G)$. As we have seen, $G = G_1/H$ in this case. Furthermore, $\mathfrak{q} = \{(X,-X) \mid X \in \mathfrak{g}\}$. The Cartan subspace $\mathfrak{a}$ is constructed by taking $\mathfrak{t}$ to be a compact Cartan subalgebra of $\mathfrak{g}$, $\mathfrak{t} \subset \mathfrak{k}$, and then setting $\mathfrak{a} = \{(X,-X) \mid X \in \mathfrak{t}\}$. Let $\alpha \in \Delta(\mathfrak{g}_{\mathbb{C}}, \mathfrak{t}_{\mathbb{C}})$ and let $X_{\pm\alpha} \in (\mathfrak{g}_{\mathbb{C}})_{\pm\alpha}$. Then

$$[(X,-X),(X_\alpha, X_{-\alpha})] = \alpha(X)(X_\alpha, X_{-\alpha})$$

for every $(X,-X) \in \mathfrak{a}$. In this way we get a bijective map,

$$\Delta(\mathfrak{g}_{\mathbb{C}}, \mathfrak{t}_{\mathbb{C}}) \ni \alpha \mapsto (\alpha, -\alpha) \in \Delta \subset \mathfrak{a}_{\mathbb{C}}^*.$$

We see also that the root spaces are exactly $(\mathfrak{g}_{\mathbb{C}})_\alpha \times (\mathfrak{g}_{\mathbb{C}})_{-\alpha}$. In particular, $Z_1^0 = (Z^0, -Z^0)$ and the space $\mathfrak{p}_1^+$, where $\mathfrak{p}_1 = \mathfrak{p} \times \mathfrak{p} \subset \mathfrak{g}_1$, is given by $\mathfrak{p}_1^+ = \mathfrak{p}^+ \times \mathfrak{p}^-$. The bounded realization of G_1/K_1 is $G_1/K_1 = G/K \times \overline{G/K}$, where $\overline{}$ means the opposite complex structure. So a holomorphic function on G_1/K_1 is the same as a function $f : \Omega_+ \times \Omega_+ \to \mathbb{C}$ that is holomorphic in the first variable and antiholomorphic in the second variable.

If π_1 is an irreducible unitary representation of $K_1 = K \times K$, then π_1 is of the form $\pi \otimes \delta$, where π and δ are irreducible representations of K. In particular, $\mathbf{V}_{\pi_1} \simeq \mathbf{V}_\pi \otimes \mathbf{V}_\delta$. Assume that u_o is a nonzero $\mathrm{diag}(K) = (K_1 \cap H)$-invariant element in $\mathbf{V}_{\pi_1}$. Define $\varphi : \mathbf{V}_\delta \to \mathbf{V}_\pi^*$ by

$$< u, \varphi(v) >= (u \otimes v | u_o),$$

where $(\cdot|\cdot)$ is the K_1-invariant inner product on $\mathbf{V}_{\pi_1}$. Then, for $k \in K$,

$$\begin{aligned} < u, \varphi(\rho(k)v) > &= (u \otimes \rho(k)v|u_o) \\ &= (\pi_1(k,k)(\pi(k^{-1})u \otimes v)|u_o) \\ &= (\pi(k^{-1})u \otimes v|u_o) \\ &= < \pi(k^{-1})u, \varphi(v) > \\ &= < u, \pi^\vee(k)\varphi(v) > \end{aligned}$$

Thus $\varphi : \mathbf{V}_\delta \to \mathbf{V}_\pi^*$ is a K-intertwining operator. As both spaces are irreducible, it follows that φ is an isomorphism. In particular, we have the following lemma.

Lemma 7.3.8 *Let π_1 be an irreducible representation of K_1 with a nonzero diag(K)-fixed vector. Then there exists an irreducible representation π of K such that $\pi_1 \simeq \pi \otimes \pi^\vee$.* □

Now $\mathbf{V}_\pi \otimes \mathbf{V}_\pi^* \simeq \mathrm{Hom}_\mathbb{C}(\mathbf{V}_\pi, \mathbf{V}_\pi)$ and the representation is carried over to

$$\pi_1(k,h)T = \pi(k)T\pi(h)^{-1}.$$

The invariant inner product on $\mathrm{Hom}_\mathbb{C}(\mathbf{V}_\pi, \mathbf{V}_\pi)$ is $(T|S) = \mathrm{Tr}\, TS^*$. In this realization the invariant element u_o is (up to constant) the identity id and

$$(T|u_o) = \mathrm{Tr}(T).$$

Let μ be the highest weight of π. Then $< \pi + \rho, \alpha >< 0$ for every $\alpha \in \Delta(\mathfrak{p}^+, \mathfrak{t}_\mathbb{C})$. Thus π corresponds to a unique holomorphic discrete series representation $\mathbf{E}_\pi$ of G. We have (cf. [135]) the following theorem.

Theorem 7.3.9 *The holomorphic discrete series $\mathbf{E}_{\pi_1}$ is canonically isomorphic to $\mathbf{E}_\pi \otimes \mathbf{E}_\pi^*$.* □

7.4 Classical Hardy Spaces

In this section we explain the construction of the Hardy space related to a regular cone field on a compactly causal symmetric space. We start with a short overview of the classical theory as it can be found, e.g., in the book by Stein and Weiss [160].

Let C be a regular cone in $\mathbb{R}^n$ and let $\Omega = C^o$. Let $\Xi(\Omega) := \mathbb{R}^n + i\Omega$. Then $\Xi(\Omega)$ is an open subset of $\mathbb{C}^n$. Let $\mathcal{O}(\Omega)$ be the space of holomorphic functions on $\Xi(\Omega)$. If $f \in \mathcal{O}(\Omega)$ and $u \in \Omega$, then

$$\mathbb{R}^n \ni x \mapsto f_u(x) := f(x + iu) \in \mathbb{C}$$

is well defined. We define the *Hardy norm* of f with respect to C be

$$\|f\|_2^2 := \sup_{u\in\Omega} \|f_u\|^2_{\mathbf{L}^2(\mathbb{R}^n)} = \sup_{u\in\Omega} \int_{\mathbb{R}^n} |f(x+iu)|^2\, dx\,.$$

We can now define the *Hardy space* $\mathcal{H}_2(C)$ by

$$\mathcal{H}_2(C) = \{f \in \mathcal{O}(\Xi(\Omega)) \mid \|f\|_2 < \infty\}. \tag{7.2}$$

Define the *boundary value map* $\beta : \mathcal{H}_2(C) \to \mathbf{L}^2(\mathbb{R}^n)$ by

$$\beta(f) = \lim_{\Omega\ni u\to 0} f(\cdot + iu)$$

Then β is an isometry into $\mathbf{L}^2(\mathbb{R}^n)$. To describe the image of β let

$$\mathcal{F}: \mathbf{L}^2(\mathbb{R}^n) \to \mathbf{L}^2(\mathbb{R}^n)$$

be the Fourier transform, i.e.,

$$\mathcal{F}(f)(v) = \frac{1}{(2\pi)^{n/2}} \int_{\mathbb{R}^n} f(x) e^{-i(x|v)}\, dx, \quad \forall f \in C_c(\mathbb{R}^n)\,.$$

To simplify the notation we define the function e_u, $u \in \mathbb{C}^n$, by

$$\mathbb{C}^n \ni x \mapsto e_u(x) = e^{(x|u)}$$

Theorem 7.4.1 *Let* $\mathbf{E} = \{f \in \mathbf{L}^2(\mathbb{R}^n) \mid \mathrm{Supp}(\mathcal{F}(f)) \subset C^*\} \simeq \mathbf{L}^2(C^*)$. *Then* $\mathrm{Im}(\beta) = \mathbf{E}$. □

Let us sketch the construction of the map $\mathbf{E} \to \mathrm{Im}(\beta)$. Consider $f \in \mathbf{E}$, $F = \mathcal{F}(f)$ and let $u \in \Omega$. As $\mathrm{Supp}(F) \subset C^*$, it follows that $|Fe_{-u}| \le |F|$. Hence $Fe_{-u} \in \mathbf{L}^2(\mathbb{R}^n)$, and we may define $g : \Xi(\Omega) \to \mathbb{C}$ by

$$g(x+iu) = \mathcal{F}^{-1}(Fe_{-u})(x)\,.$$

Formally, this is

$$\begin{aligned} g(x+iu) &= \frac{1}{(2\pi)^{n/2}} \int_{\mathbb{R}^n} F(\lambda) e^{i(x+iu|\lambda)}\, d\lambda \\ &= \frac{1}{(2\pi)^{n/2}} \int_{C^*} \left[F(\lambda) e^{-(u|\lambda)}\right] e^{i(x|\lambda)}\, d\lambda \end{aligned}$$

By construction, $g \in \mathbf{H}_2(\Omega)$, and one has to show that $\beta(g) = f$.

We define the *Cauchy kernel* associated with the tube domain $\Xi(\Omega)$ by

$$\begin{aligned} K(x+iu) &= \frac{1}{(2\pi)^{n/2}}\int_{\Omega^*} e^{i(x+iu|\lambda)}\,d\lambda \\ &= \mathcal{F}^{-1}(\chi_{\Omega^*}e_{-u}), \end{aligned}$$

where χ_A denotes the characteristic function of a set $A \subset \mathbb{R}^n$ and $x+iu \in \Xi(\Omega)$. Then $\|K(\cdot+iu)\|^2 = K(2iu) < \infty$. Thus $K(\cdot+iu) \in \mathbf{L}^2(\mathbb{R}^n)$. Furthermore, we get the following theorem.

Theorem 7.4.2 *Let $F \in \mathcal{H}_2(C)$. Then*

$$F(z) = \int_{\mathbb{R}^n} f(x)K(z-x)\,dx$$

for all $z \in \Xi(\Omega)$, where $f = \beta(F)$. □

We define the *Poisson kernel* by

$$P(x,y) = \frac{|K(x+iy)|^2}{K(2iy)}, \quad x+iy \in \Xi(\Omega)$$

Theorem 7.4.3 *Let $f \in \mathbf{H}_2(\Omega)$. Then*

$$F(x+iy) = \int_{\mathbb{R}^n} P(x-t,y)f(t)\,dt,$$

where $f = \beta(F)$. □

7.5 Hardy Spaces

In this section G/H is a compactly causal symmetric space. Let $C \in \mathrm{Cone}_H(\mathfrak{q})$ be such that $C^o \cap \mathfrak{k} \neq \emptyset$. There are two different ways to generalize the tube domain $\Xi(\Omega)$ from the last section to this setting. First, we may construct a local tube domain in $T(\mathcal{M})_{\mathbb{C}}$ by $G \times_H iC^o$. Second, we may view $\Xi(\Omega)$ as the orbit of $0 \in \mathbb{R}^n$ under the semigroup $\mathbb{R}^n + i\Omega$. The corresponding construction in this setting is to consider the semigroup $S = G \exp iD^o$, where $D \in \mathrm{Cone}_G(\mathfrak{g})$. Then $\Xi(\Omega) = S^{-1}\mathbf{o} \subset G_{\mathbb{C}}/H_{\mathbb{C}}$. Here we have to assume that $G \subset G_{\mathbb{C}}$, where $G_{\mathbb{C}}$ is a complex Lie group with Lie algebra $\mathfrak{g}_{\mathbb{C}}$. The inverse is necessary, as we want the semigroup to act on functions on $\Xi(C^o)$.

We will use the second approach (cf. [64, 63]). In particular, we will assume that $G \subset G_{\mathbb{C}}$, where $G_{\mathbb{C}}$ is a complex Lie group with Lie algebra $\mathfrak{g}_{\mathbb{C}}$ such that τ integrates to an involution on $G_{\mathbb{C}}$. We assume that $H = G^\tau$.

Let $H_{\mathbb{C}} = G_{\mathbb{C}}^{\tau}$ and define $\mathcal{M}_{\mathbb{C}} := G_{\mathbb{C}}/H_{\mathbb{C}}$. Then $\mathcal{M} \subset \mathcal{M}_{\mathbb{C}}$, and $\mathcal{M}_{\mathbb{C}}$ is a complexification of $\mathcal{M}$. Let $C \in \mathrm{Cone}_H(\mathfrak{q})$ and D be an extension of iC to a G-invariant closed cone in $i\mathfrak{g}$ such that $D \cap i\mathfrak{q} = \mathrm{pr}_{i\mathfrak{q}} D = iC$, cf. Section 4.5. Let $S(D)$ be the Ol'shanskii semigroup $G \exp D$. Then $S(D)^o = S(D^o) := G \exp D^o \simeq G \times D^o \neq \emptyset$. Define an open complex submanifold $\Xi(C^o) \subset \mathcal{M}_{\mathbb{C}}$ by

$$\Xi(C^o) := S(D^o)^{-1}\mathbf{o}. \tag{7.3}$$

Then $\Xi(C) := S(D)^{-1}\mathbf{o} \subset \overline{\Xi(C^o)}$, where the bar denotes the closure in $\mathcal{M}_{\mathbb{C}}$. Furthermore, $\mathcal{M} \subset \partial\Xi(C)$. For simplicity we will write S for $S(C)$, S^o for $S(C^o)$, Ξ for $\Xi(C)$, and Ξ^o for its interior $\Xi(C^o)$. That Ξ and Ξ^o depend only on C and not on the extension as indicated in the above notation follows from the next lemma, which shows that Ξ^o locally is a tube domain.

Lemma 7.5.1 *The manifold $\Xi(C)$ is independent of the extension D, and $\Xi(C) \simeq G \times_H -iC^o$.* □

As $S^oS \subset S^o$, it follows that $(S^o)^{-1}\Xi \subset \Xi^o$. In particular,

$$\gamma^{-1}\mathcal{M} \subset \Xi^o, \quad \forall \gamma \in S(C^o) \tag{7.4}$$

Thus, if f is a function on Ξ^o and $s \in S^o$, we can define a function $s \cdot f$ on Ξ by

$$[s \cdot f](x) = f(s^{-1}x).$$

Let $f \in \mathcal{O}(\Xi^o)$ and let $s \in S$. By (7.4) the function $s \cdot f|_{\mathcal{M}}$ is well defined as long as $s \in S^o$. In particular, $\|s \cdot f\|$ is well defined, where $\|\cdot\|$ stands for the $\mathbf{L}^2$-norm on $\mathcal{M}$. Define the *Hardy norm* of a holomorphic function f on Ξ^o by

$$\|f\|_2 := \sup_{s \in S^o} \|s \cdot f\|$$

We define the *Hardy space* $\mathcal{H}_2(C)$ by

$$\mathcal{H}_2(C) := \{f \in \mathcal{O}(\Xi^o) \mid \|f\|_2 < \infty\}. \tag{7.5}$$

As we are using a G-invariant measure on $\mathcal{M}$ and $GS^o \subset S^o$, it follows that

$$\|s \cdot f\|_2 \leq \|f\|_2 .$$

Define the "boundary value map" $\beta : \mathcal{H}_2(C) \to \mathbf{L}^2(\mathcal{M})$ by

$$\beta(f) = \lim_{s \to 1}(s \cdot f)|_{\mathcal{M}},$$

where the limit is taken in $\mathbf{L}^2(\mathcal{M})$.

Theorem 7.5.2 *Let the notation be as above. Then the following hold:*

1) *$\mathcal{H}_2(C)$ is a Hilbert space, and the action of G is unitary.*

2) *The map $\beta : \mathcal{H}_2(C) \to \mathbf{L}^2(\mathcal{M})$ is an isometric G-intertwing operator.*

3) *Let ρ_π be the holomorphic discrete series representation of G in $\mathbf{E}_\pi$. Then*

$$\operatorname{Im}\beta = \bigoplus_{\rho_\pi \in \mathcal{A}(D)} \mathbf{E}_\pi,$$

where D is some extension of C to a G-invariant cone in $\mathfrak{g}$. If $C = C_{\min}$, then the sum is over the full holomorphic discrete series. □

Assume now that G/K is a tube domain and that $\tau = \tau_{iY_o}$. We also assume that G is contained in the simply connected group $G_\mathbb{C}$. We know from Section 2.6 that

$$G/H = G \cdot (E, -E) = \{\xi \in \mathcal{S}_1 \mid \Psi_m(\xi) \neq 0\}.$$

We also have the following lemma.

Lemma 7.5.3 *The G-invariant measure on G/K is given by*

$$\int_{G/H} f(g \cdot (E, -E))\, d\dot{g} = \int_{\mathcal{S}_1} f(\xi)|\Psi_2(\xi)|^{-1}\, d\mu(\xi)$$

where $d\mu$ is a suitably normalized $K \times K$-invariant measure on $\mathcal{S}_1$. □

By the right choice of the cone $C = C_k$ we also know that the semigroup S is just the contraction semigroup of the bounded domain $G/K \simeq \Omega_+$. In particular, $\Xi^o \subset \Omega_+ \times \Omega_+$. By construction Ψ_m is holomorphic on $\Omega_+ \times \Omega_+$. We then have the following theorem.

Theorem 7.5.4 $\Xi^o = \{\xi \in \Omega_+ \times \Omega_+ \mid \Psi_m(\xi) \neq 0\}$. □

The classical Hardy space $\mathcal{H}_2$ can, via the biholomorphic Cayley transform $\mathbf{C}_h$, be viewed as space of holomorphic function on $\Omega_+ \times \Omega_+$. The semigroup acts on this space via

$$s \cdot f(z, w) = j(s^{-1}, z)^{\rho_n} j(s^{-1}, w)^{\rho_n} f(s^{-1} \cdot z, s^{-1} \cdot w)$$

which is well defined if $-\rho_n$ is the lowest weight of a holomorphic representation of $G_\mathbb{C}$. In that case we also have a holomorphic square rooth Ψ_1 of Ψ_2. The result is Theorem 7.5.5.

Theorem 7.5.5 (Isomorphism of Hardy Spaces) *Denote by $\mathcal{H}_2(\Omega_+ \times \Omega_+)$ the classical Hardy space on the bounded symmetric domain $\Omega_+ \times \Omega_+$. Assume that $\mathfrak{g}$ is not isomorphic to $\mathfrak{sp}(2n, \mathbb{R})$ or $\mathfrak{so}(2, 2k+1)$, $k, n \geq 1$. Then*

$$\mathcal{H}_2(\Omega_+ \times \Omega_+) \ni f \mapsto \frac{f}{\Psi_1} \in \mathcal{H}_2(C)$$

is an isometric G-isomorphism. □

For the remaining two cases one has to construct a double covering of Ξ^o, $\mathcal{M}$, G, and S and define the corresponding Hardy space. The classical Hardy space is then isomorphic to the space of odd functions in that Hardy space. Refer to [136] for the exact statements.

7.6 The Cauchy–Szegö Kernel

For $w \in \Xi(C^o)$ the linear form $f \mapsto f(w)$ is continuous. Hence there exists an element $K_w \in \mathcal{H}_2(C)$ such that for $f \in \mathcal{H}_2(C)$ we have $f(w) = (f|K_w)$. Let

$$K(z, w) := K_w(z). \tag{7.6}$$

The kernel $(z, w) \mapsto K(z, w)$ is called the *Cauchy–Szegö kernel.* We note that $K(z, w)$ depends on the cone C used in constructing the Hardy space. By Definition 7.1.4 we have

$$(\rho_\pi(s)f|g) = (f|\rho_\pi(s^*)g)$$

for $s \in S(C)$ and $f, g \in \mathcal{H}_2(C)$. This gives

Lemma 7.6.1 *Let $z, w \in \Xi(C^o)$ and let $s \in S(C)$. Then*

$$K(s^{-1}z, w) = K(z, \sigma(s)w).$$ □

We collect further properties of the Cauchy–Szegö kernel together in the following theorem.

Theorem 7.6.2 *Let $\mathcal{H}_2(C)$ be the Hardy space corresponding to an invariant regular cone $C \subset \mathfrak{q}$. Let $K(z, w)$ denote the corresponding Cauchy–Szegö kernel. Then the following hold:*

1) *$K(z, w) = \overline{K(w, z)}$.*

2) *For fixed $z \in \Xi(C^o)$, the map*

$$\Xi(C^o) \ni w \mapsto K(z, w) \in \mathbb{C}$$

extends to a smooth map on $\Xi(C)$. If $x \in \mathcal{M}$, then

$$K(z, x) = \overline{\beta(K_z)}(x).$$

3) Let $K(z) := K(z, \mathbf{o})$, $z \in \Xi(C^o)$. Then K is holomorphic and

$$K(s_1\mathbf{o}, s_2\mathbf{o}) = K(s_2^\sharp s_1)$$

for every $s_1, s_2 \in S(C^o)^{-1}$.

4) The map $\beta^{-1} : \operatorname{Im}\beta \to \mathcal{H}_2(C)$ is given by

$$(\beta^{-1}f)(z) = \int_{\mathcal{M}} f(m)K(z,m)\, dm. \qquad \square$$

Let $z \in \Xi(C^o)$. Then for a suitable $u \in \mathbf{V}_\pi$ we have $\varphi_u(z) \neq 0$. Assume that $K(z,z) = 0$. As

$$K(z,w) = \int_{\mathcal{M}} K(m,w)K(z,m)\, dm,$$

it follows that

$$K(z,z) = \int_{\mathcal{M}} |K(m,z)|^2\, dm \geq 0,$$

and $K(z,z) = 0$ if and only if $K(m,z) = 0$ for every $m \in \mathcal{M}$. But then

$$\varphi_u(z) = \int_{\mathcal{M}} \varphi_u(m)K(z,m)\, dm = 0,$$

a contradiction. Thus we obtain Lemma 7.6.3.

Lemma 7.6.3 *Let $z \in \Xi(C^o)$. Then $K(z,z) \neq 0$.* $\square$

We can now define the *Poisson kernel* by

$$P(z,m) := \frac{|K(z,m)|^2}{K(z,z)}. \tag{7.7}$$

Theorem 7.6.4 *Let f be a continuous function on $\Xi(C)$ which is holomorphic on $\Xi(C^o)$. Then*

$$f(z) = \int_{\mathcal{M}} P(z,m)f(m)dm$$

for every $z \in \Xi(C^o)$. In particular,

$$\int_{\mathcal{M}} P(z,m)dm = 1. \qquad \square$$

Notes for Chapter 7

Holomorphic representations of the semigroup $S(D) = G \exp iD$ were introduced by Ol'shanskii [137]. The theory was generalized to the situation we present here in [64]. For the general theory of holomorphic representations of Ol'shankii semigroups, refer to the work of K.-H. Neeb, [122] and [121, 123], where the Theorems 7.1.9, 7.1.10, and 7.1.11 were proved (cf. also [119]).

There is an extensive literature on highest-weight modules and the classification of highest-weight modules. Our exposition follows closely the work of Davidson and Fabec [18]. More general results were obtained by Neeb in [119]. The classification of unitary highest-weight modules can be found in [22, 71]. For the connection between positive-definite operator-valued kernels and unitary representations, refer to [92].

The articles [38] and [39], where Harish-Chandra constructed the holomorphic discrete series of the group, were the starting point of the analysis on bounded symmetric domains, unitary highest-weight modules, and the discrete series of the group. The analytic continuation was achieved by Wallach in [167].

Most of Section 7.3 is taken from [133] and [135] except for the "only if" part in Theorem 7.3.4, which is from [63]. The construction of $\varphi(\pi, v)$ in [135] was by using the dual representation $\pi^{\vee}$. The construction here is taken from [82].

The general theory of the discrete series on $\mathcal{M}$ was initiated by the seminal work of M. Flensted-Jensen [32], where he used the Riemannian dual of $\mathcal{M}$ to construct "most" of the discrete series. The complete construction was done by Oshima and Matsuki in [102, 142, 144]. The first construction of what is now called the holomorphic discrete series can be found in [103], where S. Matsumoto used the method of Flensted-Jensen to construct those representations.

The material in Section 7.4 is standard and can be found, e.g., in [160]. Most of Sections 7.5 and 7.6 are from [63]. The introduction of the Poisson kernel is new. The part on Cayley-type spaces is taken from [136]. Theorem 7.5.4 was also proved in [15]. Further results on the H-invariant distribution character of the holomorphic discrete series representations can be found in [132]. An overview of the theory of Hardy spaces in the group case can be found in a set of lecture notes by J. Faraut ([27]). In these notes the definition of the Poisson kernel was given and Theorem 7.6.4 was proved for the group case. A shorter overview can also be found in [30].

Chapter 8

Spherical Functions on Ordered Spaces

In this chapter we describe the theory of spherical functions and the spherical Laplace transform on noncompactly causal symmetric spaces as developed in [28] and [131]. The theory is motivated by the classical theory which we explain in the first section. The second motivation is the Harish-Chandra–Helgason theory of spherical functions on Riemannian symmetric spaces (cf. [40, 41, 45]).

8.1 The Classical Laplace Transform

Before we talk about the Laplace transform on ordered symmetric spaces, let us briefly review the classical situation. Let $\mathcal{M} = \mathbb{R}^n$ and let C be a closed, regular cone in $\mathbb{R}^n$, e.g., the light cone. As explained in Example 2.2.3, $\mathbb{R}^n$ becomes an ordered space by defining

$$x \geq y \Leftrightarrow x - y \in C\,.$$

Let

$$\mathcal{M}_{\leq} = \{(x, y) \in \mathcal{M} \mid x \leq y\}\,.$$

$\mathcal{M}_{\leq}$ is closed in $\mathcal{M} \times \mathcal{M}$. A *causal kernel* or *Volterra kernel* is a map $K : \mathcal{M} \times \mathcal{M} \to \mathbb{C}$ such that K is continuous on $\mathcal{M}_{\leq}$ and zero outside $\mathcal{M}_{\leq}$. Let $\mathcal{V}(\mathcal{M})$ be the vector space of causal kernels. For $F, G \in \mathcal{V}(\mathcal{M})$, define

$$F\#G(x, y) := \int_{[x,y]} F(x, z)G(z, y)\, dz \tag{8.1}$$

With this product $\mathcal{V}(\mathcal{M})$ becomes an algebra. The kernel K is *invariant* if, for all $x, y, z \in \mathcal{M}$,

$$K(x+z, y+z) = K(x,y).$$

An invariant causal kernel corresponds to a function of one variable supported on C via $F(x) = K(0,x)$, $K(x,y) = F(y-x)$. The above product is then given by the usual convolution of functions, $F\#G = F * G$.

Associated to the Volterra algebra is the Volterra integral equation of the second kind,

$$A(x,y) = B(x,y) + \int_{[x,y]} K(x,z)A(z,y)\,dz,$$

where B and K are given. If A and B are invariant kernels such that a and b are the corresponding functions, this reads

$$a(x) = b(x) + \int_{[x,y]} K(x,z)a(z)\,dz.$$

Theorem 8.1.1 (M. Riesz) *The Volterra equation has a unique solution given by*

$$A = B + R\#B,$$

where R is the resolvent $R = \sum_{k=1}^{\infty} K^{(n)}$, with $K^{(n+1)} = K^{(n)}\#K$. The series defining R converges uniformly on bounded sets and $R \in \mathcal{V}(\mathcal{M})$. □

For a, b and $K = k$ invariant, the Volterra equation is

$$a = b + k * a.$$

Recall the exponential functions $e_{-\lambda}$ from p. 215 and assume that $fe_{-\lambda}$ is bounded for $\lambda \in C + i\mathbb{R}^n$. Define the *Laplace transform* of f by

$$\mathcal{L}(f)(\lambda) := \int e^{-(\lambda|x)} f(x)\,dx \quad \text{for} \quad \lambda \in C + i\mathbb{R}^n.$$

If f has compact support, then $\mathcal{L}(f)$ is defined for every λ in $\mathbb{R}^n$. Write $\lambda = u + iy$, $u \in C$. Then $\mathcal{L}(f)(u+iy) = (2\pi)^{n/2}\mathcal{F}(fe_u)(y)$. Hence

$$f(x)e^{-(u|x)} = \frac{1}{(2\pi)^n} \int_{\mathbb{R}^n} \mathcal{L}f(u+iy)e^{i(y|x)}\,dx$$

or

$$f(x) = \frac{1}{(2\pi)^n} \int_{\mathbb{R}^n} \mathcal{L}f(u+iy)e^{(u+iy|x)}\,dx\,.$$

Furthermore, the Laplace transform has the following two properties.

1) If $p(D)$ is a differential operator with constant coefficients, then

$$\mathcal{L}(p(D)f)(\lambda) = p(\lambda)\mathcal{L}(f)(\lambda).$$

2) The Laplace transform is a homomorphism:

$$\mathcal{L}(f * g) = \mathcal{L}(f)\mathcal{L}(g).$$

From 2) one sees that the Laplace transform transforms the Volterra equation into

$$\mathcal{L}(a) = \mathcal{L}(b) + \mathcal{L}(k)\mathcal{L}(a),$$

which gives

$$\mathcal{L}(a) = \frac{\mathcal{L}(b)}{1 - \mathcal{L}(k)}.$$

8.2 Spherical Functions

In order to generalize the notion of the Laplace transform to ordered symmetric spaces, we need to find the functions corresponding to the exponential function e_u. These will be the ***spherical functions.*** As one already sees in the case of Riemannian symmetric spaces (cf. [45]), there are different ways to define a spherical function.

1) The differential equation: The spherical functions are the normalized, $\varphi(\mathbf{o}) = 1$, eigenfunctions of the commutative algebra $\mathbb{D}(G/K)$ of invariant differential operators on G/K.

2) The integral equation: Spherical functions satisfy the integral equation

$$\int_K \varphi(xky)\, dk = \varphi(x)\varphi(y).$$

3) The algebraic property: Denote the algebra of compactly supported, K-bi-invariant function on G by $C_c^\infty(G//K)$. Let φ be a K-bi-invariant function on G. Then the map

$$C_c^\infty(G//K) \ni f \mapsto \int_G f(x)\varphi(x)\, dx \in \mathbb{C}.$$

is a homomorphism if and only if φ is a spherical function.

4) The integral representation: For $\lambda \in \mathfrak{a}_C^*$ and ρ half the sum over the positive roots, we define

$$\varphi_\lambda^K(x) = \int_K e_\lambda^K(kx)\, dk, \tag{8.2}$$

where $e_\lambda^K(x) = e^{<\lambda-\rho, a_K(x)>}$ (cf. (5.18) on p. 136), with H replaced by K. Then φ_λ^K is spherical function, and every spherical function has an integral representation of this form for a suitable λ. Furthermore, $\varphi_\lambda^K = \varphi_\mu^K$ if and only if there is a w in the Weyl group of A such that $\lambda = w\mu$.

We remark here that there is no hope in general of using 4) to define a function on $\mathcal{M}$ if we replace K by H. This is due to the fact that $G \neq HAN$ and H is not compact, so the integral does not converge for arbitrary x.

One of the reasons for the fact that all those different definitions give the same class of functions in the Riemannian case is that $\mathbb{D}(G/K)$ contains an elliptic differential operator. Thus every joint eigendistribution is automatically an analytic eigenfunction. As this does not hold for the non-Riemannian symmetric spaces, the different definitions may lead to different classes of functions, distributions, or hyperfunctions.

Let $\mathcal{M} = G/H$ be an irreducible, noncompactly causal symmetric space with $G \subset G_\mathbb{C}$. We recall some basic structure theory. Let $\mathfrak{a}$ be maximal abelian in $\mathfrak{p}$ contained in $\mathfrak{q}_\mathfrak{p}$ and $\Delta = \Delta_0 \cup \Delta_+ \cup \Delta_-$ be the set of roots of $\mathfrak{a}$ in $\mathfrak{g}$. Choose a positive system Δ^+ in Δ such that $\Delta^+ = \Delta_0^+ \cup \Delta_+$. As usual, we set

$$\rho = \frac{1}{2} \sum_{\alpha \in \Delta^+} [\dim \mathfrak{g}_\alpha]\, \alpha\,.$$

Let $C = C(\Delta^+)$ be the positive open Weyl chamber in $\mathfrak{a}$ corresponding to Δ^+ (cf. p. 116). Fix a cone-generating element $X^0 \in \mathfrak{a}$ such that $\Delta_+ = \{\alpha \in \Delta \mid \alpha(X^0) = 1\}$ and $\Delta_0 = \{\alpha \in \Delta \mid \alpha(X^0) = 0\}$. Let $C = C_{\max}(X^0)$, (cf. (4.19) on p. 98). Finally, let $S = S(C) = H \exp C$. Then S is a closed semigroup and $S \subset HAN$ (cf. Theorem 5.4.7). In particular, the function

$$e_\lambda(s) = e_\lambda^H(s) := e^{<\lambda-\rho, a_H(s)>}\,, \quad \lambda \in \mathfrak{a}_C^*, \tag{8.3}$$

is well defined on S.

Definition 8.2.1 A *spherical function* is a H–bi-invariant function φ defined on the interior of S such that for all $s, t \in S^o$, the function

$$h \mapsto \varphi(sht)$$

is integrable over H and

$$\int_H \varphi(sht)\, dh = \varphi(s)\varphi(t). \qquad \square$$

We will often, without further comment, identify the H–bi-invariant function on S or S^o with the H-invariant function on $\mathcal{M}_+$ and $\mathcal{M}_+^o$, respectively. Thus we may view spherical functions as functions on $\mathcal{M}$.

Let us fix a Haar measure on G and other groups before we go on. We normalize the Haar measures on A and $\mathfrak{a}^*$ such that they are dual to each other, i.e., the Fourier inversion formula for the abelian group A holds without constants:

$$\hat{f}(\lambda) := \int_A f(a) a^{-i\lambda}\, da \Rightarrow f(a) = \int_{\mathfrak{a}^*} \hat{f}(\lambda) a^{i\lambda}\, d\lambda$$

Further, we normalize the measures dn and $dn^\sharp = \tau(dn)$ on N and $N^\sharp$ such that

$$\int_{N^\sharp} e^K_{-2\rho}(n^\sharp)\, dn^\sharp = 1\,.$$

We normalize the Haar measure on N_0 and $N_0^\sharp$ in the same way by using $\rho_0 = \frac{1}{2}\sum_{\alpha\in\Delta_0^+} \dim \mathfrak{g}_\alpha\, \alpha$ instead of ρ. We choose dn_+ on N_+ such that

$$dn = dn_+\, dn_0\,.$$

Let $dn_- = \tau(dn_+)$. We choose the measure dX on $\mathfrak{n}_-$ such that for all $f \in C_c(N_-)$,

$$\int_{N_-} f(n_-)\, dn_- = \int_{\mathfrak{n}_-} f(\exp X)\, dX\,.$$

Then we relate the measures on G and H by

$$\int_G f(x)\, dx = \int_H \int_A \int_N f(han) a^{2\rho}\, dh\, da\, dn$$

for every $f \in C_c(G)$ with $\mathrm{Supp}(f) \subset HAN$.

Theorem 8.2.2 *Let* $\mathcal{E} = \{\lambda \in \mathfrak{a}_C \mid \forall\alpha \in \Delta_+ : \mathrm{Re}(\lambda + \rho|\alpha) < 0\}$. *Let* $\lambda \in \mathcal{E}$ *and let* $s \in S^o$. *Then* $H \ni h \mapsto e_\lambda(sh) \in \mathbb{C}$ *is integrable and*

$$\varphi_\lambda(s) := \int_H e_\lambda(sh)\, dh \in \mathbb{C}$$

is a spherical function. $\square$

The idea of the proof is to use the integral formulas in [128] to rewrite the integral defining φ_λ as an integral over K:

$$\varphi_\lambda(s) = \int_{K\cap HAN} e_\lambda(sk)e_{-\lambda}(k)\,dk.$$

As the semigroup S^o acts by compressions on $H/(H\cap K)$, it follows that $HAN\cap K \ni k \mapsto e_\lambda(sk)$ is actually bounded. A simple $\mathfrak{sl}_2$ reduction shows that the exstension of $K\cap HAN \ni k \mapsto e_{-\lambda}(k)$ by zero outside $K\cap HAN$ is continuous if $\lambda \in \mathcal{E}$. The proof actually shows that φ_λ is a well-defined spherical function on the set

$$\mathcal{E}' := \left\{\lambda \in \mathfrak{a}_C^* \,\middle|\, \int_{K\cap HAN} e_{-\operatorname{Re}\lambda}(k)\,dk < \infty\right\}. \tag{8.4}$$

which in general is bigger than $\mathcal{E}$.

For $g \in G_0$, the decomposition $g = h(g)a(g)n(g)$ is just the usual Iwasawa decomposition. Let

$$\rho_+ = \frac{1}{2}\sum_{\alpha\in\Delta_+} m_\alpha\alpha. \tag{8.5}$$

Then $\rho = \rho_0 + \rho_+$. Let φ^0_λ denote the spherical function on the Riemannian symmetric space $G_0/(K\cap H)$:

$$\varphi^0_\lambda(x) = \int_{K\cap H} e_{\lambda+\rho_+}(xk)\,dk.$$

Denote the G_0-component of $g \in HG_0N$ by $g_0(x)$.

Lemma 8.2.3 *Let $s \in S^o$ and $\lambda \in \mathcal{E}$. Then*

$$\varphi_\lambda(s) = \int_{H\cap K\backslash H} \varphi^0_{\lambda-\rho_+}(g_0(sh))\,d\dot{h},$$

where $d\dot{h}$ denotes a suitable normalized invariant measure on $H/(H\cap K)$. In particular, $\varphi_{w\lambda} = \varphi_\lambda$ for every $w \in W_0$. □

Let $\mathbb{D}(\mathcal{M})$ be the algebra of invariant differential operators on $\mathcal{M}$. To $\lambda \in \mathfrak{a}_C^*$ there corresponds a homomorphism $\chi_\lambda : \mathbb{D}(\mathcal{M}) \to \mathbb{C}$. In short, this homomorphism can be constructed by choosing an element u in the universal enveloping algebra $U(\mathfrak{g})$ that corresponds to D by right differentiation. Then project u to $U(\mathfrak{a}) \simeq S(\mathfrak{a}^*)$ along $\mathfrak{h}U(\mathfrak{g}) \oplus U(\mathfrak{g})\mathfrak{n}$ and evaluate at $\lambda - \rho$.

Lemma 8.2.4 *Let $\lambda \in \mathcal{E}$. Let $A(C) := \exp C \subset A$. Then $\varphi_\lambda|_{HA(C)H}$ is analytic and $D\varphi_\lambda = \chi_\lambda(D)\varphi_\lambda$ on $HA(C)H$.* □

8.3 The Asymptotics

In this section we describe the asymptotic behavior of φ_λ on $A(C)$. For that we need three c-functions. Define

$$c_\Omega(\lambda) = \int_\Omega e_{-\lambda}(\exp X)\, dX\,, \quad c_0(\lambda) = \int_{\overline{N}_0} e_{-\lambda}(n^\sharp)\, dn^\sharp,$$

and

$$c(\lambda) = c_\Omega(\lambda) c_0(\lambda)$$

The function $c_0(\lambda)$ is the usual Harish-Chandra c-function for the symmetric space $\mathcal{M}_0 = G_0/(K \cap H)$ and has thus a well-known product formula $c_0(\lambda) = \prod_{\alpha \in \Delta_0^+} c_\alpha(\lambda_\alpha)$, cf. [35] or [45]. On the other hand, the function $c_\Omega(\lambda)$ is known only for some special cases, cf. [26]. The integral defining $c_\Omega(\lambda)$ converges exactly for $\lambda \in \mathcal{E}'$. The integral defining $c_0(\lambda)$ converges for $\lambda \in \mathfrak{a}_C^*$ such that $\mathrm{Re}(\lambda|\alpha) > 0$ for every $\alpha \in \Delta_0^+$. One should note that one can replace λ and ρ by $\lambda|_{\mathfrak{a} \cap [\mathfrak{g}_0, \mathfrak{g}_0]}$ and ρ_0, respectively, in all calculations involving $c_0(\lambda)$.

Lemma 8.3.1 *Let $\lambda \in \mathcal{E}$ be such that* $\mathrm{Re}(\lambda|\alpha) > 0$ *for every $\alpha \in \Delta_0^+$. Then*

$$c(\lambda) = \int_{N^\sharp \cap HAN} e_{-\lambda}(n^\sharp)\, dn^\sharp. \qquad \square$$

Let us introduce the notation

$$a \xrightarrow{A(C)} \infty$$

for the fact that $a \in A(C)$ and for all $\alpha \in \Delta^+$ we have

$$\lim a^\alpha = \infty.$$

Rewriting the integral defining φ_λ as an integral over $\overline{N} \cap HAN$, we get the following theorem.

Theorem 8.3.2 *Let $\lambda \in \mathcal{E}$ be such that* $\mathrm{Re}(\lambda|\alpha) > 0$ *for every $\alpha \in \Delta_0^+$. Then*

$$\lim_{a \xrightarrow{A(C)} \infty} a^{\rho-\lambda} \varphi_\lambda(a) = c(\lambda)\,.$$

Furthermore,

$$\lim_{t\to\infty} e^{t<\rho-\lambda, X^0>} \varphi_\lambda(a \exp tX^0) = c_\Omega(\lambda) \varphi^0_{\lambda+\rho_+}(a). \qquad \square$$

Example 8.3.3 (The Hyperboloids) Let $G = \mathrm{SO}_o(1,n)$ and let $H = \mathrm{SO}_o(1,n-1)$, $n \geq 2$, cf. Section 1.5. Let

$$\mathfrak{a} = \left\{ a_t = \begin{pmatrix} \cosh t & 0 & \sinh t \\ 0 & I_{n-1} & 0 \\ \sinh t & 0 & \cosh t \end{pmatrix} \middle| t \in \mathbb{R} \right\}.$$

Let $X^0 = E_{1,n+1} + E_{n+1,1}$. Then $\mathfrak{a} = \mathbb{R}X^0$. We choose the positive root such that $\alpha(X^0) = 1$ and identify $\mathfrak{a}_C^*$ with $\mathbb{C}$ by $z \mapsto -z\alpha$. Then $\rho = -(n-1)/2$. The spherical function φ_λ is given by

$$\varphi_\lambda(a_t) = \int_0^\infty (\cosh t + \sinh t \cosh\theta)^{-\lambda-(n-1)/2} (\sinh\theta)^{n-2} d\theta$$

Let Q_ν^μ be the usual Legendre function of the second kind and let ${}_2F_1$ be the hypergeometric function. From [23] we get Theorem 8.3.4

Theorem 8.3.4 *The integral defining $\varphi_\lambda(a_t)$ converges for $t > 0$ and* $\mathrm{Re}\,\lambda < -(n-3)/2$. *Furthermore,*

$$\begin{aligned} \varphi_\lambda(a_t) &= \gamma_n \frac{\Gamma(\lambda - \frac{n-3}{2})}{\Gamma(\lambda + \frac{n-1}{2})(\sinh t)^{\frac{n}{2}-1}} Q^{\frac{n}{2}-1}_{\lambda-\frac{1}{2}}(\cosh t) \\ &= 2^{n-2}\Gamma\left(\frac{n-1}{2}\right) \frac{\Gamma(\lambda - \frac{n-3}{2})}{\Gamma(\lambda+1)} (2\cosh t)^{-\lambda-\frac{n-1}{2}} \cdot \\ &\quad \cdot {}_2F_1\left(\frac{\lambda + \frac{n+1}{2}}{2}, \frac{\lambda + \frac{n-1}{2}}{2}, \lambda+1, \frac{1}{(\cosh t)^2}\right) \end{aligned}$$

where γ_n is a constant depending only on n. □

In particular, for $n = 2$,

$$\varphi_\lambda(a_t) = \frac{1}{\lambda} \frac{1}{\sinh t} e^{-\lambda t}.$$

In this case $\Delta^+ = \Delta_+$. Thus $c(\lambda) = c_\Omega(\lambda)$ and

$$\begin{aligned} c_\Omega(\lambda) &= 2^{n-2}\Gamma\left(\frac{n-1}{2}\right) \frac{\Gamma(\lambda - \frac{n-3}{2})}{\Gamma(\lambda+1)} \\ &= 2^{n-2} B\left(\lambda - \frac{n-3}{2}, \frac{n-1}{2}\right) \end{aligned}$$

Furthermore,

$$\frac{1}{c_\Omega(\lambda)} \varphi_\lambda(a_t) = {}_2F_1\left(\frac{\lambda + \frac{n+1}{2}}{2}, \frac{\lambda + \frac{n-1}{2}}{2}, \lambda+1, \frac{1}{\cosh^2 t}\right)$$

which extends to a meromorphic function on $\mathfrak{a}_C^*$ holomorphic for $\mathrm{Re}\,\lambda > -1$.

□

Example 8.3.5 (The Group Case) Let G be a connected semisimple Lie group such that G_C/G is ordered. Let $\mathfrak{a}$ be a maximal abelian subalgebra of $\mathfrak{q} = i\mathfrak{g}$ contained in $\mathfrak{p} \oplus i\mathfrak{k}$. Note that $i\mathfrak{a}$ is a compact Cartan subalgebra of $\mathfrak{g}$. The Weyl group W_0 is given as $N_K(\mathfrak{a})/Z_K(\mathfrak{a})$, which is the Weyl group of $\mathfrak{a}$ in K. Let $\epsilon(w) = \det w$, $w \in W$.

Theorem 8.3.6 *The spherical function φ_λ is given by*

$$\varphi_\lambda(\exp X) = c\frac{\sum_{w\in W_o} \epsilon(w)e^{-<w\lambda,X>}}{\prod_{\alpha\in\Delta^+} <\alpha,\lambda> \prod_{\alpha\in\Delta^+} \sinh <\alpha, X>}$$

for $X \in C^o \cap \mathfrak{a}$. □

We get for suitable constants γ, γ_0, and γ_1 such that $\gamma_0\gamma_1 = \gamma$,

$$c_0(\lambda) = \frac{\gamma_0}{\prod_{\alpha\in\Delta_0^+} <\lambda,\alpha>},$$

$$c_\Omega(\lambda) = \frac{\gamma_1}{\prod_{\alpha\in\Delta_+} <\lambda,\alpha>},$$

and

$$c(\lambda) = \frac{\gamma}{\prod_{\alpha\in\Delta^+} <\lambda,\alpha>}.$$

Furthermore,

$$c(\lambda)^{-1}\varphi_\lambda(\exp X) = \frac{\sum_{w\in W_o} \epsilon(w)e^{-<w\lambda,X>}}{\prod_{\alpha\in-\Delta^+} \sinh <\alpha, X>}$$

for $X \in C^o \cap \mathfrak{a}$ and as a function of λ this function extends to a holomorphic function on $\mathfrak{a}_\mathbb{C}$. □

8.4 Expansion Formula for the Spherical Functions

Let us recall the case of spherical functions on the Riemannian symmetric space G/K, cf. [40, 41, 45], before we talk about the causal symmetric spaces. Let $\varphi_\lambda(x) = \int_K e_\lambda(xk)\,dk$ as before. Let $c^r(\lambda) = \int_{N^\sharp} e_{-\lambda}(n^\sharp)\,dn^\sharp$ be the usual Harish-Chandra c-function for the Riemannian symmetric space G/K, which is isomorphic to the r-dual space $G^r/K^r = \mathcal{M}^r$. Notice that we can view A as a subset of $\mathcal{M}^r$. If $D \in \mathbb{D}(\mathcal{M}^r)$, then the radial part $\Delta_{\mathcal{M}^r}(D)$ of D is a differential operator on $A(C)$, such that for every K–bi-invariant function F on G we have

$$(DF)|_{A(C)} = \Delta_{\mathcal{M}^r}(D)(F|_{A(C)}).$$

In particular, this holds for the spherical functions $\varphi_\lambda^{K^r}$.

$$\Delta_{\mathcal{M}^r}(D)(\varphi_\lambda^{K^r}|_{A(C)}) = \chi_\lambda(D)(\varphi_\lambda^{K^r}|_{A(C)}).$$

Let $\Lambda = \mathbb{N}\Delta^+$ and construct the function Γ_μ, $\mu \in \Lambda$ recursively by

$$\begin{aligned} \Gamma_0 &= 1, \\ &[(\mu|\mu) - 2(\mu|\lambda)]\Gamma_\mu(\lambda) \\ &= 2\sum_{\alpha\in\Delta^+} m_\alpha \sum_{k\geq 0} \Gamma_{\mu-2k\alpha}(\lambda)\left((\mu+\rho-2k\alpha|\alpha) - (\alpha|\lambda)\right). \end{aligned}$$

Define for $a \in A(C)$:

$$\Phi_\lambda(a) := a^{\lambda-\rho}\sum_{\mu\in\Lambda}\Gamma_\mu(\lambda)a^{-\mu} = a^{\lambda-\rho}\left(1 + \sum_{\mu\in\Lambda\setminus\{0\}}\Gamma_\mu(\lambda)a^\mu\right). \tag{8.6}$$

From [40, 45] we see

Theorem 8.4.1 *Let W be the Weyl group of Δ. Then there exists an open dense set $U \subset \mathfrak{a}_C^*$ such that for $\lambda \in U$, $\{\Phi_{s\lambda} \mid s \in W\}$ is a basis of the space of functions on $A(C)$ satisfying the differential equation*

$$\Delta_{\mathcal{M}^r}(D)\Phi = \chi_\lambda(D)\Phi, \quad \forall D \in \mathbb{D}(\mathcal{M}^r). \qquad \square$$

For an H-bi-invariant function on $HA(C)H$, define a K^r-bi-invariant function f^γ on $K^rA(C)K^r \subset G^r$ by $f^\gamma(k_1ak_2) = f(a)$. Denote the natural isomorphism $\mathbb{D}(\mathcal{M}) \simeq \mathbb{D}(\mathcal{M}^r)$ by γ. Then $(Df)^\gamma = D^\gamma f^\gamma$. This, together with Theorem 8.4.1 and the asymptotics for the spherical function φ_λ (cf. [130]), gives Theorem 8.4.2

Theorem 8.4.2 *Let $\lambda \in \mathcal{E} \cap U$ and $a \in A(C)$. Then*

$$\varphi_\lambda(a) = c_\Omega(\lambda)\sum_{w\in W_0} c_0(w\lambda)\Phi_{w\lambda}(a). \qquad \square$$

As a corollary of this we get the following.

Corollary 8.4.3 *The functions*

$$\mathcal{E} \times A(C) \ni (\lambda, a) \mapsto \frac{1}{c_\Omega(\lambda)}\varphi_\lambda(a),\ \frac{1}{c(\lambda)}\varphi_\lambda(a)$$

extend to $\mathfrak{a}_C^* \times A(C)$ as meromorphic functions in λ. $\square$

We will denote the H–bi-invariant extension $h_1 a h_2 \mapsto [1/c_\Omega(\lambda)]\varphi_\lambda(a)$ by the same symbol. Similarly we denote the H–bi-invariant function on $HA(C)H$ (or $HA(C)\mathrm{o}$) that extends Φ_λ by the same symbol, Φ_λ. Then the product formula for c shows that

$$c^r(\lambda) = c_+(\lambda)c_0(\lambda)\,,$$

where $c_+(\lambda)$ is the part of the product coming from the roots in Δ_+. As Δ_+ is W_0-invariant, it follows that $c_+(\lambda)$ is W_0-invariant. Thus

$$\frac{c^r(\lambda)}{c(\lambda)} = \frac{c_+(\lambda)}{c_\Omega(\lambda)},$$

and $c_+(\lambda)/c_\Omega(\lambda)$ is W_0-invariant. Let

$$\varphi^r_\lambda(x) = \sum_{w\in W} c^r(w\lambda)\Phi_{w\lambda}(x)$$

be the spherical function on $\mathcal{M}^r$ for the parameter λ.

Theorem 8.4.4 *Let the notation be as above. Then*

$$\varphi^r_\lambda = \sum_{w\in W_\circ\backslash W} \frac{c(w\lambda)}{c(w\lambda)}\varphi_{w\lambda} = \sum_{w\in W_\circ\backslash W} \frac{c_+(w\lambda)}{c_\Omega(w\lambda)}\varphi_{w\lambda}. \qquad \square$$

8.5 The Spherical Laplace Transform

A *causal* kernel or *Volterra kernel* on $\mathcal{M}$ is a function on $\mathcal{M}\times\mathcal{M}$ which is continuous on $\{(x,y) \mid x \leq y\}$ and zero outside this set. We compose two such kernels F and G via the formula

$$\begin{aligned} F\#G(x,y) &= \int_{\mathcal{M}} F(x,m)G(m,y)\,dm \\ &= \int_{[x,y]} F(x,m)G(m,y)\,dm\,. \end{aligned}$$

This definition makes sense because $\mathcal{M}$ is globally hyperbolic (cf. Theorem 5.3.5). With respect to this multiplication, the space of Volterra kernels $V(\mathcal{M})$ becomes an algebra, called the *Volterra algebra* of $\mathcal{M}$. A Volterra kernel is said to be *invariant* if

$$F(gx,gy) = F(x,y) \quad \forall g\in G\,.$$

The space $V(\mathcal{M})^\sharp$ of all invariant Volterra kernels is a commutative subalgebra of $V(\mathcal{M})$, cf. [25], Théorème 1. An invariant kernel is determined by the function

$$f(m) = F(\mathbf{o}, m), \quad m \in \mathcal{M}_+,$$

which is continuous on $\mathcal{M}_+$ and H-invariant. On the other hand, for f a continuous H-invariant function on $\mathcal{M}_+$, we can define an invariant Volterra kernel F by

$$F(a \cdot \mathbf{o}, b \cdot \mathbf{o}) = f(a^{-1}b \cdot \mathbf{o}).$$

Under this identification the product $\#$ corresponds to the "convolution"

$$f\#g(m) = \int_{G/H} f(x \cdot \mathbf{o})g(x^{-1} \cdot m)\, d\dot{x}.$$

So the algebra $V(\mathcal{M})^\sharp$ becomes the algebra of continuous H-invariant functions on $S \cdot \mathbf{o}$ with the above "convolution" product.

The *spherical Laplace transform* of an invariant Volterra kernel F is defined by

$$\mathcal{L}F(\lambda) = \int_{\mathcal{M}} F(\mathbf{o}, m)e_\lambda(m)\, dm\,.$$

Here, by abuse of notation, we view the H-invariant function e_λ as a function on $\mathcal{M}_+$. The corresponding formula for the H-invariant function on $S \cdot x_o$ is

$$\mathcal{L}(f)(\lambda) = \int_{\mathcal{M}} f(x)e_\lambda(x)\, dx.$$

Let $\mathcal{D}(f)$ be the set of λ for which the integral converges absolutely. Using Fubini's theorem, we get Lemma 8.5.1.

Lemma 8.5.1 *Let $f, g \in V(\mathcal{M})^\sharp$ be invariant causal kernels. Then $\mathcal{D}(f) \cap \mathcal{D}(g) \subset \mathcal{D}(f\#g)$. For $\lambda \in \mathcal{D}(f) \cap \mathcal{D}(g)$ we have*

$$\mathcal{L}(f\#g)(\lambda) = \mathcal{L}f(\lambda)\mathcal{L}g(\lambda). \qquad \square$$

Let $M = Z_{H\cap K}(A)$. In "polar coordinates" on $\mathcal{M}$,

$$H/M \times A(C) \ni (hM, a) \mapsto ha \cdot \mathbf{o} \in \mathcal{M}_+,$$

we have, for $f \in C_c(S/H)$,

$$\int_{S/H} f(x)\, dx = c \int_H \int_{A(C)} f(ha\mathbf{o})\delta(a)\, dX\, dh\,,$$

where c is some positive constant depending only on the normalization of the measures, and

$$\delta(a) = \prod_{\alpha\in\Delta^+} (\sinh < \alpha, \log a >)^{m_\alpha} .$$

Theorem 8.5.2 *Let $c > 0$ be the constant defined above. Let $f : S/H \to \mathbb{C}$ be continuous and H-invariant. If $\lambda \in \mathcal{D}(f)$, then φ_λ exists and*

$$\mathcal{L}f(\lambda) = c \int_{A(C)} f(a)\varphi_\lambda(a)\delta(a)\, da. \qquad \square$$

To invert the Laplace transform, we define the *normalized* spherical function $\tilde{\varphi}_\lambda$ by

$$\tilde{\varphi}_\lambda(x) := \frac{1}{c(\lambda)}\varphi_\lambda(x)$$

and the *normalized* Laplace transform by

$$\tilde{\mathcal{L}}(f)(\lambda) := \int_{A(C)} f(a)\tilde{\varphi}_\lambda(a)\delta(a)\, da.$$

Then

$$\tilde{\mathcal{L}}(f)(\lambda) = \sum_{w\in W_o} \frac{c_0(w\lambda)}{c_0(\lambda)} \int_{A(C)} f(a)\Phi_{w\lambda}(a)\delta(a)\, da.$$

as $c_\Omega(\lambda)$ is W_0-invariant. Note that the unknown function $c_\Omega(\lambda)$ disappears in this equation.

Let $\lambda \in i\mathfrak{a}^*$. Then

$$c_0(-\lambda) = \overline{c_0(\lambda)}$$

and

$$c_0(-\lambda)c_0(\lambda) = c_0(-w\lambda)c_0(w\lambda)\,. \tag{8.7}$$

Thus

$$\left|\frac{c_0(w\lambda)}{c_0(\lambda)}\right| = \left|\frac{c_0(-\lambda)}{c_0(-w\lambda)}\right| = \left|\frac{c_0(\lambda)}{c_0(w\lambda)}\right|$$

and

$$\left|\frac{c_0(w\lambda)}{c_0(\lambda)}\right| = 1,$$

so $\lambda \mapsto c_0(w\lambda)/c_0(\lambda)$ has no poles on $i\mathfrak{a}^*$. Let $C_c^\infty(HA(C)H//H)$ be the space of H–bi-invariant functions with compact support in $H\backslash HA(C)H/H$.

Theorem 8.5.3 *Let $f \in C_c^\infty(HA(C)H//H)$. Then $\mathcal{E} \ni \lambda \mapsto \tilde{\mathcal{L}}(f)(\lambda) \in \mathbb{C}$ extends to a meromorphic function on $\mathfrak{a}_C^*$ with no poles on $i\mathfrak{a}^*$.* $\square$

From the functional equation for φ_λ we get

$$\sum_{w\in W_o\backslash W} c^r(w\lambda)\tilde{\mathcal{L}}(f)(w\lambda) = c_1\mathcal{F}(f^\gamma)(\lambda),$$

where c_1 is a positive constant independent of f and λ, and $\mathcal{F}$ denotes the spherical Fourier transform on $\mathcal{M}^r$:

$$\mathcal{F}(f^\gamma)(\lambda) = \int_{G^r} f^\gamma(x)\varphi_\lambda^r(x)\,dx\,.$$

Let $E_\lambda(h_1ah_2) := \varphi_\lambda^r(a)$ be the H–bi-invariant function on G with the same restriction to $A(C)$ as φ_λ^r. By the inversion formula for the spherical Fourier transform, [45], we have for some constant depending on the normalization of measures and $w = |W|$,

$$\begin{aligned} cf(a) &= \frac{1}{w}\int_{ia^*} \mathcal{F}(f^\gamma)(\lambda)\frac{E_{-\lambda}(a)}{c^r(\lambda)c^r(-\lambda)}\,d\lambda \\ &= \frac{1}{w}\sum_{w\in W_o\backslash W}\int_{ia^*}\tilde{\mathcal{L}}(f)(w\lambda)\frac{c^r(w\lambda)}{c^r(\lambda)c^r(-\lambda)}E_{-\lambda}(a)\,d\lambda \\ &= \frac{1}{w}\sum_{w\in W_o\backslash W}\int_{ia^*}\tilde{\mathcal{L}}(f)(w\lambda)\frac{1}{c^r(-w\lambda)}E_{-\lambda}(a)\,d\lambda \\ &= \frac{1}{w}\sum_{w\in W_o\backslash W}\int_{ia^*}\tilde{\mathcal{L}}(f)(w\lambda)\frac{E_{-w\lambda}(a)}{c^r(-w\lambda)}d\lambda \\ &= \frac{1}{w_o}\int_{ia^*}\tilde{\mathcal{L}}(f)(\lambda)\frac{E_{-\lambda}(a)}{c^r(-\lambda)}d\lambda\,. \end{aligned}$$

This proves the inversion formula at least for $f \in C_c^\infty(HA(C)H//H)$:

Theorem 8.5.4 *Let $\tilde{E}_\lambda := E_\lambda/c^r(\lambda)$. Let $f \in C_c^\infty(HA(C)H//H)$. Then there exists a positive constant c that depends only on the normalization of the measures involved such that for every $a \in A(C)$,*

$$cf(a) = \frac{1}{w_o}\int_{ia^*}\tilde{\mathcal{L}}(f)(\lambda)\tilde{E}_{-\lambda}(a)d\lambda. \qquad \square$$

8.6 The Abel Transform

As in the Riemannian case, we can define the Abel transform and relate that to the spherical Laplace transform. The main difference is that in this case $\mathcal{A}(f)$ does not have compact support even if the support of f is compact. On the contrary, by the nonlinear convexity theorem, the support of $\mathcal{A}(f)$ can be described as a "cone" with a base constructed out of the support of f.

Let f be an H-invariant function on $\mathcal{M}_+$. We define the *Abel transform*, $\mathcal{A}f\colon A \to \mathbb{C}$ of f, by

$$\mathcal{A}(f)(a) = a^{\rho} \int_N f(an)\, dn$$

whenever the integral exists. Using the nonlinear convexity theorem, p. 151, we prove the following lemma.

Lemma 8.6.1 *Let f be a continous H-invariant function on $\mathcal{M}_+$ (extended by zero outside $\mathcal{M}_+$) such that $n \mapsto f(an)$ is integrable on N for all $a \in A$. Let $L \subset c_{\max}$ be the convex hull of* $\log(\mathrm{Supp}(f|_{S\cap A}))$. *Then*

$$\log(\mathrm{Supp}(\mathcal{A}f)) \subset L + c_{\min}. \qquad \square$$

We rewrite now the Integral over $\mathcal{M}_+ \subset \mathcal{M}$ as an integral over AN, cf. [128], to get Theorem 8.6.2.

Theorem 8.6.2 *Let f be an H-invariant function S/H and $\lambda \in \mathcal{D}(f)$. Then*

$$\mathcal{L}(f)(\lambda) = \int_{\exp c_{\max}} a^{\lambda}\mathcal{A}f(a)\, da = \mathcal{L}_A(\mathcal{A}f)(-\lambda),$$

where $\mathcal{L}_A$ is the Euclidean Laplace transform on A with respect to the cone $c_{\max}$. $\square$

The Abel transform can be split up further according to the semidirect product decomposition $N = N_+N_0$. Set

$$\mathcal{A}_+f(g_0) = a^{\rho_+} \int_{N_+} f(g_0 n_+)\, dn_+$$

for $g_0 \in G_0$. Then obviously $\mathcal{A}_+(f)$ is K_0–bi-invariant and

$$\mathcal{A}f(a) = a^{\rho_0} \int_{N_0} \mathcal{A}_+(f)(an_0)\, dn_0\,.$$

Denote by $\mathcal{A}_0$ the Abel transform with respect to the *Riemannian* symmetric space $G_0/K \cap H$. Then we have

$$\mathcal{A}f(a) = \mathcal{A}_0(\mathcal{A}_+f)(a)$$

for all continuous, H-invariant functions $f : S/H \to \mathbb{C}$ such that the above integrals make sense and all $a \in A$. As it is well known how to invert the transform $\mathcal{A}_0$, at least for "good" functions, the inversion of the Abel transform associated to the ordered space reduces to inverting $\mathcal{A}_+f$.

Theorem 8.6.3 *If $f : \mathcal{M}_+ \to \mathbb{C}$ is continuous, H-invariant, and such that the Abel transform exists, then its Abel transform is invariant under W_0, i.e.,*

$$\mathcal{A}f(wa) = \mathcal{A}f(a) \quad \forall a \in A, w \in W_0 . \qquad \square$$

8.7 Relation to Representation Theory

The spherical functions are related to the representation theory on both G and the dual group G^c. Here we explain the relation to the representation theory of G.

Let M be the centralizer of A in K and let P be the minimal parabolic subgroup MAN. For $\lambda \in \mathfrak{a}^*_{\mathbb{C}}$, let $(\pi(\lambda), \mathbf{H}(\lambda))$ be the principal series representation induced from the character $\chi_\lambda : man \mapsto a^\lambda$ of P. The space of smooth vectors in $\mathbf{H}(\lambda)$ is given by

$$\mathcal{C}^\infty(\lambda) = \{f \in \mathcal{C}^\infty(G) \mid \forall man \in MAN, \forall g \in G : f(gman) = a^{-(\lambda+\rho)}f(g)\}$$

and

$$[\pi(\lambda)(g)f](x) = f(g^{-1}x) .$$

The bilinear form

$$\mathcal{C}^\infty(\lambda) \times \mathcal{C}^\infty(-\lambda) \ni (f, g) \mapsto < f, g > := \int_K f(k)g(k)\, dk$$

defines an invariant pairing $\mathcal{C}^\infty(\lambda) \times \mathcal{C}^\infty(-\lambda)$. Extend e_λ to be zero outside HAN. Using the above pairing, we find that $e_\lambda \in \mathcal{C}^{-\infty}(\lambda)^H$ for $\lambda \in -\mathcal{E}$, where $\mathcal{C}^{-\infty}(\lambda)$ is the continuous dual of $\mathcal{C}^\infty(\lambda)$. Furthermore,

$$< f, e_\lambda > = \int_K f(k)e_\lambda(k)\, dk = \int_H f(h)\, dh$$

for $f \in \mathcal{C}^\infty(\lambda)$ and $\lambda \in -\mathcal{E}$. The linear form $f \mapsto < f, e_\lambda >$ has a meromorphic continuation to all of $\mathfrak{a}^*_C$ as an H-invariant element in $\mathcal{C}^{-\infty}(\lambda)$, cf. [2, 3, 128, 140, 141, 145]. Let $f \in C_c^\infty(G)$. Then $\pi_\lambda^{-\infty}(f)e_\lambda \in \mathcal{C}^\infty(-\lambda)$. Hence

$$f \mapsto \Theta_\lambda(f) = < \pi_\lambda^{-\infty}(f)e_\lambda, e_{-\lambda} >$$

is well defined.

Definition 8.7.1 A distribution Θ on G is called *H-spherical* if

1) Θ is H–bi-invariant.

2) There exists a character $\chi : \mathbb{D}(\mathcal{M})$ such that $D(\Theta) = \chi(D)\Theta$. □

Theorem 8.7.2 *Θ_λ is an H-spherical function and $D\Theta_\lambda = \chi_\lambda(D)\Theta_\lambda$ for every $D \in \mathbb{D}(\mathcal{M})$.* □

The relation between the spherical distribution Θ_λ and the spherical function φ_λ is given by Theorem 8.7.3.

Theorem 8.7.3 *If $\lambda \in \mathcal{E}$ and* $\mathrm{Supp}(f) \subset (S^o)^{-1}$, *then*

$$\Theta_\lambda(f) = \int_G f(x)\varphi_{-\lambda}(x^{-1})\, dx.$$ □

Notes for Chapter 8

The material in Section 8.1 is standard, but usually Γ is defined as the set $\{(x, y) \in \mathcal{M} \mid x \geq y\}$. In that case the product for invariant kernels becomes $F\#G = G*F$.

Spherical functions on symmetric spaces of the form $G_\mathbb{C}/G$ were introduced by J. Faraut in [24], where they were used to diagonalize certain integral equations with symmetry and causality conditions. Most of the material in Section 8.2, Section 8.3, and Section 8.5 is from [28]. The proof of Theorem 8.3.6 in [63] used the relation to the principal series in Section 8.7 and the formula of the H-spherical character due to P. Delorme [19]. Examples 8.3.3 and 8.3.5 are from [28]. Lemma 8.2.4 was proved in [131]. A more explicit formula for the spherical function and the $c_\Omega(\lambda)$ function for Cayley-type spaces was obtained by J. Faraut in [26]. The inversion formula for the Laplace transform was proved in [28] by using the explicit formula for the spherical functions. In the same article, an inversion formula was proved for the rank 1 spaces by using the Abel transform. The general inversion formula presented here was proved in [131].

The first main results on spherical functions on Riemannian symmetric spaces are in [40, 41]. The theory was further developed by S. Helgason, cf. [43, 45]. The isomorphism $\gamma : \mathbb{D}(\mathcal{M}) \to \mathbb{D}(\mathcal{M}^r)$ was first constructed by M. Flensted-Jensen in [32].

A Laplace transform associated with the Legendre functions of the second kind was introduced by [17]. In [164] this transform was related to harmonic analysis of the unit disc. A more general Laplace-Jacobi transform associated with the Jacobi functions of the second kind was studied by M. Mizony in [105, 106].

There is by now an extensive literature on the function e_λ and its generalizations. Its importance in harmonic analysis on $\mathcal{M}$ comes from the generalized Poisson transformation, i.e., the embedding of generalized principal series representations into spaces of eigenfunctions on $\mathcal{M}$. A further application is the construction of the spherical distributions Θ_λ. We refer to [2, 3], [12, 13], [128], [140, 141, 145] and [161], to mention just a few.

Chapter 9

The Wiener-Hopf Algebra

The classical Wiener-Hopf equation is an equation of the form

$$[I + W(f)]\xi = \eta,$$

where $\eta \in \mathbf{L}^2(\mathbb{R}^+)$, ξ is an unknown function on $\mathbb{R}^+$, and

$$[W(f)\xi](s) = \int_0^\infty f(s-t)\xi(t)\, dt,$$

where $f \in L^1(\mathbb{R})$. Equations of this type can be studied using C^*-algebra techniques because the C^*-algebra $\mathcal{W}_{\mathbb{R}^+}$ generated by these operators has a sufficiently tractable structure. It contains the ideal $\mathcal{K}$ of compact operators on $\mathbf{L}^2(\mathbb{R}^+)$ and the quotient $\mathcal{W}_{\mathbb{R}^+}/\mathcal{K}$ is isomorphic to $C_0(\mathbb{R})$.

There is a natural generalization of these Wiener-Hopf operators to operators acting on square integrable functions defined on the positive domain in an ordered homogeneous space (cf. [54, 55, 108]). In this section we consider Wiener-Hopf operators on the positive domain $\mathcal{M}_+$ of a non-compactly causal symmetric space $\mathcal{M} = G/H$. We recall the ordering $\leq_S$ that has been used in the definition of the order compactification $\mathcal{M}^{cpt}$ and the corresponding positive domain $\mathcal{M}_+ \subset \mathcal{M}$. Next we consider an invariant measure $\mu_{\mathcal{M}}$ on M and the corresponding unitary action of G on $\mathbf{L}^2(\mathcal{M})$ as well as the integrated representation $\pi_{\mathcal{M}}$ of the group algebra $\mathbf{L}^1(G)$ on $\mathbf{L}^2(\mathcal{M})$, i.e.,

$$[\pi_{\mathcal{M}}(f)\phi](x) = \int_G f(g)\phi(g^{-1}\cdot x)\, d\mu_G(g) \qquad \phi \in L^2(M), f \in \mathbf{L}^1(G).$$

The *Wiener-Hopf algebra* $\mathcal{W}_{\mathcal{M}_+}$ is defined to be the C^*-algebra generated by the compressions of the operators $\pi_{\mathcal{M}}(f)$ to the subspace $\mathbf{L}^2(\mathcal{M}_+)$. Our intention in this chapter is to describe the structure of the Wiener-Hopf algebra in terms of composition series which are determined by the G-orbit structure of $\mathcal{M}^{cpt}$. Further, we explain how the Wiener-Hopf algebra can be obtained as the homomorphic image of the C^*-algebra of a groupoid.

We start by showing the connection of the order compactification as described in Section 2.4 with the functional analytic compactification using the weak-star topology in $L^\infty(G)$ and characteristic functions, which is commonly used in the context of Wiener-Hopf algebras.

Lemma 9.1.1 *The mapping*

$$\Psi : \mathcal{F}_\downarrow(G) \to L^\infty(G), \quad A \mapsto \chi_A,$$

where χ_A is the characteristic function of A, is a continuous injection.

Proof: 1) Ψ is injective: Let $A, B \in \mathcal{F}_\downarrow(G)$ with $\chi_A = \chi_B$ in $L^\infty(G)$, i.e., almost everywhere. Let $a \in \text{Int}(A)$. If $x \notin B$, then there exists a neighborhood U of a in A such that $U \cap B = \emptyset$. Therefore $\mu(U) \leq \mu(A \setminus B) = 0$, a contradiction. We conclude that $\text{Int}(A) \subset B$. In view of Lemma 2.4.7, this implies that $A \subset B$. The inclusion $B \subset A$ follows by symmetry.

2) Ψ is continuous: Let $A_n \to A$ in $\mathcal{F}_\downarrow(G)$. We have to show that $\lim \chi_{A_n} = \chi_A$ almost everywhere. We proceed in two steps:

a) $\chi_{\text{Int}(A)} \leq \text{liminf}\chi_{A_n}$: Let $a \in \text{Int}(A)$. Then there exists $b \in \text{Int}(A) \cap \text{Int}(\uparrow a)$. We choose a neighborhood U of b in $\uparrow a$ which is contained in A. Then there exists $n_U \in \mathbb{N}$ such that $A_n \cap U \neq \emptyset$ for all $n \geq n_U$. Let $b_n \in A_n \cap U$. Then $b_n \in \uparrow a$ and therefore $a \in \downarrow b_n \subset \downarrow A_n = A_n$. This shows that $\text{liminf}\chi_{A_n}(a) = 1$.

b) $\text{limsup}\chi_{A_n} \leq \chi_A$: Let $a \in G$ with $\text{limsup}\chi_{A_n}(a) = 1$. We have to show that $\chi_A(a) = 1$, i.e., $a \in A$. To see this, let U be an arbitrary neighborhood of a in G. Then the condition $\text{limsup}\chi_{A_n}(a) = 1$ implies for every $n_0 \in \mathbb{N}$ the existence of $n \geq n_0$ with $a \in A_n$. It follows in particular that $A_n \cap U \neq \emptyset$. Hence $a \in \text{limsup}A_n = A$. □

Lemma 9.1.2 *Let $B := \{f \in L^\infty(G) \mid \|f\|_\infty \leq 1\}$ and set*

$$g \cdot f := f \circ \lambda_{g^{-1}} \quad \textit{for} \quad g \in G, f \in B,$$

where λ_g is left multiplication by g in G. Then the mapping $G \times B \to B, (g, f) \mapsto g \cdot f$ is continuous, i.e., G acts continuously on the compact space B.

Proof: Let $g_0 \in G, f_0 \in B$ and $h \in L^1$. We have to show that the function $(g, f) \mapsto \langle g \cdot f, h \rangle$ is continuous at (g, f). For $g \in G$ and $f \in B$ we have

$$\begin{aligned}\langle g \cdot f, h \rangle - \langle g_0 \cdot f_0, h \rangle &= \langle g \cdot f - g_0 \cdot f, h \rangle + \langle g_0 \cdot f - g_0 \cdot f_0, h \rangle \\ &= \langle f, g^{-1} \cdot h - g_0^{-1} \cdot h \rangle + \langle f - f_0, g_0^{-1} \cdot h \rangle\end{aligned}$$

and both summands tend to 0 for $g \to g_0$ and $f \to f_0$ because $g_0^{-1} \cdot h \in L^1$ and $\|g^{-1} \cdot h - g_0^{-1} \cdot h\|_1 \to 0$. □

Lemma 9.1.3 *1) The mapping*

$$\Psi : \mathcal{F}_\downarrow(G) \to B, \qquad A \mapsto \chi_A$$

is a G-equivariant continuous and monotone mapping of compact G-pospaces.

2) The mapping

$$\mathcal{M}^{cpt} \to L^\infty(G), \qquad A \mapsto \chi_A$$

is a homeomorphism onto its image.

Proof: 1) follows from Lemma B.1.2, Proposition 9.1.1, and Lemma 9.1.2.
2) is a direct consequence of 1) and the compactness of $\mathcal{M}^{cpt}$ □

This lemma shows that there is no essential difference between the compactifications in $L^\infty(G)$ and in $\mathcal{F}(G)$.

Recall from Theorem 2.4.6 that $\mathcal{M}$ is open in $\mathcal{M}^{cpt}$. Thus it is reasonable to extend $\mu_{\mathcal{M}}$ to $\mathcal{M}^{cpt}$ via $\mu_{\mathcal{M}}(\mathcal{M}^{cpt} \backslash \mathcal{M}) = 0$. In particular, we can identify $\mathcal{M}_+$ and $\mathcal{M}_+^{cpt}$ as measure spaces. We write $\mu_{\mathcal{M}_+}$ for $\mu_M|_{\mathcal{M}_+}$. If now

$$p : \mathbf{L}^2(\mathcal{M}, \mu_{\mathcal{M}}) \to \mathbf{L}^2(\mathcal{M}_+, \mu_{\mathcal{M}_+})$$

and

$$j : \mathbf{L}^2(\mathcal{M}_+, \mu_{\mathcal{M}_+}) \to \mathbf{L}^2(\mathcal{M}, \mu_{\mathcal{M}})$$

are given by restriction and extension by 0, respectively, the *Wiener-Hopf operator*

$$W_{\mathcal{M}_+}(f) : \mathbf{L}^2(\mathcal{M}_+, \mu_{\mathcal{M}_+}) \to \mathbf{L}^2(\mathcal{M}_+, \mu_{\mathcal{M}_+})$$

associated to the *symbol* $f \in L^1(G)$ is given by $W_{\mathcal{M}_+}(f) := p \circ \pi_M(f) \circ j$, i.e.,

$$[W_{\mathcal{M}_+}(f)\phi](x) = \int_{\overline{\eta}(x)} f(g)\phi(g^{-1} \cdot x) \, d\mu_G(g),$$

where μ_G is a suitable Haar measure on G (cf. Lemma 2.4.2). If we multiply two such Wiener-Hopf operators we obtain for $f_1, f_2 \in C_c(G)$

$$\begin{aligned}&\left[W_{\mathcal{M}_+}(f_1) \circ W_{\mathcal{M}_+}(f_2)\phi\right](x)\\ &= \int_{\overline{\eta}(x)} \left[\int_{\overline{\eta}(x)} f_1(g) f_2(g^{-1}a)\, d\mu_G(g)\right] \phi(a^{-1}\cdot x)\, d\mu_G(a).\end{aligned}$$

Thus the product of two such Wiener-Hopf operators is again some sort of Wiener-Hopf operator with a two-variable symbol. To construct a domain on which these symbols are functions, we need the notion of a *groupoid*. A *locally compact groupoid* is a locally compact topological space $\mathcal{G}$ together with a pair of continuous mappings satisfying the following axioms. The domain of the first mapping is a subset $\mathcal{G}^2 \subset \mathcal{G} \times \mathcal{G}$ called the set of *composable pairs*, and the image if $(x, y) \in \mathcal{G}^2$ is denoted xy. The second mapping is an involution on $\mathcal{G}$ denoted by $x \mapsto x^{-1}$. The axioms are as follows.

1) $(x, y), (y, z) \in \mathcal{G}^2 \Longrightarrow (xy, z), (x, yz) \in \mathcal{G}^2$ and $(xy)z = x(yz)$;

2) $(x, x^{-1}), (x^{-1}, x) \in \mathcal{G}^2$ for all $x \in \mathcal{G}$;

3) $(x, y), (z, x) \in \mathcal{G}^2 \Longrightarrow x^{-1}(xy) = y$ and $(zx)x^{-1} = z$.

The maps

$$d: \mathcal{G} \to \mathcal{G},\ x \mapsto x^{-1}x, \qquad r: \mathcal{G} \to \mathcal{G},\ x \mapsto xx^{-1}$$

are called the *domain* and *range* maps, respectively. They have a common image $\mathcal{G}^0$ called the *unit space* of $\mathcal{G}$. Note that a pair (x, y) belongs to $\mathcal{G}^2$ if and only if $d(x) = r(y)$, hence $\mathcal{G}^2$ is closed in $\mathcal{G} \times \mathcal{G}$. Also, since $u \in \mathcal{G}^0$ if and only if $(u, u) \in \mathcal{G}^2$ and $u^2 = u$, it follows that $\mathcal{G}^0$ is closed in $\mathcal{G}$. Let $u \in \mathcal{G}^0$. Then $r^{-1}(u) \cap d^{-1}(u)$ is called the *isotropy group* in u. It carries the structure of a locally compact group. We say that two points u, v in $\mathcal{G}^0$ lie in the same *orbit* if $d^{-1}(u) \cap r^{-1}(v) \neq \emptyset$. Note that this defines a partition of $\mathcal{G}^0$ into orbits because $d(x^{-1}) = r(x)$ for $x \in \mathcal{G}$. Finally, we note that $r(xy) = r(x)$ and that $d(xy) = d(y)$ for $x, y \in \mathcal{G}$.

We consider the right action of G on $\mathcal{M}^{cpt}$ defined by $A \cdot g := g^{-1}A$. Our groupoid will be the reduction of the transformation group $\mathcal{M}^{cpt} \times G \to \mathcal{M}^{cpt}$ to $\mathcal{M}_+^{cpt}$ (cf. [108], 2.2.4 and 2.2.5). This amounts to the following. We set

$$\mathcal{G} = \{(x, g) \in \mathcal{M}_+^{cpt} \times G \mid g \in x\}.$$

The domain of the multiplication is

$$\mathcal{G}^2 = \{((x, g), (y, h)) \in \mathcal{G} \times \mathcal{G} \mid x \cdot g = y\},$$

the multiplication map $\mathcal{G}^2 \to \mathcal{G}$ is

$$((x,g),(y,h)) \mapsto (x,g)(y,h) = (x,gh),$$

and the involution $\mathcal{G} \to \mathcal{G}$ is defined by

$$(x,g) \mapsto (x,g)^{-1} := (x \cdot g, g^{-1}).$$

The domain and range maps for this groupoid are given by

$$d(x,g) = (x \cdot g, 1) \quad \text{and} \quad r(x,g) = (x,1).$$

Thus the unit space $\mathcal{G}^0$ of $\mathcal{G}$ is equal to $\mathcal{M}_+^{cpt} \times \{1\}$. We note that it is instructive to visualize the elements (x,g) of the groupoid as arrows from $d(x,g)$ to $r(x,g)$. Then the involution corresponds to inversion of arrows and the composable elements are the arrows which fit together. Moreover, multiplication is composition of arrows. We endow $\mathcal{G}$ with the locally compact topology inherited from $\mathcal{M}_+^{cpt} \times G$.

Next we define a convolution product on the set $C_c(\mathcal{G})$ of all compactly supported complex valued functions on $\mathcal{G}$ by

$$\begin{aligned} F_1 * F_2(x,g) &= \int_G F_1(x,ga)F_2(x \cdot ga, a^{-1})\chi_{\mathcal{M}_+^{cpt}}(x \cdot ga)\, d\mu_G(a) \\ &= \int_{\overline{\eta}(x)} F_1(x,a)F_2(a^{-1} \cdot x, a^{-1}g)\, d\mu_G(a) \end{aligned}$$

which is obviously related to the product formula of our Wiener-Hopf operators. We define a map $C_c(G) \to C_c(\mathcal{G})$ via

$$f \mapsto \tilde{f}, \quad \tilde{f}(x,g) := f(g).$$

Note that the compactness of $\mathcal{M}_+^{cpt}$ implies that $\tilde{f} \in C_c(\mathcal{G})$. Then $C_c(\mathcal{G})$ is a $*$-algebra with respect to the involution

$$F^*(x,g) = \overline{F((x,g)^{-1})}\Delta_G(g)^{-1} = \overline{F(g^{-1} \cdot x, g^{-1})}\Delta_G(g)^{-1}$$

(cf. [54], III.6). We define norms on $C_c(\mathcal{G})$ via (cf. [108], 2.7)

$$\|F\|_0 := \sup_{x \in \mathcal{M}_+^{cpt}} \|F(x,\cdot)\|_1 \quad \text{and} \quad \|F\|_1 := \max\{\|F\|_0, \|F^*\|_0\}.$$

Now we write $L^1(\mathcal{G})$ for the normed $*$-algebra obtained form $C_c(\mathcal{G})$ by completion with respect to the norm $\|\cdot\|_1$ (cf. [150], p. 51). We also recall from [109], 2.11, that $C_c(\mathcal{G})$ and therefore $L^1(\mathcal{G})$ admits a two sided approximate identity. Thus we obtain a *universal enveloping C^*-algebra* $C^*(\mathcal{G})$ as the subalgebra of $C^*(L^1(\mathcal{G})_1)$ generated by $L^1(\mathcal{G})$. The following result is proven in [54].

Theorem 9.1.4 *For $F \in C_c(\mathcal{G})$, the prescription*

$$[W_{\mathcal{M}_+}(F)\phi](x) = \int_{\overline{\eta}(x)} F(x,g)\phi(g^{-1}\cdot x)\sqrt{s(g^{-1},x)}\,d\mu_G(g),$$

where $s(g^{-1},x)$ is a cocycle depending on the measure chosen on $\mathcal{M}$, defines a norm-contractive $*$-representation

$$W_{\mathcal{M}_+} : C_c(\mathcal{G}) \to B(\mathbf{L}^2(\mathcal{M}_+))$$

and the extension

$$W_{\mathcal{M}_+} : C^*(\mathcal{G}) \to B(\mathbf{L}^2(\mathcal{M}_+))$$

is a C^*-representation with image $\mathcal{W}_{\mathcal{M}_+}$. □

The preceding theorem shows that it is reasonable to try to describe the ideal structure of the algebra $\mathcal{W}_{\mathcal{M}_+}$ of Wiener-Hopf operators via the C^*-algebra $C^*(\mathcal{G})$ of the groupoid. But it should be noted here that it is an open problem to determine the kernel of the representation $W_{\mathcal{M}_+}$.

Whenever a subset $\mathcal{M}_+^{cpt'}$ of $\mathcal{M}_+^{cpt}$ is invariant in the groupoid sense, i.e., satisfies

$$d(x,g) \in \mathcal{M}_+^{cpt'} \quad \Leftrightarrow \quad r(x,g) \in \mathcal{M}_+^{cpt'},$$

we set

$$\mathcal{G}_{\mathcal{M}_+^{cpt'}} := r^{-1}(\mathcal{M}_+^{cpt'}) = d^{-1}(\mathcal{M}_+^{cpt'}).$$

If $\mathcal{M}_+^{cpt'}$ is locally compact in $\mathcal{M}_+^{cpt}$, then $\mathcal{G}_{\mathcal{M}_+^{cpt'}}$ is again a locally compact groupoid and one can talk about its groupoid C^*-algebra. Now one can show (cf. [55]) the following.

Theorem 9.1.5 *Let U be an open invariant subset of $\mathcal{M}_+^{cpt}$. Then we have a short exact sequence of C^*-algebras,*

$$0 \to C^*(\mathcal{G}_U) \overset{j_U}{\to} C^*(\mathcal{G}) \overset{\beta_U}{\to} C^*(\mathcal{G}_{\mathcal{M}_+^{cpt}\setminus U}) \to 0.$$

For $F \in C_c(\mathcal{G})$ the map β_U is given by restriction to $r^{-1}(\mathcal{M}_+^{cpt} \setminus U)$. The sequence splits if U is open and closed in $\mathcal{M}_+^{cpt}$. □

Recall that there are only finitely many different G-orbits in $\mathcal{M}^{cpt}$ which are in addition locally closed, i.e., each orbit is open in its closure, and we also know that all other orbits in the closure of a given orbit have lower dimension.

Let $\mathcal{M}_k$ denote the set of all orbits of dimension $(\dim \mathcal{M}) - k$ in $\mathcal{M}^{cpt}$ and $(\mathcal{M}_+^{cpt})_k := \mathcal{M}_+^{cpt} \cap \mathcal{M}_k$ (cf. Section 6.4). Let

$$I_k := C^*(\mathcal{G}_{\bigcup_{j=0}^{k}(\mathcal{M}_+cpt)_j})$$

for $k \in \mathbb{N}_0$ and consider this set as an ideal of $C^*(\mathcal{G})$ according to Theorem 9.1.5. Then one finds (cf. [55]) the following theorem.

Theorem 9.1.6 *Let $x_{k,i}$, $i = 1, \dots, i_k$ be a set of representatives for the different invariant subset of $(\mathcal{M}_+^{cpt})_k$ and $H_{k,i}$ the stabilizer of $x_{k,i}$ in G. Then the C^*-algebra $C^*(\mathcal{G})$ has a composition series*

$$I_0 \subset \dots \subset I_{\dim M} = C^*(\mathcal{G})$$

with

$$I_k/I_{k-1} \cong \bigoplus_{i=1}^{i_k} \left(C^*(H_{i,k}) \otimes \mathcal{K}\big(\mathbf{L}^2([\mathcal{M}_+^{cpt}]_k, \mu_k)\big)\right)$$

for $k \geq 1$ and

$$I_0 \cong C^*(H) \otimes \mathcal{K}\big(\mathbf{L}^2([\mathcal{M}_+^{cpt}]_0, \mu_0)\big),$$

where $\mathcal{K}\big(\mathbf{L}^2([\mathcal{M}_+cpt]_k, \mu_k)\big)$ denotes the C^*-algebra of compact operators on the Hilbert space $\mathbf{L}^2([\mathcal{M}_+^{cpt}]_k, \mu_k)$ and μ_k is a positive Radon measure on $[\mathcal{M}_+^{cpt}]_k$.

The main result of [54] is the following theorem, which contains all the information one has on the first ideal in the composition series of the Wiener-Hopf algebra.

Theorem 9.1.7 *Let $\mathcal{K}\left(\mathbf{L}^2(\mathcal{M}_+)\right)$ be the ideal of compact operators on $\mathbf{L}^2(\mathcal{M}_+)$. Then $\mathcal{K}\left(\mathbf{L}^2(\mathcal{M}_+)\right) \subset \mathcal{W}_{\mathcal{M}_+}$. More precisely, this ideal is the image of $C^*(\mathcal{G}_{\mathcal{M}_+})$ under the Wiener-Hopf representation.*

Notes for Chapter 9

The groupoid approach to C^*-algebras generated by Wiener-Hopf operators goes back to Muhly and Renault [108] (cf. also [126, 109]). They also gave definitions of Wiener-Hopf operators for arbitrary ordered homogeneous spaces, but treated only the case of vector spaces ordered by polyhedral or homogeneous cones in detail. First attempts to also deal with nonabelian symmetry groups are due to Nica [127]. A more systematic approach is given in [54] (cf. also [53]). The case of noncompactly causal symmetric spaces was developed in [55].

Appendix A

Reductive Lie Groups

In this appendix, on the one hand we collect the basic notation for semisimple and reductive Lie theory used throughout the book. On the other hand we quote, and in part prove, various results which are purely group theoretical but not readily available in the textbooks on Lie groups.

A.1 Notation

Let $\mathbb{K}$ be either $\mathbb{R}$ or $\mathbb{C}$ and $\mathfrak{b}$ a finite-dimensional abelian Lie algebra over the field $\mathbb{K}$. Given a finite-dimensional $\mathfrak{b}$-module $\mathbf{V}$, we denote the corresponding representation of $\mathfrak{b}$ on $\mathbf{V}$ by π. If $\lambda \in \mathbb{K}$ and $T \in \mathrm{End}(\mathbf{V})$, we define

$$\mathbf{V}(\lambda, T) := \{v \in \mathbf{V} \mid Tv = \lambda v\}. \tag{A.1}$$

For $X \in \mathfrak{b}$ we set

$$\mathbf{V}(\lambda, X) := \mathbf{V}(\lambda, \pi(X)). \tag{A.2}$$

For $\alpha \in \mathfrak{b}^*$ we define $\mathbf{V}(\alpha, \mathfrak{b})$ by

$$\mathbf{V}(\alpha, \mathfrak{b}) := \{v \in \mathbf{V} \mid \forall X \in \mathfrak{b} : X \cdot v = \alpha(X)v\} = \bigcap_{X \in \mathfrak{b}} \mathbf{V}(\alpha(X), X). \tag{A.3}$$

Furthermore,

$$\mathbf{V}^{\mathfrak{b}} := \mathbf{V}(0, \mathfrak{b}). \tag{A.4}$$

If the role of $\mathfrak{b}$ is obvious, we abbreviate $\mathbf{V}(\alpha, \mathfrak{b})$ by $\mathbf{V}_\alpha$. Define

$$\Delta(\mathbf{V}, \mathfrak{b}) := \{\alpha \in \mathfrak{b}^* \setminus \{0\} \mid \mathbf{V}_\alpha \neq \{0\}\} \tag{A.5}$$

and

$$\mathbf{V}(\Gamma) := \bigoplus_{\alpha\in\Gamma} \mathbf{V}_\alpha, \quad \emptyset \neq \Gamma \subset \Delta(\mathbf{V}, \mathfrak{b}). \tag{A.6}$$

The elements of $\Delta(\mathbf{V}, \mathfrak{b})$ are called *weights*, as is the zero functional if $\mathbf{V}^{\mathfrak{b}} \neq \{0\}$.

If $\mathbf{V}$ is a real vector space, we set $\mathbf{V}_{\mathbb{C}} := \mathbf{V}\otimes\mathbb{C}$. We define the *conjugation of* $\mathbf{V}_{\mathbb{C}}$ *relative to* $\mathbf{V}$ by

$$\overline{u+iv} := u - iv, \qquad u, v \in \mathbf{V}. \tag{A.7}$$

Sometimes we also denote this involution by σ.

We denote Lie groups by capital letters G, H, K, etc., and the associated Lie algebras by the corresponding German lowercase letter. For complexifications and dual spaces we use the subscript $_{\mathbb{C}}$, respectively the superscript *. If $\varphi : L \to K$ is a homomorphism of Lie groups, we use φ for the corresponding homomorphism of Lie algebras and complexified Lie algebras. In particular, we have $\varphi(\exp_L(X)) = \exp_K(\varphi(X))$ for all $X \in \mathfrak{l}$.

We call a real Lie group *semisimple* if its Lie algebra is semisimple. A Lie algebra is called reductive if it is the direct sum of a semisimple and an abelian Lie algebra. In contrast to the Lie algebra case, there is no generally agreed-on definition for real reductive *groups*. Therefore we make explicitly clear that *we* call a real Lie group *reductive* if its Lie algebra is reductive.

Let G be a semisimple connected Lie group with Cartan involution θ. Let $K = G^\theta$ be the corresponding group of θ-fixed points in G. The Lie algebra of K is given by

$$\mathfrak{k} = \mathfrak{g}(1,\theta) = \{X \in \mathfrak{g} \mid \theta(X) = X\}.$$

Let $\mathfrak{p} = \mathfrak{g}(-1,\theta)$. Then we have the *Cartan decomposition*

$$\mathfrak{g} = \mathfrak{k} \oplus \mathfrak{p}. \tag{A.8}$$

We denote the Killing form of $\mathfrak{g}$ by B , i.e. , $B(X,Y) = \mathrm{Tr}(\mathrm{ad}\, X \circ \mathrm{ad}\, Y)$. For $X, Y \in \mathfrak{g}$ set

$$(X \mid Y) := (X \mid Y)_\theta := -B(X, \theta(Y)). \tag{A.9}$$

Then $(\cdot \mid \cdot)_\theta$ is an inner product on $\mathfrak{g}$. We will denote the corresponding norm by $|\cdot|_\theta$ or simply by $|\cdot|$. With respect to this inner product, the transpose $\mathrm{ad}(X)^\top$ of $\mathrm{ad}(X)$ is given by $-\mathrm{ad}(\theta(X))$ for all $X \in \mathfrak{g}$. In particular, $\mathrm{ad}(X)$ is symmetric if $X \in \mathfrak{p}$.

Let $\mathfrak{a}$ be a maximal abelian subalgebra of $\mathfrak{p}$ and $\mathfrak{m}$ be the centralizer of $\mathfrak{a}$ in $\mathfrak{k}$, i.e.,

$$\mathfrak{m} = \mathfrak{z}_{\mathfrak{k}}(\mathfrak{a}). \tag{A.10}$$

As $\{\operatorname{ad} X \mid X \in \mathfrak{a}\}$ is a commutative family of symmetric endomorphisms of $\mathfrak{g}$, we get

$$\mathfrak{g} = \mathfrak{m} \oplus \mathfrak{a} \oplus \bigoplus_{\alpha \in \Delta(\mathfrak{g},\mathfrak{a})} \mathfrak{g}_\alpha .$$

The elements of $\Delta = \Delta(\mathfrak{g}, \mathfrak{a})$ are called *restricted roots*. For $X \in \mathfrak{a}$ such that $\alpha(X) \neq 0$ for all $\alpha \in \Delta$, one can define a set of *positive restriced roots* to be

$$\Delta^+ := \{\alpha \in \Delta \mid \alpha(X) > 0\}. \tag{A.11}$$

Set

$$\mathfrak{n} = \bigoplus_{\alpha \in \Delta^+} \mathfrak{g}_\alpha = \mathfrak{g}(\Delta^+).$$

Then $\mathfrak{n}$ is a nilpotent Lie algebra and we have the following *Iwasawa decomposition* of $\mathfrak{g}$:

$$\mathfrak{g} = \mathfrak{k} \oplus \mathfrak{a} \oplus \mathfrak{n}. \tag{A.12}$$

Let $N = \exp \mathfrak{n}$ and $A = \exp \mathfrak{a}$. Then N and A are closed subgroups of G and we have the *Iwasawa decomposition* of G:

$$K \times A \times N \ni (k, a, n) \mapsto kan \in G. \tag{A.13}$$

This map is an analytic diffeomorphism. For a subset L of G and a subset $\mathfrak{b}$ of $\mathfrak{g}$, we denote by $Z_L(\mathfrak{b})$ *the centralizer of* $\mathfrak{b}$ *in* L:

$$Z_L(\mathfrak{b}) = \{b \in L \mid \forall X \in \mathfrak{b} \ : \ \operatorname{Ad}(b)X = X\}. \tag{A.14}$$

Similarly, we define the *normalizer of* $\mathfrak{b}$ *in* L by

$$N_L(\mathfrak{b}) = \{b \in L \mid \forall X \in \mathfrak{b} \ : \ \operatorname{Ad}(b)X \in \mathfrak{b}\}. \tag{A.15}$$

We fix the notation $M = Z_K(\mathfrak{a})$, $M^* = N_K(\mathfrak{a})$ and $W := W(\mathfrak{a}) := M^*/M$. The group W is called the *Weyl group* of Δ. We will use the following facts that hold for *any* connected semisimple Lie group (cf. [44]).

1) If $\mathfrak{a}$ and $\mathfrak{b}$, are maximal abelian in $\mathfrak{p}$, then there exists a $k \in K$ such that

$$\operatorname{Ad}(k)\mathfrak{b} = \mathfrak{a}. \tag{A.16}$$

2) If $k \in M^*$ and $\alpha \in \Delta$, then $k \cdot \alpha = \alpha \circ \operatorname{Ad}(k^{-1})$ is again a root.

3) Let Δ^+ be a set of positive roots in Δ. For $k \in M^*$ we have $k \in M$ if and only if $k \cdot \Delta^+ = \Delta^+$.

4) If $\Lambda \subset \Delta$ is a set of positive roots, then there exists a $k \in M^*$ such that $k \cdot \Lambda = \Delta^+$.

Let $P_{\min} := MAN$. Then $P_{\min}$ is a group. The groups conjugate to $P_{\min}$ are called *minimal parabolic subgroups* of G. A subgroup of G containing a minimal parabolic subgroup is called a *parabolic subgroup.*

If $\lambda \in \mathfrak{a}_{\mathbb{C}}^*$ we define a character $a \mapsto a^\lambda$ on A by

$$a^\lambda := \exp \lambda(X), \; a = \exp X.$$

By restriction, $(\cdot \mid \cdot)_\theta$ defines an inner product on $\mathfrak{a}$ and then also on $\mathfrak{a}^*$ by duality. Choose a Cartan subalgebra $\mathfrak{t}$ containing $\mathfrak{a}$. Then $\mathfrak{t}$ is θ-stable and $\mathfrak{t} = \mathfrak{t}_k \oplus \mathfrak{a}$, where $\mathfrak{t}_k = \mathfrak{t} \cap \mathfrak{k} \subset \mathfrak{m}$. The elements of $\Delta(\mathfrak{g}_{\mathbb{C}}, \mathfrak{t}_{\mathbb{C}})$ are called *roots* . Similar to the case of restricted roots, one can choose a system of positive roots. This can be done in such a way that $\Delta^+(\mathfrak{g}, \mathfrak{a}) = \{\alpha|_\mathfrak{a} \mid \alpha \in \Delta^+(\mathfrak{g}_{\mathbb{C}}, \mathfrak{t}_{\mathbb{C}}), \alpha|_\mathfrak{a} \neq 0\}$.

A.2 Finite-Dimensional Representations

Definition A.2.1 Let $\mathbb{K}$ be $\mathbb{R}$ or $\mathbb{C}$ and $\mathbf{V}$ a $\mathbb{K}$-vector space. Given a connected real Lie group G and a subgroup $L \subset G$, a representation π of G on $\mathbf{V}$ is called *L-spherical* if there exists a nonzero L-fixed vector $v \in \mathbf{V}$ generating $\mathbf{V}$ as an L-module. We define

$$\mathbf{V}^L = \{v \in \mathbf{V} \mid \forall a \in L : \pi(a)v = v\}. \tag{A.17}$$

Similarly, if $\mathfrak{l}$ is a subalgebra of the real Lie algebra $\mathfrak{g}$ and π is a representation of $\mathfrak{g}$ on $\mathbf{V}$, then π is called $\mathfrak{l}$-spherical if there is a nonzero $v \in \mathbf{V}$ such that $\pi(\mathfrak{l})v = \{0\}$. □

If G is semisimple and $L = K$, then we call π *spherical* if it is K-spherical. For the following theorem, see [45], p. 535.

Theorem A.2.2 *Let G be a connected semisimple Lie group and π an irreducible complex representation of G in the finite-dimensional Hilbert space $\mathbf{V}$.*

1) $\pi(K)$ has a nonzero fixed vector if and only if $\pi(M)$ leaves the highest-weight vector of π fixed.

2) Let λ be a linear form on $i\mathfrak{t}_k \oplus \mathfrak{a}$, where $\mathfrak{t}_k = \mathfrak{t} \cap \mathfrak{k}$. Then λ is the highest weight of an irreducible finite-dimensional spherical representation π of G if and only if

$$\lambda|_{i\mathfrak{t}_k} = 0 \qquad \text{and} \qquad \forall \alpha \in \Delta^+(\mathfrak{g}, \mathfrak{a}) \; : \; \frac{(\lambda|\alpha)}{(\alpha|\alpha)} \in \mathbb{Z}^+. \tag{A.18}$$

Here $(\cdot \mid \cdot)$ is the inner product on $(i\mathfrak{t}_k + \mathfrak{a})^*$ abtained from the Killing form by duality. □

Lemma A.2.3 *Let G be a connected semisimple Lie group with Cartan involution θ and $(\pi, \mathbf{V})$ be a finite-dimensional (complex or real) representation of G. Then there exists an inner product on $\mathbf{V}$ such that $\pi(x)^* = \pi(\theta(x)^{-1})$. In particular, $\pi(x)$ is unitary for all $x \in K$ and symmetric for $x \in \exp \mathfrak{p}$.*

Proof: If $\mathbf{V}$ is real, we replace $\mathbf{V}$ by $\mathbf{V}_{\mathbb{C}}$. Since the real part of an inner product on $\mathbf{V}_{\mathbb{C}}$ defines an inner product on $\mathbf{V}$ by restriction, we may assume that $\mathbf{V}$ is complex. The corresponding representation of $\mathfrak{g}_{\mathbb{C}}$ is again denoted by π. Consider the compact real form $\mathfrak{u} = \mathfrak{k} \oplus i\mathfrak{p}$ of $\mathfrak{g}_{\mathbb{C}}$ and let U be a simply connected Lie group with Lie algebra $\mathfrak{u}$. Then U is compact. As U is simply connected, there exists a representation π_U of U whose derived representation is $\pi|_{\mathfrak{u}}$. Let $(\cdot \mid \cdot)_0$ be an inner product on $\mathbf{V}$. Since U is compact, we can define a new inner product by

$$(w \mid v) = \int_U (\pi_U(u)w \mid \pi_U(u)v)_0 \, du \, .$$

Relative to this inner product, π_U is unitary. From the definition of $\mathfrak{u}$, it now follows easily that for any $x \in G$ we have

$$\pi(x)^* = \pi\left(\theta(x)^{-1}\right), \tag{A.19}$$

and the lemma follows. □

Lemma A.2.4 *Let G be a group and $\mathbf{V}$ an irreducible finite-dimensional real G-module. Then the following statements are equivalent:*

1) The complexified module $\mathbf{V}_{\mathbb{C}}$ is irreducible.

2) $\mathbf{V}$ carries no complex structure making it a complex G-module.

3) The commutant of G in $\mathrm{End}(\mathbf{V})$ *is* $\mathbb{R} \cdot \mathrm{Id}_{\mathbf{V}}$.

Proof: Suppose that $\mathbf{V}_{\mathbb{C}}$ is not irreducible. Then we can find an irreducible complex submodule $\mathbf{W}$ of $\mathbf{V}_{\mathbb{C}}$. If $\mathbf{W} \cap \mathbf{V} \neq \{0\}$, then $\mathbf{W} \cap \mathbf{V}$ is an invariant nonzero subspace of $\mathbf{V}$. As $\mathbf{V}$ is irreducible, it follows that $\mathbf{W} \cap \mathbf{V} = \mathbf{V}$. But then $\mathbf{W} = \mathbf{V}_{\mathbb{C}}$ and $\mathbf{V}_{\mathbb{C}}$ would be irreducible. Thus we have $\mathbf{W} \cap \mathbf{V} = \{0\}$. Now we have $\mathbf{W} \cap \overline{\mathbf{W}} = \{0\}$ because otherwise there would be a $w \in \mathbf{W}$, $w \neq 0$, such that $\overline{w} \in \mathbf{W}$. Then one of the vectors $\overline{w} + w$, $i(\overline{w} - w)$ is nonzero and

$$\overline{w} + w,\ i(\overline{w} - w) \in \mathbf{V} \cap \mathbf{W},$$

which contradicts $\mathbf{V} \cap \mathbf{W} = \{0\}$. This implies that the $\mathbb{R}$-linear map

$$\mathbf{W} \ni w \mapsto \frac{1}{2}(w + \overline{w}) \in \mathbf{V}$$

is injective and G-equivariant. On the other hand, the image is nonzero. Therefore the map is an $\mathbb{R}$-linear G-isomorphism $\mathbf{W} \simeq \mathbf{V}$. Thus $\mathbf{V}$ is a complex G-module.

Conversely, if I is a complex structure on $\mathbf{V}$ for which $\mathbf{V}$ is a complex G-module, then the complex linear extension of I to $\mathbf{V}_{\mathbb{C}}$ is an intertwiner on $\mathbf{V}_{\mathbb{C}}$ which is not a multiple of the identity. Thus Schur's lemma implies that $\mathbf{V}_{\mathbb{C}}$ is not irreducible.

Clearly, 3) implies 2). Conversely, if 1) holds, the commutant of G in $\mathrm{End}_{\mathbb{C}}(\mathbf{V}_{\mathbb{C}})$ is $\mathbb{C} \cdot \mathrm{Id}_{\mathbf{V}_{\mathbb{C}}}$. Since $\mathrm{End}_{\mathbb{R}}(\mathbf{V}) \cap \mathbb{C} \cdot \mathrm{Id}_{\mathbf{V}_{\mathbb{C}}} = \mathbb{R} \cdot \mathrm{Id}_{\mathbf{V}}$, we obtain 3). □

Lemma A.2.5 *Let G be a connected semisimple Lie group and $\mathbf{V}$ a finite-dimensional irreducible real G-module.*

1) $\dim \mathbf{V}^K \leq 1$.

2) If $\dim \mathbf{V}^K = 1$, *then* $\mathbf{V}_{\mathbb{C}}$ *is irreducible.*

Proof: 1) We apply Theorem A.2.2 to $\mathbf{V}_{\mathbb{C}}$. Let $\lambda \in \mathfrak{a}^*$ be a maximal weight of $\mathbf{V}_{\mathbb{C}}$ and let v be a nonzero weight vector of weight λ. Multiplying by a suitable scalar, we may assume that $\overline{v} \neq -v$. Then

$$(man) \cdot v = a^{\lambda} v, \qquad m \in M,\ a \in A,\ n \in N.$$

In particular,

$$P \cdot v = \mathbb{R}^+ v. \tag{A.20}$$

It follows that $(man) \cdot \overline{v} = a^{\lambda} \overline{v}$. Thus $u := v + \overline{v}$ is in $\mathbf{V}$ and $\pi(man)u = a^{\lambda} u$. Let now $v_K \in \mathbf{V}^K \setminus \{0\}$. As $G = KAN$ and $(\pi(kan)u \mid v_K) = a^{\lambda}(u \mid v_K)$, we find that $(u \mid v_K) \neq 0$. If $\dim \mathbf{V}^K > 1$, we can choose a K-fixed vector w_K which is orthogonal to v_K. But then $(u \mid w_K)v_K - (u \mid v_K)w_K \in \mathbf{V}^K \setminus \{0\}$, whereas

$$(u \mid (u \mid w_K)v_K - (u \mid v_K)w_K) = 0,$$

which is impossible by the above argument. Thus $\dim \mathbf{V}^K \leq 1$.

2) This is an immediate consequence of Lemma A.2.4 ,since a complex G-module has even real dimension and $\mathbf{V}^K$ is a complex subspace of $\mathbf{V}$ if $\mathbf{V}$ carries a complex structure commuting with G. □

Lemma A.2.6 *Let G be a connected semisimple Lie group acting irreducibly on $\mathbf{V}$. Assume that $\mathbf{V}$ is spherical so that $\mathbf{V}_\mathbb{C}$ is irreducible. Let $\lambda \in \mathfrak{a}^*$ be the highest weight of $\mathbf{V}_\mathbb{C}$, u be a highest-weight vector, and $v_K \in \mathbf{V}^K$ be such that $(u \mid v_K) > 0$. Further, let $P = MAN$ be the minimal parabolic subgroup of G with $\Delta^+(\mathfrak{g}, \mathfrak{a}) = \Delta(\mathfrak{n}, \mathfrak{a})$. Then $u \in \overline{P \cdot \mathbb{R}^+ v_K}$.*

Proof: Write $v_K = \sum v_\mu$ with $v_\mu \in (\mathbf{V}_\mathbb{C})_\mu \setminus \{0\}$. As different weight spaces are orthogonal and $\dim \mathbf{V}_\lambda = 1$, it follows that λ occurs in the above sum and that $v_\lambda = cu$ with $c > 0$. We can write $\mu = \lambda - \sum n_\alpha \alpha$ with $n_\alpha \geq 0$ and $\alpha \in \Delta^+(\mathfrak{g}_\mathbb{C}, \mathfrak{t}_\mathbb{C})$. But as v_K is M-fixed, it follows that all the v_μ's are M-fixed. Thus $\alpha|_{\mathfrak{t}_k} = 0$. Now choose $H \in \mathfrak{a}$ such that $\alpha(H) > 0$ for all $\alpha \in \Delta^+(\mathfrak{g}, \mathfrak{a})$. Let $a_t := \exp tH \in P$. Then

$$\lim_{t \to \infty} e^{-t\lambda(H)} a_t \cdot v_K = \lim_{t \to \infty} \left[cu + \sum_{\mu \neq \lambda} \exp(-t \sum n_\alpha \alpha(H)) v_\mu \right] = cu$$

which proves the claim. □

Remark A.2.7 Let L be a connected reductive Lie group with Lie algebra $\mathfrak{l}$ and $G := L'$ the commutator subgroup of L. Then G is a semisimple connected Lie group, but not necessarily closed in L. However, if L is linear, then G is closed (cf. [66]). □

Lemma A.2.8 *Let L be a connected reductive Lie group and π a representation on the finite-dimensional real vector space $\mathbf{V}$. Suppose that the restriction of π to the commutator subgroup of L is irreducible and spherical. Then the center of L acts by real multiples of $\mathrm{Id}_\mathbf{V}$.*

Proof: Since the restriction of π to $G := L'$ is spherical, Lemma A.2.5 implies that $\mathbf{V}_\mathbb{C}$ is an irreducible G-module. Therefore the commutatant of $\pi|_G$ is $\mathbb{R} \cdot \mathrm{Id}_\mathbf{V}$ by Lemma A.2.4. Let $Z := Z(L)$ be the center of L. Then $\pi(Z)$ of L is contained in the commutatant of $\pi|_G$ and this implies the claim. □

A.3 Hermitian Groups

A *Hermitian* group is a real connected simple Lie group G for which the corresponding Riemannian symmetric space G/K is a complex bounded symmetric domain. In terms of group theory, this is equivalent to

$$\mathfrak{z}(\mathfrak{k}) \neq \{0\}, \tag{A.21}$$

where $\mathfrak{z}(\mathfrak{k})$ is the center of $\mathfrak{k}$ and $\mathfrak{g} = \mathfrak{k} + \mathfrak{p}$ is a fixed but arbitrary Cartan decomposition. We call a Lie algebra *Hermitian* if it is the Lie algebra of a Hermitian group. For the results in this section, we refer to [44, 83, 84, 155].

If $\mathfrak{g}$ is Hermitian, then $\mathfrak{g}$ and $\mathfrak{k}$ have the same rank, i.e., there exists a Cartan subalgebra $\mathfrak{t}$ of $\mathfrak{g}$ contained in $\mathfrak{k}$. Moreover, $\mathfrak{z}(\mathfrak{k})$ is one dimensional and every $\operatorname{ad} Z|_{\mathfrak{p}}$, $Z \in \mathfrak{z}(\mathfrak{k}) \setminus \{0\}$ is regular (see [44], Chapter 6, Theorem 6.1 and Proposition 6.2). Further, there exists an element $Z^0 \in \mathfrak{z}(\mathfrak{k})$ with eigenvalues $0, i, -i$ such that the zero-eigenspace is $\mathfrak{k}$, and $\operatorname{ad}_{\mathfrak{p}} Z^0$ is a complex structure on $\mathfrak{p}$. The $\pm i$-eigenspace $\mathfrak{p}^{\pm}$ of $\operatorname{ad} Z^0|_{\mathfrak{p}_{\mathbb{C}}}$ is an abelian algebra.

Let $\mathfrak{g}$ be Hermitian and $G_{\mathbb{C}}$ a simply connected complex Lie group with Lie algebra $\mathfrak{g}_{\mathbb{C}}$. Then the analytic subgroup G of $G_{\mathbb{C}}$ with Lie algebra $\mathfrak{g}$ is Hermitian and closed in $G_{\mathbb{C}}$. Let $P^{\pm}$ be the analytic subgroup of $G_{\mathbb{C}}$ with Lie algebra $\mathfrak{p}^{\pm}$. The group $P^{\pm}$ is simply connected, and $\exp : \mathfrak{p}^{\pm} \to P^{\pm}$ is an isomorphism of Lie groups. Denote the inverse of $\exp|_{\mathfrak{p}^+}$ by $\log : P^+ \to \mathfrak{p}^+$. Then $G \subset P^+ K_{\mathbb{C}} P^-$, where $K_{\mathbb{C}}$ is the analytic subgroup of $G_{\mathbb{C}}$ with Lie algebra $\mathfrak{k}_{\mathbb{C}}$, and we have the following embeddings of complex manifolds:

$$G/K \subset P^+/(K_{\mathbb{C}}P^-) \subset G_{\mathbb{C}}/(K_{\mathbb{C}}P^-). \tag{A.22}$$

Since $P^+ \ni p \mapsto p/K_{\mathbb{C}}P^-$ is injective and holomorphic, we get an embedding

$$G/K \hookrightarrow P^+/(K_{\mathbb{C}}P^-) \simeq P^+ \overset{\log}{\hookrightarrow} \mathfrak{p}^+. \tag{A.23}$$

We denote this composed map by $m \mapsto \zeta(m)$ and call it the *Harish-Chandra embedding*. Then $\zeta(G/K)$ is a bounded symmetric domain in $\mathfrak{p}^+$. Furthermore, the map

$$P^+ \times K_{\mathbb{C}} \times P^- \ni (p, k, q) \mapsto pkq \in G_{\mathbb{C}}$$

is a diffeomorphism onto an open dense submanifold of $G_{\mathbb{C}}$ and $G \subset P^+ K_{\mathbb{C}} P^-$. If $g \in G$, then $g = p^+(g) k_{\mathbb{C}}(g) p^-(g)$ uniquely with $p^+(g) \in P^+$, $k(g) \in K_{\mathbb{C}}$ and $p^-(g) \in P^-$.

Let $\mathfrak{g}$ be a Hermitian algebra and $G \subset G_{\mathbb{C}}$ a Hermitian group with Lie algebra $\mathfrak{g}$ and simply connected $G_{\mathbb{C}}$. We denote the complex conjugation of $\mathfrak{g}_{\mathbb{C}}$ w.r.t. $\mathfrak{g}$ and the corresponding complex conjugation of $G_{\mathbb{C}}$ by σ and note that $G = G_{\mathbb{C}}^{\sigma}$ (cf. Theorem 1.1.11). Let $\mathfrak{t}$ be a Cartan subalgebra of $\mathfrak{g}$ containing Z^0. Then $\mathfrak{t} \subset \mathfrak{k}$. Let $\Delta = \Delta(\mathfrak{g}_{\mathbb{C}}, \mathfrak{t}_{\mathbb{C}})$, $\Delta_k = \Delta(\mathfrak{k}_{\mathbb{C}}, \mathfrak{t}_{\mathbb{C}})$ and $\Delta_n = \Delta(\mathfrak{p}_{\mathbb{C}}, \mathfrak{t}_{\mathbb{C}})$. The elements of Δ_k are called *compact roots*, whereas the elements of Δ_n are called *noncompact roots*. Choose a basis $Z_1, Z_2, \ldots, Z_t$

of $\mathfrak{t} \cap [\mathfrak{k}, \mathfrak{k}]$. The *lexicographic ordering* of $i\mathfrak{t}^*$ with respect to this basis is now defined by $\lambda > \mu$ if the first nonzero number in the sequence $(\lambda - \mu)(iZ^0), (\lambda - \mu)(iZ_1), (\lambda - \mu)(iZ_2), \ldots, (\lambda - \mu)(iZ_t)$ is positive. Denote as usual by a superscript $^+$ the corresponding set of positive roots. Note that the ordering is chosen such that $\Delta_n^+ = \Delta(\mathfrak{p}^+, \mathfrak{t}_\mathbb{C})$. For $\alpha \in \Delta$ we choose $E_\alpha \in (\mathfrak{g}_\mathbb{C})_\alpha$ such that $E_{-\alpha} = \sigma(E_\alpha)$. The normalization of the E_α can be chosen such that the element $H_\alpha = [E_\alpha, E_{-\alpha}] \in i\mathfrak{t}$ satisfies $\alpha(H_\alpha) = 2$ (cf. [44], p. 387).

Two roots $\alpha, \beta \in \Delta$ are called ***strongly orthogonal*** if $\alpha \pm \beta \notin \Delta$. Note that strongly orthogonal roots are in fact orthogonal w.r.t. the inner product on $i\mathfrak{t}^*$ induced by the Killing form. We recall the standard construction of a maximal system of strongly orthogonal roots: Let r be the rank of $D = G/K$, i.e., the dimension of a maximal abelian subalgebra of $\mathfrak{p}$. Let $\Gamma_r := \Delta(\mathfrak{p}^+, \mathfrak{t}_\mathbb{C})$ and γ_r be the highest root in Γ_r. If we have defined $\Gamma_r \supset \Gamma_{r-1} \supset \cdots \supset \Gamma_k \neq \emptyset$ and $\gamma_j \in \Gamma_j$, $j = k, \ldots r$, we define Γ_{k-1} to be the set of all γ in $\Gamma_k \setminus \{\gamma_k\}$ that are strongly orthogonal to γ_k. If Γ_{k-1} is not empty (or, equivalently, $k > 1$), we let γ_{k-1} be the highest root in Γ_{k-1}. Set $\Gamma := \{\gamma_1, \ldots, \gamma_r\}$. Then Γ is a maximal set of strongly orthogonal roots in $\Delta(\mathfrak{p}^+, \mathfrak{t}_\mathbb{C})$. We get a maximal set $\Gamma := \{\gamma_1, \ldots, \gamma_r\}$ of strongly orthogonal roots. Let $E_{\pm j} := E_{\pm\gamma_j}$ and $H_j := H_{\gamma_j}$. Further, we set

$$X_j := -i(E_j - E_{-j}) \in \mathfrak{p}, \quad X_o := \frac{1}{2}\sum_{j=1}^r X_j, \tag{A.24}$$

$$Y_j := E_j + E_{-j} \in \mathfrak{p}, \quad Y_o := \frac{1}{2}\sum_{j=1}^r Y_j, \tag{A.25}$$

$$E_o := \sum_{j=1}^r E_j \in \mathfrak{p}^+ \tag{A.26}$$

and

$$Z_o := \frac{i}{2}\sum_{j=1}^r H_j \in \mathfrak{t}.$$

Furthermore, we let

$$\mathfrak{a} := \bigoplus_{j=1}^r \mathbb{R}X_j \subset \mathfrak{p} \cap \sum_{\gamma\in\Gamma}((\mathfrak{g}_\mathbb{C})_\gamma + (\mathfrak{g}_\mathbb{C})_{-\gamma}), \tag{A.27}$$

$$\mathfrak{a}_h := \bigoplus_{j=1}^r \mathbb{R}Y_j \subset \mathfrak{p} \cap \sum_{\gamma\in\Gamma}((\mathfrak{g}_\mathbb{C})_\gamma + (\mathfrak{g}_\mathbb{C})_{-\gamma}), \tag{A.28}$$

and

$$\mathfrak{t}^- := i\bigoplus_{j=1}^{r}\mathbb{R}H_j .$$

The strong orthogonality of the roots γ_i, γ_j, $i \neq j$ implies that the inner automorphisms, $\mathbf{C}_j \in \mathrm{Aut}(\mathfrak{g}_\mathbb{C})$, $j = 1, \ldots, r$, defined by

$$\mathbf{C}_j := \mathrm{Ad}(c_j), \quad \text{with} \quad c_j = \exp\frac{\pi i}{4}X_j,$$

commute. They are called *partial Cayley transforms* . Their product $\mathbf{C} := \mathbf{C}_1 \circ \ldots \circ \mathbf{C}_r \in \mathrm{Aut}(\mathfrak{g}_\mathbb{C})$ is called the *Cayley transform* . It satisfies

$$\mathbf{C} = \mathrm{Ad}(c) \quad \text{with} \quad c = \exp\frac{\pi i}{2}X_o, \tag{A.29}$$

so in particular we have $c^8 = 1$.

Example A.3.1 For $\mathfrak{g} = \mathfrak{su}(1,1)$ we get

$$H_1 = \begin{pmatrix} 1 & 0 \\ 0 & -1 \end{pmatrix}, \; E_1 = \begin{pmatrix} 0 & 1 \\ 0 & 0 \end{pmatrix}, \; E_{-1} = \begin{pmatrix} 0 & 0 \\ 1 & 0 \end{pmatrix}$$

and

$$Z_o = \frac{i}{2}\begin{pmatrix} 1 & 0 \\ 0 & -1 \end{pmatrix}, \quad X_o = \frac{1}{2}\begin{pmatrix} 0 & -i \\ i & 0 \end{pmatrix}, \quad \text{and} \quad Y_o = \frac{1}{2}\begin{pmatrix} 0 & 1 \\ 1 & 0 \end{pmatrix}.$$

This gives

$$c = \frac{1}{\sqrt{2}}\begin{pmatrix} 1 & 1 \\ -1 & 1 \end{pmatrix}. \qquad \square$$

Define $\varphi_j : \mathfrak{sl}(2,\mathbb{C}) \to \mathfrak{g}_\mathbb{C}$ by

$$\varphi_j \;:\; H_1 \mapsto H_j, \quad E_1 \mapsto E_j, \quad \text{and} \quad E_{-1} \mapsto E_{-j}.$$

Then $[\mathrm{Im}\,\varphi_j, \mathrm{Im}\,\varphi_k] = 0$ for $j \neq k$ and $\varphi_j \circ \sigma = \sigma \circ \varphi_j$. Thus, in particular, $\varphi_j(\mathfrak{su}(1,1)) \subset \mathfrak{g}$. As $\mathrm{SL}(2,\mathbb{C})$ is simply connected, φ_j integrates to a homomorphism $\varphi_j : \mathrm{SL}(2,\mathbb{C}) \to G_\mathbb{C}$ such that $\varphi_j(\mathrm{SU}(1,1)) \subset G$.

The use of the φ_j allows to reduce many problems to calculations in $\mathrm{SL}(2,\mathbb{C})$. We give an application of this principle.

Lemma A.3.2 *The elements Z_o, X_o, and Y_o span a three dimensional subalgebra of $\mathfrak{g}$ isomorphic to $\mathfrak{su}(1,1)$. Set $\tilde{c} := \exp([\pi/2]Z_o)$, $c_h := \exp([i\pi/2)Y_o)$, $\tilde{\mathbf{C}} := \mathrm{Ad}(\tilde{c})$, and $\mathbf{C}_h := \mathrm{Ad}(c_h)$. Then we have*

1) $\mathbf{C}(H_j) = -Y_j$, $\mathbf{C}(X_j) = X_j$, $\mathbf{C}(Y_j) = H_j$.

2) $\tilde{\mathbf{C}}(H_j) = H_j$, $\tilde{\mathbf{C}}(X_j) = Y_j$, $\tilde{\mathbf{C}}(Y_j) = -X_j$.

3) $\mathbf{C}_h(H_j) = X_j$, $\mathbf{C}_h(X_j) = -H_j$, $\mathbf{C}_h(Y_j) = Y_j$.

Proposition A.3.3 *$\mathfrak{a}$ and $\mathfrak{a}_h$ are maximal abelian subalgebras of $\mathfrak{p}$. Furthermore,*

$$\mathbf{C}_h(\mathfrak{t}^-) = i\mathfrak{a} \qquad \textit{and} \quad \mathbf{C}(\mathfrak{t}^-) = i\mathfrak{a}_h. \qquad \square$$

The following theorem of C. C. Moore [110] describes the set of restricted roots $\Delta(\mathfrak{g}, \mathfrak{a}_h)$. A similar statement is also true for $\mathfrak{a}$ instead of $\mathfrak{a}_h$.

Theorem A.3.4 (Moore) *Define $\alpha_j = \gamma_j \circ \mathbf{C}^{-1}$. Then the set of roots of $\mathfrak{a}$ in $\mathfrak{g}$ is given either by*

$$\Delta(\mathfrak{g}, \mathfrak{a}) = \pm\{\alpha_j, \frac{1}{2}(\alpha_i \pm \alpha_k) \mid 1 \leq i, j, k \leq r; i < k\}$$

or by

$$\Delta(\mathfrak{g}, \mathfrak{a}) = \pm\{\frac{1}{2}\alpha_j, \alpha_j, \frac{1}{2}(\alpha_i \pm \alpha_k) \mid 1 \leq i, j, k \leq r; i < k\}.$$

Taking + gives a positive system of roots. The dimensions of the root spaces for the roots $\pm\frac{1}{2}(\alpha_i \pm \alpha_k)$ all agree. $\square$

Theorem A.3.5 (Korányi–Wolf) *The following properties are equivalent:*

1) The Cayley transform $\mathbf{C} \in \mathrm{Aut}(\mathfrak{g}_{\mathbb{C}})$ has order 4.

2) $Z^0 = Z_o$.

3) $\mathrm{ad}\, X_o$ *has only the eigenvalues* 0 *and* ± 1.

4) The restricted root system is reduced.

If these conditions are satisfied, $\Omega = \mathrm{Ad}(\mathbf{C}_h(L))(iE_o)$ is an open convex cone in $\mathfrak{g}(1, X_o)$, where $L := K_{\mathbb{C}} \cap (cGc^{-1})$. Moreover, the Cayley transform induces a biholomorphic map

$$\mathbf{C}_h \circ c_h : \zeta(G/K) \to \mathfrak{g}(1, X_o) + i\Omega$$

defined by $Z \mapsto \mathbf{C}_h^{-1}(c_h \cdot Z)$ (cf. [155], p. 137*).* $\square$

If the conditions of Theorem A.3.5 hold, the Riemannian symmetric space G/K is called a *tube domain* .

Appendix B

The Vietoris Topology

In this appendix we describe some topological properties of the set of closed subsets of a locally compact space. This material is needed to study compactifications of homogeneous ordered spaces.

Let (K, d) be a compact metric space. We write $\mathcal{C}(K)$ for the set of compact subsets of K and $\mathcal{C}_0(K)$ for the set of nonempty compact subsets. For $A \in \mathcal{C}_0(K)$ and $b \in K$ we set

$$d(A, b) = d(b, A) := \min\{d(a, b) \mid a \in A\} \tag{B.1}$$

and for $A, B \in \mathcal{C}_0(K)$ we define the *Hausdorff distance*:

$$d(A, B) := \max\left\{\max\{d(a, B) \mid a \in A\}, \max\{d(b, A) \mid b \in B\}\right\}. \tag{B.2}$$

This metric defines a compact topology, called *Vietoris topology* , on $\mathcal{C}_0(K)$ ([11], Ch. II, §1, Ex. 15). We set

$$d(A, \emptyset) = d(\emptyset, A) := \infty. \tag{B.3}$$

For two open subsets $U, V \subset K$ we set

$$K(U, V) := \{F \in \mathcal{C}(K) \mid F \subset U, F \cap V \neq \emptyset\}. \tag{B.4}$$

Then the sets $K(U, V)$ form a subbase for the Vietoris topology ([14], p. 162).

Let X be a locally compact space which is metrizable and σ-compact. We write $\mathcal{F}(X)$ for the set of closed subsets of X and $\mathcal{C}(X)$ for the set of compact subsets. To get a compact topology on $\mathcal{F}(X)$, we consider the one-point compactification $X^\omega := X \cup \{\omega\}$ and identify X with the corresponding subset of X^ω. Note that our assumption on X implies that X^ω is metrizable. We define the mapping

$$\beta : \mathcal{F}(X) \to \mathcal{C}_0(X^\omega), \quad F \mapsto F \cup \{\omega\}. \tag{B.5}$$

Then β is one-to-one and we identify $\mathcal{F}(X)$, via β, with the closed subset $\operatorname{Im}\beta = \{K \in \mathcal{C}_0(X^\omega) \mid \omega \in K\}$. As a closed subspace of $\mathcal{C}_0(X^\omega)$, the space $\mathcal{F}(X)$ is a compact metrizable topological space. For a sequence $(A_n)_{n\in\mathbb{N}}$ in $\mathcal{F}(X)$, we define

$$\operatorname{liminf} A_n := \{x \in X \mid \forall m \in \mathbb{N}(\exists n_m), \forall n \geq n_m : d(x, A_n) < \tfrac{1}{m}\} \tag{B.6}$$

and

$$\operatorname{limsup} A_n := \{x \in X \mid \forall m \in \mathbb{N}, \forall n_0 \in \mathbb{N}, \exists n \geq n_0 : d(x, A_n) < \tfrac{1}{m}\}. \tag{B.7}$$

We note that these sets are always closed.

Lemma B.1.1 *The following assertions hold:*

1) *If $U \subset X$, then $\{F \in \mathcal{F}(X) \mid F \cap U \neq \emptyset\}$ is open; and if $A \subset X$ is closed, then $\{F \in \mathcal{F}(X) \mid F \subset A\}$ is closed.*

2) *Let A_n be a sequence in $\mathcal{F}(X)$. Then A_n converges to $A \in \mathcal{F}(X)$ if and only if*
$$A = \operatorname{limsup} A_n = \operatorname{liminf} A_n.$$
In this case A consists of the set of limit points of sequences (a_n) with $a_n \in A_n$.

3) *If $A \subset X$ is closed, then $\{F \in \mathcal{F}(X) \mid A \subset F\}$ is closed.*

4) *If A_n is a sequence of connected sets, $A_n \to A \neq \emptyset$, and for every $n \in \mathbb{N}$ the set $\bigcup_{m\geq n} A_m$ is not relatively compact, then every connected component of A is noncompact.*

5) *The relation $\subset$ is a closed subset of $\mathcal{F}(X) \times \mathcal{F}(X)$.*

Proof. 1) The first assertion follows from the observation that

$$\{F \in \mathcal{F}(X) \mid F \cap U \neq \emptyset\} = \mathcal{F}(X) \cap \{F \in \mathcal{C}(X^\omega) \mid F \cap U \neq \emptyset\}$$

because U is open in X^ω. The second assertion follows by applying the first one with $U := X \setminus A$.

2) $\Rightarrow$: Let $x \in A$. Then there exist numbers $n_m \in \mathbb{N}$ such that A_n intersects the $(1/m)$-ball around a if $n \geq n_m$. W.l.o.g. we may assume that

the sequence n_m is increasing. For $k = n_m + 1, ..., n_{m+1}$ we choose $a_k \in A_k$ with distance less than $1/m$ from a. Then $a = \lim_{k\to\infty} a_k$ and therefore $a \in \text{liminf} A_n$. If, conversely, $a \in \text{limsup} A_n$, then there exists a subsequence $n_m \in \mathbb{N}$ with $n_m \geq m$ and elements $a_m \in A_{n_m}$ with $d(a, a_m) < 1/m$. We conclude that $a = \lim a_m \in \lim A_{n_m} = A$. This proves that

$$\text{limsup} A_n \subset A \subset \text{liminf} A_n.$$

($\Leftarrow$): We use the compactness of $\mathcal{F}(X)$. Let A_{n_m} be a convergent subsequence of A_n. Then the first part implies that $\lim A_{n_m} = \text{liminf} A_{n_m}$. Moreover, we have that

$$\text{liminf} A_n \subset \text{liminf} A_{n_m}$$

and

$$\text{limsup} A_{n_m} \subset \text{limsup} A_n.$$

Hence

$$A = \text{liminf} A_n \subset \text{liminf} A_{n_m} \subset \text{limsup} A_{n_m} \subset \text{limsup} A_n = A$$

implies that $\lim A_{n_m} = A$. Finally, the arbitrariness of the subsequence of A_n entails that $A_n \to A$.

3) This is an immediate consequence of 2).

4) Let $a \in A$ be arbitrary and $C(a)$ the connected component of a in A. We have to show that $C(a)$ is noncompact. Suppose this is false. Then $C(a)$ is a compact subset of X and there exists a relatively compact open neighborhood V of $C(a)$ in X such that $V \cap A$ is closed ([11], Ch. II, §4.4, Cor.]) and therefore compact. Let $\delta := \min\{d(a', b) \mid a' \in V \cap A, b \in X \setminus V\} > 0$ and $\epsilon := \min\{\delta/4, d(\omega, V)/3\}$. Since $a \in \lim A_n$, there exists $n_V \in \mathbb{N}$ such that $d(A_n, a) < \epsilon$ for all $n \geq n_V$. Moreover, we may assume that $d(A_n, A \cup \{\omega\}) < \epsilon$. Let $a_n \in A_n$ with $d(a_n, V \cap A) < \epsilon$. Then $d(a_n, \omega) > \epsilon$ because $d(\omega, V \cap A) > 3\epsilon$ and therefore $d(a_n, A) < \epsilon$. Let $b \in A$ such that $d(a_n, b) < \epsilon$. Then $d(b, V \cap A) < 3\epsilon < \delta$. Hence $b \in V \cap A$ and this entails that $d(a_n, V \cap A) < \epsilon$. Thus

$$\{c \in A_n \mid d(c, A \cap V) \in]\epsilon, 2\epsilon[\} = \emptyset.$$

Since A_n is connected, this implies that $A_n \subseteq V$. This contradicts the assumption that $\bigcup_{m \geq n} A_m$ is not relatively compact.

5) Let $(A_n, B_n) \to (A, B)$ with $A_n \subset B_n$. Then 2) shows that $A \subset B$. □

We recall that a Hausdorff space Y endowed with a closed partial order $\leq$ is called a *pospace*. Thus we see that $(\mathcal{F}(X), \subset)$ is a compact pospace.

Proposition B.1.2 *Let $\mu : G \times X \to X$ be a continuous action of a locally compact group G on X. Then*

$$\mathcal{F}(\mu) : G \times \mathcal{F}(X) \to \mathcal{F}(X), \quad (g, F) \mapsto g(F)$$

defines a continuous action of G on $\mathcal{F}(X)$.

Proof: Let $(g, F) = \lim(g_n, F_n)$ with $g_n \in G$ and $F_n \in \mathcal{F}(X)$. We have to show that $g_n(F_n) \to g(F)$. To see this, we have to show that each convergent subsequence converges to $g(F)$. So we may assume that $g_n(F_n)$ is convergent. Let $f \in F$. Then there exists a sequence $f_n \in F_n$ with $f_n \to f$. Hence $g_n \cdot f_n \to g \cdot f$ implies that $g \cdot f \in \lim g_n(F_n)$. This proves that $g(F) \subset \lim g_n(F_n)$. If, conversely, $f' = \lim g_n \cdot f'_n \in \lim g_n(F_n)$, then

$$f'_n = g_n^{-1} \cdot (g_n \cdot f'_n) \to g^{-1} \cdot f'$$

entails that $f' = \lim g \cdot f'_n \in \lim g(F_n) = g(\lim F_n) = g(F)$. □

Proposition B.1.3 *Suppose that the locally compact group G acts continuously on X. Further suppose that $\mathcal{O} \subset X$ is open.*

1) The set $S := S(\mathcal{O}) = \{g \in G \mid g \cdot \mathcal{O} \subset \mathcal{O}\}$ is a closed subsemigroup of G.

2) If X is a homogeneous G-space, then the interior of S is given by

$$S^o = \{g \in G \mid g \cdot \overline{\mathcal{O}} \subset \mathcal{O}\},$$

where $\overline{\mathcal{O}}$ is the closure of $\mathcal{O}$ in X.

Proof: 1) According to Lemma B.1.1 the set $\{F \in \mathcal{F}(X) \mid \mathcal{X} \setminus \mathcal{O} \subset F\}$ is closed. Therefore Proposition B.1.2 implies that also

$$S = \{g \in G \mid X \setminus \mathcal{O} \subset g(X \setminus \mathcal{O})\}$$

is closed. Since S obviously is a semigroup, this implies the claim.

2) Let $g \in S^o$ and U be a neighborhood of 1 in G such that $Ug \subset S$. Then

$$g \cdot \overline{\mathcal{O}} = \overline{g \cdot \mathcal{O}} \subset (Ug) \cdot \mathcal{O} \subset S \cdot \mathcal{O} \subset \mathcal{O}.$$

Conversely, if $g \cdot \overline{\mathcal{O}} \subset \mathcal{O}$, then there exists a neighborhood U of 1 in G such that $(Ug) \cdot \overline{\mathcal{O}} \subset \mathcal{O}$. In particular, we find $Ug \subset S$, whence $g \in S^o$. □

Proposition B.1.4 *Let G be a Lie group. Set $\mathcal{F}(G)^H := \{F \in \mathcal{F}(G) \mid \forall h \in H : Fh = F\}$. Then the mapping*

$$\pi^* : \mathcal{F}(G/H) \to \mathcal{F}(G)^H, \quad F \mapsto \pi^{-1}(F)$$

is a homeomorphism.

Proof: That π^* is a bijection follows from the fact that the quotient mapping $\pi : G \to G/H$ is continuous. So it remains to show that π^* is continuous. Let $F_n \to F$ in $\mathcal{F}(G/H)$ and set $E_n := \pi^{-1}(F_n)$ and $E := \pi^{-1}(F)$. We may assume that $E_n \to E'$. Then we have to show that $E = E'$.

Let $e' \in E'$ and $e_n \in E_n$ with $e_n \to e'$. Then $\pi(e_n) \to \pi(e') \in \lim F_n = F$. Hence $e' \in E$ and therefore $E' \subset E$. If, conversely, $e \in E$, then $\pi(e) \in F$ and there exists a sequence $f_n \in F_n$ with $f_n \to \pi(e)$. Using a local cross section $\sigma : U \to G$, where U is a neighborhood of $\pi(e)$ and $\sigma(\pi(e)) = e$, we find that $\sigma(f_n) \to e \in \lim E_n = E'$. □

Lemma B.1.5 *Let G be a Lie group acting on a locally compact space $\mathcal{Y}$ and $X \in \mathfrak{g}$. For $p \in \mathcal{Y}$ let $\mathfrak{g}^p$ be the Lie algebra of the group $G^p = \{g \in G \mid g \cdot p = p\}$. If $q = \lim_{t\to\infty} \exp(tX) \cdot p$ then*

$$\operatorname{limsup}_{t\to\infty} e^{\operatorname{ad} tX} \mathfrak{g}_p \subset \mathfrak{g}_q.$$

Proof: Suppose $Z \in \operatorname{limsup}_{n\to\infty} e^{\operatorname{ad} nX} \mathfrak{g}_p$. Then $Z = \lim_{k\to\infty} e^{\operatorname{ad} n_k X} Y_k$ for suitable sequences $Y_k \in \mathfrak{g}_p$ and $(n_k)_{k\in\mathbb{N}}$. But then

$$\begin{aligned}
\exp Z \cdot q &= \lim_{k\to\infty} (\exp(n_k X) \exp(Y_k) \exp(-n_k X)) \lim_{k\to\infty} \exp(n_k X) \cdot p \\
&= \lim_{k\to\infty} \exp(n_k X) \exp(Y_k) \exp(-n_k X) \exp(n_k X) \cdot p \\
&= \lim_{k\to\infty} \exp(n_k X) \exp(Y_k) \cdot p \\
&= \lim_{k\to\infty} \exp(n_k X) \cdot p = q
\end{aligned}$$

which proves the asertion. □

Notation

Chapter 1

H, stabilizer of a base point: 2
τ, a nontrivial involution on G and $\mathfrak{g}$: 1
$G^\tau = \{a \in G \mid \tau(a) = a\}$: 2
$\mathfrak{g}(1,\tau)$, the $+1$-eigenspace of τ: 2
$\mathfrak{h} = \mathfrak{g}^\tau$, the Lie algebra of H: 2
$\mathfrak{q}$, the (-1)-eigenspace of τ: 2
σ, the conjugation $X + iY \mapsto X - iY$ relative to $\mathfrak{g}$: 3
$\eta = \tau \circ \sigma$: 6
$\mathfrak{g}(-1,\tau)$, (-1)-eigenspace of τ: 2
$\mathrm{ad}_{\mathfrak{q}}(X) = \mathrm{ad}(X)|_{\mathfrak{q}}$, restriction of $\mathrm{ad}\, X$, $X \in \mathfrak{h}$, to $\mathfrak{q}$: 2
ℓ_a, left translation by a: 4
$\mathcal{M}$, the symmetric space G/H: 4
$\tilde{\mathcal{M}}$, universal covering space of $\mathcal{M}$: 5
$\mathfrak{h}_k, \mathfrak{h}_p, \mathfrak{q}_k, \mathfrak{q}_p$: 7
$\check{G}$, the analytic subgroup for $\mathfrak{g}$ in the simply connected complexification: 6
$\check{G}^c$, the same with $\mathfrak{g}$ replaced by $\mathfrak{g}^c$: 6
$\check{H}$, the τ-fixed group in $\check{G}$: 6
$\check{\mathcal{M}} = \check{G}/\check{H}$: 6
$\check{\mathcal{M}}^c = \check{G}/\check{H}$: 6
$\tau_X = \mathrm{Ad}(\exp \pi X) = e^{\pi\, \mathrm{ad}\, X}$: 8
$\varphi_X = \mathrm{Ad}\,(\exp([\pi/2]X)) = e^{[\pi/2]\, \mathrm{ad}\, X}$: 8
$\mathfrak{g}^c = \mathfrak{h} \oplus i\mathfrak{q}$, the c-dual Lie algebra: 6
$\mathfrak{k}^c$, the maximal compactly embedded algebra $\mathfrak{h}_k \oplus i\mathfrak{q}_p$ in $\mathfrak{g}^c$: 8
$\mathfrak{p}^c = \mathfrak{h}_p \oplus i\mathfrak{q}_k \subset \mathfrak{g}^c$: 8
$\mathfrak{q}^c = i\mathfrak{q}$, the -1 eigenspace of $\tau|_{\mathfrak{g}^c}$: 8
$\tau^a = \tau \circ \theta$, the associated involution: 9
$\mathfrak{h}^a = \mathfrak{g}(+1,\tau^a) = \mathfrak{h}_k \oplus \mathfrak{q}_p$: 9
$\mathfrak{q}^a = \mathfrak{g}(-1,\tau^a) = \mathfrak{q}_k \oplus \mathfrak{h}_p$: 9

$\mathcal{M}^a$, a symmetric space locally isomorphic to G/H^a: 9
$\theta^r = \tau|_{\mathfrak{g}^r}$ the Cartan involution on the Riemannian dual Lie algebra $\mathfrak{g}^r$: 9
$\mathfrak{g}^r = \mathfrak{h}_k \oplus i\mathfrak{h}_p \oplus i\mathfrak{q}_k \oplus \mathfrak{q}_p$, the Riemannian dual Lie algebra: 9
G^r, a Lie group with Lie algebra $\mathfrak{g}^r$: 9
$\mathfrak{k}^r = \mathfrak{h}_k \oplus i\mathfrak{h}_p \subset \mathfrak{g}^r$: 9
$K^r = \exp \mathfrak{k}^r$, the maximal almost compact subgroup of G^r: 9
$\mathcal{M}^r$, the Riemannian dual space G^r/K^r: 9
$\mathfrak{q}^{H\cap K} = \{X \in \mathfrak{q} \mid \forall k \in H \cap K : \mathrm{Ad}(k)X = X\}$: 12
$\mathfrak{q}^{H_o \cap K}$, the same as above with H replaced by H_o: 12
$\mathfrak{z}_{\mathfrak{l}}(\mathfrak{a}) = \{Y \in \mathfrak{l} \mid \forall X \in \mathfrak{a} : [Y, X] = 0\}$, the centralizer of $\mathfrak{a}$ in $\mathfrak{l}$: 14
$\mathfrak{z}(\mathfrak{l}) = \mathfrak{z}_{\mathfrak{l}}(\mathfrak{l})$, the center of $\mathfrak{l}$: 14
Y^0, a central element in $\mathfrak{h}_p$ such that $\mathrm{ad}(Y^0)$ has spectrum 0, 1 and -1 and $\mathfrak{h} = \mathfrak{g}(0, Y^0)$: 21
$Q_{p,q}$, the bilinear form $x_1y_1 + \ldots + x_py_p - x_{p+1}y_{p+1} - \ldots - x_ny_n$: 24
$Q_{\pm r} = \{x \in \mathbb{R}^n \mid Q_{\pm r}(x, x) = \pm r\}$: 24
$I_{p,q} = \begin{pmatrix} I_p & 0 \\ 0 & -I_q \end{pmatrix}$, where I_k is the $(k \times k)$-identity matrix: 25
$M(l \times m, \mathbb{K})$, the $\mathbb{K}$-vector space of $(k \times m)$-matrices: 25
$Z(G)$, the center of G: 14
$\mathfrak{q}^+ = \mathfrak{g}(+1, Y^0)$, $\mathfrak{q}^- = \mathfrak{q}(-1, Y^0)$, the irreducible components of $\mathfrak{q}$ for Cayley type spaces: 13

Chapter 2

$\mathbf{V}^C = C \cap -C$, the largest vector space contained in the closed cone C: 29
$< C > := C - C$, the vector space generated by the cone C: 29
$C^* = \{u \in \mathbf{V} \mid \forall v \in C,\ v \neq 0 : (u|v) > 0\}$, dual cone: 29
$U^\perp = \{v \in \mathbf{V} \mid \forall u \in U : (v|u) = 0\}$: 30
$C^o = \mathrm{int}(C)$, the interior of C: 30
$\mathrm{algint}(C)$, the interior of C in $< C >$, the algebraic interior of C: 30
$\mathrm{cone}(S)$, the cone generated by S: 30
$\mathrm{Cone}(\mathbf{V})$, the set of regular, closed convex cones in $\mathbf{V}$: 30
$\overline{\Omega} = cl(\Omega)$, the closure of Ω: 30
$\mathrm{Fa}(C)$, the set of faces of the cone C: 31
$\mathrm{op}(F) = F^\perp \cap C$, the face of C^* opposite to the face F of C: 32
$P_V^W(C)$, the orthogonal projection of the cone C into $\mathbf{V} \subset \mathbf{W}$: 32
$I_V^W(C) = C \cap \mathbf{V}$, the intersection of C with $\mathbf{V}$: 32
$\mathrm{Aut}(C) = \{a \in \mathrm{GL}(\mathbf{V}) \mid a(C) = C\}$, the automorphism group of C: 33
$\mathrm{Cone}_G(\mathbf{V})$: G-invariant cones in $\mathrm{Cone}(\mathbf{V})$: 33,38
$E_{V,N}^{W,L}(C) = \overline{\mathrm{conv}\,(L \cdot C)}$, the minimal L-invariant extension of C: 34
$M(m, \mathbb{K}) = M(m \times m, \mathbb{K})$: 34

$\mathrm{H}(m,\mathbb{K}) = \{X \in M(m,\mathbb{K}) \mid X^* = X\}$, space of Hermitian matrices over $\mathbb{K}$: 34
$\mathrm{H}^+(m,\mathbb{K})$, the cone of positive definite matrices in $\mathrm{H}(m,\mathbb{K})$: 34
$\mathrm{conv}(L)$, the convex hull of the set L: 36
$u_K = \int_K (k \cdot u)dk \in \mathrm{conv}(K \cdot u)$, a K-invariant vector in C obtained as the center of gravity of a K-orbit: 36
$C_{\min}$, a minimal invariant cone: 39
$C_{\max}$, a maximal invariant cone, $C_{\max} = C_{\min}^*$: 39
$\mathbf{o}$, the basepoint $\mathbf{o} = eH$ if $\mathcal{M} = G/H$: 40
$\preceq_s$, strict causality relation via connecting by causal curves: 41
$\preceq$, the closure of the relation $\preceq_s$: 41
$\uparrow A = \{y \in Y \mid \exists a \in A : a \leq y\}$: 41
$\downarrow A = \{y \in Y \mid \exists a \in A : y \leq a\}$: 41
$[m,n]_{\leq} = \{z \in \mathcal{M} \mid m \leq z \leq n\} = \uparrow m \cap \downarrow n$: 42
$S_{\leq} = \{a \in G \mid \mathbf{o} \leq a \cdot \mathbf{o}\}$, the causal semigroup: 43
$\mathbf{L}(S_{\preceq}) = \{X \in \mathfrak{g} \mid \exp \mathbb{R}^+ X \subset S_{\preceq}\}$, the tangent cone: 44, 44
$\leq_S$, $g \leq_S g'$ if $g' \in gS$: 44
$\mathrm{Mon}(S)$, the set of monotone functions: 44
$\mathcal{F}(\mathcal{M})$, the set of closed subset of $\mathcal{M}$: 45
$\mathcal{F}_\downarrow(G) = \{F \in \mathcal{F}(G) \mid \downarrow F = F\} \subset \mathcal{F}(G)^H$: 45
$\mathcal{F}_\downarrow(G/H)$: 45
$\overline{\eta} : G/H \to \mathcal{F}_\downarrow(G), \quad gH \mapsto \downarrow(gH)$, the causal compactification map: 46
$\mathcal{M}_+ = [\mathbf{o},\infty) = S \cdot \mathbf{o}$, the positive cone in $\mathcal{M}$: 46
$\mathcal{M}^{cpt} = \overline{\overline{\eta}(\mathcal{M})} = \overline{\eta(G)} \subset \mathcal{F}(G)$, the order compactification of $\mathcal{M}$: 46
$\mathcal{M}_+^{cpt} = \overline{\overline{\eta}(\mathcal{M}_+)} = \overline{\eta(S)}$: 46
$\mathcal{F}_\downarrow^\infty(G/H)$, elements of $\mathcal{F}_\downarrow(G/H)$ with noncompact connected upper sets: 48
∂A, the boundary of A: 49
$C_k = C_+ - C_- \subset \mathfrak{q}$, a cone such that $C_k^o \cap \mathfrak{k} \neq \emptyset$: 53
$C_p = C_+ + C_- \subset \mathfrak{q}$, a cone with $C_p^o \cap \mathfrak{p} \neq \emptyset$: 53
$X_\pm$, $(K \cap H)$-invariants in $\mathfrak{q}^\pm$: 52
$\mathbf{C}$, a Cayley transform commuting with τ^a: 56,255
$p^+(g), k_{\mathbb{C}}(g), p^-(g)$, the projections of $g \in P^+K_{\mathbb{C}}P^-$ onto its components: 56
Ω_+, the bounded realization of G/K: 56
$j(g,Z)$, the $K_{\mathbb{C}}$ projection of $g \exp Z$: 56
$g \cdot Z$, the P^+ component of $g \exp Z$: 56
$\mathcal{S}$, the Shilov boundary of G/K: 56
$\zeta(p) = \log(p) \in P^+$: 56
$E = \zeta(c)$, a base point in $\mathcal{S}$: 56
$\mathcal{S}_1 = \mathcal{S} \times \mathcal{S}$: 57
ρ_n, half the sum of positive noncompact roots: 58

π_m, irreducible representation of $G_{\mathbb{C}}$ with lowest weight $-m\rho_n$: 59
$\Phi_m(Z) = (\pi_m(c^{-2}\exp Z)u_0 \mid u_0)$: 59
$\Psi_m(Z,W) := \Phi_m(Z-W)$: 59

Chapter 3

X^0, a cone-generating element in $\mathfrak{q}$: 77
$\psi_k = \varphi_{Z^0}$, an isomorphism $(\mathfrak{g},\tau,\theta) \simeq (\mathfrak{g},\tau^a,\theta)$: 78
$\psi_p = \varphi_{iX^0}$, an isomorphism $(\mathfrak{g},\tau,\theta) \simeq (\mathfrak{g},\tau^a,\theta)^r$: 78
$\psi_c = \varphi_{iY^0}$, an isomorphism $(\mathfrak{g},\tau,\theta) \simeq (\mathfrak{g}^c,\tau,\tau^a)$: 78
$\Delta_0 = \{\alpha \in \Delta \mid \alpha(X^0) = 0\} = \Delta(\mathfrak{h}^a,\mathfrak{a})$: 79
$\Delta_\pm = \{\alpha \in \Delta \mid \alpha(X^0) = \pm 1\}$: 80
$\mathfrak{n}_\pm = \sum_{\alpha(X^0)=\pm 1} \mathfrak{g}_\alpha$: 80
$\mathfrak{n}_0 = \sum_{\alpha\in\Delta_0^+} \mathfrak{g}_\alpha \subset \mathfrak{g}_0$: 80

Chapter 4

$X^\lambda = \lambda/|\lambda|^2 \in \mathfrak{a}$: 92
$Y_\alpha \in \mathfrak{g}_\alpha$, such that $|Y_\alpha|^2 = \frac{2}{|\alpha|^2}$, $Y_{-\alpha} = \tau(Y_\alpha)$. Thus $[Y_\alpha, Y_{-\alpha}] = X^\alpha$: 92
$Y^\alpha = \frac{1}{2}(Y_\alpha + Y_{-\alpha}) \in \mathfrak{h}_p$: 93
$Z^\alpha = \frac{1}{2}(Y_{-\alpha} - Y_\alpha) \in \mathfrak{q}_k$: 93
$X_{\pm\alpha} = \frac{1}{2}(X^\alpha \pm Z^\alpha) \in \mathfrak{q}$: 93
φ_α, a τ-equivariant homomorphism $\mathfrak{sl}(2,\mathbb{C}) \to \mathfrak{g}_{\mathbb{C}}$: 93
$\mathfrak{s}_\alpha = \operatorname{Im}\varphi_\alpha$: 94
$\tilde{\Delta} = \Delta(\mathfrak{g}_{\mathbb{C}}, \mathfrak{t}^c_{\mathbb{C}})$: the roots of the Cartan subalgebra $\mathfrak{t}_{\mathbb{C}}$ in $\mathfrak{g}_{\mathbb{C}}$: 95
$\tilde{\Delta}_\pm = \Delta((\mathfrak{p}^c)^\pm, \mathfrak{t}^c_{\mathbb{C}})$, the set of noncompact roots: 95
$\tilde{\Delta}_0 := \Delta(\mathfrak{k}^c_{\mathbb{C}}, \mathfrak{t}^c_{\mathbb{C}})$, the set of compact roots: 95
$\tilde{\Gamma} = \{\tilde{\gamma}_1,\ldots,\tilde{\gamma}_{r^c}\}$, a maximal set of strongly orthogonal noncompact roots: 96
$\mathfrak{a}^c_h$, maximal abelian subalgebra of $\mathfrak{h}_p$: 96
$C_{\min}(X^0) = C_{\min} = \overline{\operatorname{conv}[\operatorname{Ad}(H_o)(\mathbb{R}^+X^0])}$, a minimal $\operatorname{Ad}(H_o)$-invariant cone containing the cone-generating element X^0: 98
$C_{\max}(X^0) = C_{\max}\{X \in \mathfrak{q} \mid \forall Y \in C_{\min} : B(X,Y) \geq 0\} = C_{min}(X^0)^*$, the maximal
cone containing X^0: 98
$c_{\min}(X^0) = c_{\min} = \sum_{\alpha\in\Delta_+} \mathbb{R}_0^+ X^\alpha = \sum_{\alpha\in\Delta_+} \mathbb{R}_0^+ \alpha$, the minimal Weyl group invariant regular cone in $\mathfrak{a}$: 98
$c_{\max}(X^0) = c_{\max} = \{X \in \mathfrak{a} \mid \forall \alpha \in \Delta_+ : \alpha(X) \geq 0\} = c^*_{\min}$, the corresponding maximal cone in $\mathfrak{a}$: 98
$\tilde{c}_{\min} = \sum_{\tilde{\alpha}\in\tilde{\Delta}_+} \mathbb{R}_0^+ \tilde{\alpha}$, the minimal W-invariant cone in a Cartan subalgebra

containing $\mathfrak{a}$: 102
$\tilde{c}_{\max} = \{X \in it^c \mid \forall\tilde{\alpha} \in \tilde{\Delta}_+ : \tilde{\alpha}(X) \geq 0\} = \{X \in it^c \mid \forall\tilde{\alpha} \in \tilde{\Delta}_+ : \tilde{\alpha}(X) \geq 0\}$
$= \tilde{c}^*_{\min}$: 102
$\tilde{C}_{\min}$, a minimal G^c-invariant cone in $i\mathfrak{g}^c$: 103
$\tilde{C}_{\max} = \tilde{C}^*_{\min}$, a maximal G^c-invariant cone in $i\mathfrak{g}^c$: 103
$\tilde{W}_0 = W(\tilde{\Delta}_0) = N_{K^c}(\mathfrak{t}^c)/Z_{K^c}(\mathfrak{t}^c)$: 115
$\mathcal{C}(\Delta_0^+) = \{X \in \mathfrak{a} \mid \forall\alpha \in \Delta_0^+ : \alpha(X) > 0\}$, open Weyl chamber: 116
$\tilde{W}_0(\tau) = \{w \in \tilde{W}_0 \mid \tau \circ w = w \circ \tau\}$: 117
$\tilde{W}_0^{\mathfrak{a}} = \{w \in \tilde{W}_0 \mid w|_{\mathfrak{a}} = \mathrm{id}\}$: 117

Chapter 5

$P_{\max} = H^a N_+$, a maximal parabolic subgroup in G: 121
$\sharp: G \to G, \quad g \mapsto \tau(g)^{-1}$: 121
$A_h^c = \exp \mathfrak{a}_h$, where $\mathfrak{a}_h \subset \mathfrak{h}_p$ is maximal abelian: 121
$\mathcal{O} = (G^\tau)_o \cdot o_{\mathcal{X}} \subset \mathcal{X} = G/P_{max}$: 122
$\mathcal{X} = G/P_{max}$, the real flag manifold: 122
$\kappa : \mathfrak{n}_- \to \mathcal{X}$, $\kappa(X) = (\exp X) \cdot P_{\max}$, real Harish-Chandra embedding: 123
$\Omega_- = \kappa^{-1}(\mathcal{O}) \subset \mathfrak{n}_-$, real bounded domain, isomorphic to $H/(H \cap K)$: 123
Ω_+, the bounded realization of $H/(H \cap K)$ inside $\mathfrak{n}_+$: 124
$\mathcal{X}_\mathbb{C} = G_\mathbb{C}/(P_{\max})_\mathbb{C}$, the complex flag manifold: 125
$(\Omega_\pm)_\mathbb{C} \simeq G^c/K^c$, the complexification of $H/(H \cap K)$: 125
$S(C) = H \exp C$, the closed semigroup in G with tangent cone $\mathfrak{h} \oplus C$: 129
$S(\mathcal{O})$, the compression semigroup $\{g \in G \mid g \cdot \mathcal{O} \subset \mathcal{O}\}$:133
$S(L, Q)$, the compression semigroup $\{g \in G \mid gLQ \subset LQ\} = S(LQ/Q)$: 134
$\Theta(g) = d_1\lambda_g(C_{\max} + \mathfrak{h})$, $\forall g \in N^\sharp AH$: 135
$\mu_{(t,h)} : N^\sharp AH \to N^\sharp AH, g \mapsto tgh^{-1}$: 135
I_h, the inner automorphism $g \mapsto hgh^{-1}$: 135
a_H, the causal Iwasawa projection, $g \in H \exp(a_H(g))N$: 136
$W(\mathfrak{a}) = N_K(\mathfrak{a})/Z_K(\mathfrak{a})$, the Weyl group of $\mathfrak{a}$ in G: 138
$W_\tau(\mathfrak{a}) = \{s \in W(\mathfrak{a}) \mid s(\mathfrak{a} \cap \mathfrak{h}) = \mathfrak{a} \cap \mathfrak{h}\}$: 138
$W_0(\mathfrak{a}) = N_{K \cap H}(\mathfrak{a})/Z_{K \cap H}(\mathfrak{a})$, the Weyl group of $\mathfrak{a}$ in H^a: 138
$S_A = S(G^\tau, P_{\max})^o \cap A_q$: 141
$\tilde{\Delta} = \Delta(\mathfrak{g}_\mathbb{C}, \mathfrak{t}^c_\mathbb{C})$, the set of roots of the Cartan subalgebra $\mathfrak{t}^c_\mathbb{C}$: 143
$\tilde{\mathfrak{n}} = \sum_{\tilde{\alpha} \in \tilde{\Delta}^+} (\mathfrak{g}_\mathbb{C})_{\tilde{\alpha}}$: 143
$\tilde{\mathfrak{a}} = it^c$: 143
$\tilde{A} = \exp(\tilde{\mathfrak{a}})$: 143
$\tilde{N} = \exp \tilde{\mathfrak{n}})$: 143
$L \times_U V$: 144
$I_{B^\sharp} : \mathcal{F}_\downarrow(G) \to \mathcal{F}_\downarrow(B), \quad F \mapsto F \cap B^\sharp$: 153

$\eta_{B^\sharp}:\mathcal{F}_\downarrow(G)\to\mathcal{F}_\downarrow(B),\quad F\mapsto F\cap B^\sharp$: 154
$\mathrm{Aff}(N_-)=N_-\rtimes\mathrm{End}(N_-)=N_-\rtimes\mathrm{End}(\mathfrak{n}_-)$: 159
$\mathrm{Aff}_{com}(N_-)=\{(n_-,\gamma)\in\mathrm{Aff}(N_-)\mid n_-\gamma(\overline{\Omega}_-)\subset\overline{\Omega}_-\}$: 159
$\overline{B^\sharp}$, the closure of $B^\sharp$ in $\mathrm{Aff}(N_-)$: 160
$\lambda_{e_X},\rho_{e_X}:\overline{B^\sharp}\to\overline{B^\sharp}$, left and right multiplication with e_X in $\overline{B^\sharp}$: 165
$S_A^{cpt}=\overline{S\cap A}=\overline{\exp c_{\max}}\subset\overline{B^\sharp}$, the closure of S_A^{cpt} in $\overline{B^\sharp}$: 160
$e_F:X\mapsto\begin{cases}0, & \text{if } X\in\mathfrak{g}_\alpha,\alpha\notin F\cap\Delta_-\\ X, & \text{if } X\in\mathfrak{g}_\alpha,\alpha\in F\cap\Delta_-\end{cases}$: 160
$E_X=(\mathbb{R}X-c_{\max})\cap c_{\max}$, the face of $c_{\max}$ generated by X: 162
$F_X=X^\perp\cap -c_{\max}^*=E_X^\perp\cap -c_{\max}^*$: 162
$\Delta_X=E_X^\perp\cap\Delta$: 162
W_X, the subgroup of W_0 generated by the reflections fixing X : 163
Σ_X: 163
$E_{X,0}=E_X^\perp\cap\mathrm{span}\{\alpha_1,\ldots,\alpha_k\}=E_{X,0,\mathrm{eff}}\oplus E_{X,0,\mathrm{fix}}$: 163
$E_{X,0,\mathrm{fix}}=\{Y\in E_{X,0}\mid(\forall w\in W_X)\ w\cdot Y=Y\}$: 163
$E_{X,0,\mathrm{eff}}=\mathrm{span}\{w\cdot Y-Y\mid w\in W_X,Y\in E_{X,0}\}$: 163
$U(T)$, the group of units in a monoid T: 168

Chapter 6

$\Gamma:\mathcal{F}(\mathcal{X})\to\mathcal{F}(G),\quad F\mapsto\{g\in G\mid g^{-1}\cdot F\subset\overline{\mathcal{O}}\}$, causal Galois connection: 173
$\hat{\Gamma}:2^G\to\mathcal{F}(\mathcal{X}),\quad A\mapsto\bigcap_{a\in A}a\cdot\overline{\mathcal{O}}$, dual map of Γ: 173
$\mathcal{M}^{\mathcal{O}}=\overline{\{g\cdot\overline{\mathcal{O}}\mid g\in G\}}\subset\mathcal{F}(\mathcal{X})$, causal orbit: 176
$\iota:G\to\mathcal{M}^{\mathcal{O}},g\mapsto g\cdot\overline{\mathcal{O}}$, causal orbit map: 176
$\overline{\Omega}_F=e_F\cdot\overline{\Omega}_-$, projection of $\overline{\Omega}_F$: 180
$\mathfrak{h}_X=\lim_{t\to\infty}e^{\mathrm{ad}\,tX}\mathfrak{h}=\mathfrak{z}_\mathfrak{h}(X)+\mathfrak{g}(\Delta^+\setminus X^\perp)$: 181
$H_F=\{g\in G\mid g\cdot\overline{\Omega}_F=\overline{\Omega}_F\}$, stabilizer of a projection $\overline{\Omega}_F$: 182
$\mathfrak{h}_F=\mathfrak{h}_X+\left[\mathfrak{h}_F\cap(\mathfrak{a}+\mathfrak{n}^\sharp)\right]$, the Lie algebra of H_F: 182
$\mathfrak{n}_{X,-}=\mathfrak{g}(\Delta_{X,-})$, image of the idempotent e_X: 183
$N_{X,-}=\exp(\mathfrak{n}_{X,-})$: 183
$N_\mathcal{Y}(L)=\{g\in L\mid g\cdot\mathcal{Y}=\mathcal{Y}\}$: 183
$Z_\mathcal{Y}(L)=\{g\in L\mid\forall y\in\mathcal{Y}\ :\ g\cdot y=y\}$: 183
$d(E)=\dim\ker e_F$, the degree of $E\in G\cdot\overline{\Omega}_F$: 187
$\Upsilon:\mathcal{F}\to(\mathcal{M}^{cpt}\setminus\{\emptyset\})/G,\quad F\mapsto G\cdot\overline{\Omega}_F$, classifying map for the G-orbits of $\mathcal{M}^{cpt}$: 188

Chapter 7

$\mathbf{V}^\infty$, the space of smooth vectors: 199
$C(\pi)$, the cone of negative elements: 200
$\mathcal{A}(C)$, the set of C-admissible representations: 200
$C(\mathbf{V})$, the space of contractions of $\mathbf{V}$: 200
$\widehat{S(C)}$, holomorphic dual of S: 202
Θ_π, the character of π: 202
$\mathbf{V}_K$, the K-finite elements in $\mathbf{V}$: 204
$\Phi_\pi(x) := \pi(k_H(x^{-1})^{-1})v_o$: 211
$\varphi(\pi, v)(x) = (v|\Phi_\pi(\bar{x}))$, the generating function for $\mathbf{E}_\pi \subset \mathbf{L}^2(\mathcal{M})$: 211
$\mathbf{E}_\pi$, the holomorphic discrete series: 212
$\Psi_\pi(g, x) = \pi(k_\mathbb{C}(g))\Phi_\pi(g^{-1}x)$, $g \in G$, $x \in P^+K_\mathbb{C}H_\mathbb{C}$: 213
$\mathcal{F}$, the classical Fourier transform: 215
e_u, the function $x \mapsto e^{(x|u)}$: 215
$\Xi(C), \Xi^o(C)$, the $\mathfrak{o}$-orbits of the Ol'shanskii semigroups $S(C)$ and $S(C^o)$: 217
$\mathcal{H}_2(C)$, the Hardy space corresponding to the cone C: 215,217
$\beta(f)$, the boundary value map: 217
$K(z, w)$: The Cauchy-Szegö kernel: 219
$P(z, m)$, the Poisson kernel: 220

Chapter 8

$\mathrm{M}_\leq = \{(x, y) \in \mathcal{M} \mid x \leq y\}$, the graph of the order $\leq$: 222
$\mathcal{V}(\mathcal{M})$, the Volterra algebra: 222,232
$\mathcal{V}(\mathcal{M})^\#$, the algebra of invariant Volterra kernels: 233, 233
$F\#G(x, y) = \int_{[x,y]} F(x, z)G(z, y)dz$, the product of two Volterra kernels: 222,232
$\mathbb{D}(\mathcal{M})$, the algebra of invariant differential operators on $\mathcal{M}$: 224
$C_c^\infty(G//K)$, the algebra of K-bi-invariant functions on G with compact support in $K\backslash G/K$: 224
$e_\lambda^K(x) = e^{<\lambda-\rho, a_K(x)>}$, $e_\lambda(x) = e_\lambda^H = e^{,\lambda-\rho, a_H(x)>}$: 225
$\varphi_\lambda^K(x) = \int_K e_\lambda^K(kx)\, dk$, the spherical function on G/K: 225
$\rho = \frac{1}{2}\sum_{\alpha \in \Delta^+} (\dim \mathfrak{g}_\alpha)\alpha$: 225
$\varphi_\lambda(s) = \int_H e_\lambda(sh)\, dh$, the spherical function with parameter λ: 226
$\rho_0 = \frac{1}{2}\sum_{\alpha \in \Delta_0^+} \dim \mathfrak{g}_\alpha\, \alpha$: 226
$\mathcal{E} = \{\lambda \in \mathfrak{a}_C \mid \forall \alpha \in \Delta_+ : \mathrm{Re}(\lambda + \rho|\alpha) < 0\}$, domain of definition for parameters for spherical functions: 226

$\mathcal{E}' = \{\lambda \in \mathfrak{a}_C^* \mid \int_{K\cap HAN} e_{-\mathrm{Re}\,\lambda}(k)\,dk < \infty\}$. Same as above: 227
$\rho_+ = \frac{1}{2}\sum_{\alpha\in\Delta_+} m_\alpha \alpha$: 227
$\mathbb{D}(\mathcal{M})$, the algebra of G-invariant differential operators on $\mathcal{M}$: 227
$A(C) := \exp C \subset A$: 227
$c_\Omega(\lambda) = \int_\Omega e_{-\lambda}(\exp X)\,dX$, the causal c-function: 228
$c_0(\lambda) = \int_{\overline{N}_0} e_{-\lambda}(n^\sharp)\,dn^\sharp$, classical c-function for $H^a/(H\cap K)$: 228
$c(\lambda) = c_\Omega(\lambda)c_0(\lambda)$: 228
$a \overset{A(C)}{\to} \infty$, convergence to ∞ in a Weyl chamber: 228
$c^r(\lambda) = \int_{N^\sharp} e_{-\lambda}(n^\sharp)\,dn^\sharp$, the classical c-function: 230
$\Delta_{\mathcal{M}^r}(D)$: the radial part of $D \in \mathbb{D}(\mathcal{M}^r)$: 230
Γ_μ: 231
$\Phi_\lambda(a) = a^{\lambda-\rho}\sum_{\mu\in\Lambda}\Gamma_\mu(\lambda)a^{-\mu}$: 231
$c_+(\lambda)$: 232
$\mathcal{L}F(\lambda) = \int_{\mathcal{M}} F(\mathrm{o},m)e_\lambda(m)\,dm$, the Laplace transform of a Volterra kernel: 233
$C_c^\infty(HA(C)H//H)$, the space of H-biinvariant functions on $HA(C)H$ that have compact support in $H\backslash HA(C)H/H$: 234
$\mathcal{D}(f)$, domain of definition of the Laplace transform of f: 233
$\tilde{\varphi}_\lambda(x) = [1/c(\lambda)]\varphi_\lambda(x)$, the normalized spherical function: 234
$\tilde{\mathcal{L}}(f)(\lambda) = [1/c(\lambda)]\mathcal{L}(f)(\lambda)$, the normalized Laplace transform: 234
f^γ, the K^r-bi-invariant extension: 231
$\tilde{\varphi}_\lambda^r(x) = \frac{1}{c(\lambda)}\varphi_\lambda^r(x)$: 235
$E_\lambda(h_1 a h_2) := \varphi_\lambda^r(a)$: 235
$\tilde{E}_\lambda := \frac{1}{c^r(\lambda)}E_\lambda$: 235
$\mathcal{A}(f)(a) = a^\rho \int_N f(an)dn$, the Abel transform: 236
$\mathcal{A}_+ f(g_0) = a^{\rho_+}\int_{N_+} f(g_0 n_+)dn_+$: 236

Chapter 9

$W(f)$, classical Wiener-Hopf operator with symbol f: 239
$\mathcal{W}_{\mathcal{M}_+}$, C^*-algebra generated by all Wiener-Hopf operators: 239
$W_{\mathcal{M}_+}(f)$: Wiener-Hopf operator with symbol f: 241
$\mathcal{G}$, a groupoid: 242
$\mathcal{G}^2$, composable pairs in $\mathcal{G}$: 242
$d(x)$, domain map for $\mathcal{G}$: 242
$r(x)$, range map for $\mathcal{G}$: 242
$\mathcal{G}^0$, unit space of $\mathcal{G}$: 242
$L^1(\mathcal{G})$, closure of $C_c(\mathcal{G})$: 243
$C^*(\mathcal{G})$, C^*-algebra generated by $\mathbf{L}^1(\mathcal{G})$ w.r.t. suitable norm: 243

$W_{\mathcal{M}_+}$, Wiener-Hopf representation of $C^*(\mathcal{G})$: 244

Appendixes

$\mathbf{V}(\lambda, T) = \{v \in \mathbf{V} \mid Tv = \lambda v\}$, eigenspace of T for the eigenvalue λ: 246
$\mathbf{V}(\lambda, X) = \mathbf{V}(\lambda, \pi(X))$, for a representation π: 246
$\mathbf{V}(\alpha, \mathfrak{b}) = \mathbf{V}_\alpha$, simultaneous eigenspaces: 246
$\mathbf{V}^{\mathfrak{b}} = \mathbf{V}(0, \mathfrak{b})$: 246
$\Delta(\mathbf{V}, \mathfrak{b})$, the set of weights: 246
$\mathbf{V}(\Gamma) = \bigoplus_\alpha \mathbf{V}_\alpha$, the sum of weight spaces: 247
θ, the Cartan involution: 247
$\mathfrak{k} = \mathfrak{g}^\theta$, the maximal compactly embedded subalgebra: 247
$\mathfrak{p} = \mathfrak{g}(-1, \theta)$: 247
$\mathfrak{a}$, a maximal abelian subalgebra in $\mathfrak{p}$: 248
$\mathfrak{m} = \mathfrak{k}^{\mathfrak{a}}$: 248
$\Delta = \Delta(\mathfrak{g}, \mathfrak{a})$: 248
$\mathfrak{n} = \mathfrak{g}(\Delta^+)$, the sum of positive root spaces; Iwasawa $\mathfrak{n}$: 248
$W(\mathfrak{a}) = W$, the Weyl group of $(\mathfrak{g}, \mathfrak{a})$: 248
$\mathbf{V}^L = \{v \in \mathbf{V} \mid \forall a \in L : a \cdot v = v\}$, L-fixed vectors in $\mathbf{V}$: 249
Z^0, a central element in $\mathfrak{k}$ defining a complex structure on $\mathfrak{p}$: 253
Δ_k, the set of compact roots: 253
Δ_n, the set of noncompact roots: 253
E_α, root vectors for $\alpha \in \Delta_n$ (suitably normalized): 254
H_α, co-roots for α: 254
Γ, maximal set of strongly orthogonal positive noncompact roots: 254
$E_{\pm j}$, normalized root vectors for $\gamma_j \in \Gamma$: 254
H_j, corresponding co-roots: 254
$X_j = -i(E_j - E_{-j})$: 254
$Y_j = E_j + E_{-j}$: 254
$X_o = \frac{1}{2} \sum X_j$: 254
$Y_o = \frac{1}{2} \sum Y_j$: 254
E_o: 254
Z_o: 254
$\mathfrak{t}^-$, the subspace of $\mathfrak{t}$ generated by H_j: 254
$\mathbf{C}_j$, partial Cayley transform: 255
$\mathcal{C}(K)$, the space of compact subsets of K: 257
$\mathcal{C}_0(K)$, the space of nonempty compact subset of K: 257
$d(A, b)$, the Hausdorf distance: 257
$d(A, B)$: 257
$K(U, V)$, a subbasis for the Vietoris topology: 257

$\mathcal{F}(X)$, the space of closed subsets of X: 257
X^ω, the one-point compactification of X: 257
β, the one-point compactification map: 258
liminf, the limes inferior for sets: 258
limsup, the limes superior for sets: 258
$\mathcal{F}(G)^H$, the set of H-fixed points of $\mathcal{F}(G)$ under translation: 260

Bibliography

[1] Ban, E., van den: A convexity theorem for semisimple symmetric spaces. *Pac. J. Math.* **124** (1986), 21–55.

[2] Ban, E., van den: The principal series for reductive symmetric spaces I, H-fixed distributions vectors. *Ann. Sci. École Norm. Sup.* **21** (1988), 359–412.

[3] Ban, E., van den: The principal series for a reductive symmetric space. II. Eisenstein integrals. *J. Funct. Anal.* **109** (1992), 331–441.

[4] Beem, J. K., P. E. Ehrlich,: *Global Lorentzian Geometry.* Marcel Dekker, New York, 1981.

[5] Berger, M.: Les espaces symétriques non compacts. *Ann. Sci. Ecole Norm. Sup* **74** (1957), 85–177.

[6] Bertram, W.: On the causal group of some symmetric spaces. Preprint, 1996.

[7] Betten, F.: Kausale Kompaktifizierung kompakt-kausaler Räume. Dissertation, Göttingen, 1996.

[8] Bourbaki, N.: *Groupes et algébres de Lie, Ch. 4, 5 et 6.* Masson, Paris, 1981.

[9] Bourbaki, N.: *Groupes et algèbres de Lie, Ch. 9.* Masson, Paris, 1982.

[10] Bourbaki, N.: *Groupes et algébres de Lie, Ch. 7 et 8.* Masson, Paris, 1990.

[11] Bourbaki, N.: *Topologie générale.* Hermann, Paris, 1991.

[12] Brylinski, J.-L., P. Delorme: Vecteurs distributions H-invariants pour les séries principales généralisées d'espaces symetriques reductifs et prolongement meromorphe d'intégrales d'Eisenstein. *Invent. Math.* **109** (1992), 619–664.

[13] Carmona, J., P. Delorme: Base méromorphe de vecteurs distributions H-invariants pour les séries principales généralisées d'espaces symétriques réductifs: Equation fonctionnelle. *J. Funct. Anal.* **122** (1994), 152–221.

[14] Carruth, J. N., J. A. Hildebrand, R. J. Koch: *The Theory of Topological Semigroups, Vol. II.* Marcel Dekker, New York, 1986.

[15] Chadli, M.: Domaine complexe associé à un espace symétrique de type Cayley. *C.R. Acad. Sci. Paris* **321** (1995), 1157–1161.

[16] Clerc, J.-.L.: Laplace transforms and unitary highest weight modules. *J. Lie Theory*, **5** (1995), 225-240.

[17] Cronström, C., W. H. Klink: Generalized $O(1,2)$ expansion of multiparticle amplitudes. *Ann. Phys.* **69** (1972), 218–278.

[18] Davidson, M. G., R. C. Fabec: Geometric realizations for highest weight representations. *Contemp. Math.*, in press.

[19] Delorme, P.: Coefficients généralisé de séries principales sphériques et distributions sphériques sur G_C/G_R. *Invent. Math.* **105** (1991), 305–346.

[20] Dijk, G. van: Orbits on real affine symmetric spaces. *Proc. Koninklijke Akad. van Wetenschappen* **86** (1983), 51–66.

[21] Doi, H.: A classification of certain symmetric Lie algebras. *Hiroshima Math. J.* **11** (1981), 173–180.

[22] Enright, T. J., R. Howe, N. Wallach: A classification of unitary highest weight modules. Proceedings: Representation Theory of Reductive Groups. *Progr. Math.* **40** (1983), 97–143.

[23] Erdelyi, A., et al.: *Higher Transcendental Functions*, I. McGraw-Hill, New York, 1953.

[24] Faraut, J.: Algèbres de Volterra et transformation de Laplace sphérique sur certains espaces symetriques ordonnes. *Symp. Math.* **29** (1986), 183–196.

[25] Faraut, J.: Espace symetriques ordonnés et algèbres de Volterra. *J. Math. Soc. Jpn.* **43** (1991), 133–147.

[26] Faraut, J.: Functions sphériques sur un espace symétrique ordonné du type Cayley. Preprint, 1994.

[27] Faraut, J.: Hardy spaces in non-commutative harmonic analysis. Lectures, Summer Schools, Tuczno, 1994. Preprint, Paris, 1994.

[28] Faraut, J., J. Hilgert, G. Ólafsson: Spherical functions on ordered symmetric spaces. *Ann. Inst. Fourier* **44** (1995), 927–966.

[29] Faraut, J., A. Korányi: *Analysis on Symmetric Cones.* Clarendon Press, Oxford, 1994.

[30] Faraut, J., G. Ólafsson: Causal semisimple symmetric spaces; The geometry and harmonic Analysis. In: *Semigroups in Algebra, Geometry and Analysis*, De Gruyter, 1995.

[31] Faraut, J., G. A. Viano: Volterra algebra and the Bethe-Salpeter equation. *J. Math. Phys.* **27** (1986), 840–848.

[32] Flensted-Jensen, M.: Discrete series for semisimple symmetric spaces. *Ann. Math.* **111** (1980), 253–311.

[33] Flensted-Jensen, M.: Analysis on non-Riemannian symmetric spaces. *CBMS Reg. Conf.* **61** (1986).

[34] Gel'fand, I. M., S. G. Gindikin: Complex manifolds whose skeletons are semisimple real Lie groups, and analytic discrete series of representations. *Func. Anal. Appl.* **7** (1977), 19–27.

[35] Gindikin, S.G., F.I. Karpelevich: Plancherel measure of Riemannian symmetric spaces of non-positive curvature. *Dokl. Akad. Nauk USSR* **145** (1962), 252–255.

[36] Gradshteyn, I., I. Ryzhik: *Tables of Integrals, Series and Products.* Academic Press, 1980.

[37] Günter, P.: *Huygen's Principle for Linear Partial Differential Operators of Second Order.* Birkhäuser, Boston, 1987.

[38] Harish-Chandra: Representations of semisimple Lie groups, IV. *Amer. J. Math.* **77** (1955), 743–777.

[39] Harish-Chandra: Representations of semisimple Lie groups V, VI. *Amer. J. Math.* **78** (1956), 1–41, 564–628.

[40] Harish-Chandra: Spherical functions on a semisimple Lie group, I. *Amer. J. Math.* **80** (1958), 241–310.

[41] Harish-Chandra: Spherical functions on a semisimple Lie group, II, *Amer. J. Math.* **80** (1958), 553–613.

[42] Hawking, S. W., G. F. R. Ellis: *The Large Scale Structure of Space-Time.* Cambridge University Press, Cambridge, 1973.

[43] Helgason, S.: A duality for symmetric spaces with application to group representations. *Adv. Math.* **5** (1970), 1–154.

[44] Helgason, S.: *Differential Geometry, Lie Groups and Symmetric Spaces.* Academic Press, New York, 1978.

[45] Helgason, S.: *Groups and Geometric Analysis. Integral Geometry, Invariant Differential Operators and Spherical Functions.* Academic Press, New York, 1984.

[46] Helgason, S.: *Geometric Analysis on Symmetric Spaces.* AMS Math. Surveys and Monographs **39**, Providence, R. I., 1994.

[47] Hilgert, J.: *Subsemigroups of Lie groups.* Habilitationsschrift, Darmstadt, 1987.

[48] Hilgert, J.: Invariant cones in symmetric spaces of hermitian type. Notes, 1989.

[49] Hilgert, J.: A convexity theorem for boundaries of ordered symmetric spaces. *Can. J. Math.* **46** (1994), 746–757.

[50] Hilgert, J., K. H. Hofmann, J. D. Lawson: *Lie Groups, Convex Cones and Semigroups.* Oxford University Press, 1989.

[51] Hilgert, J., K.-H. Neeb: *Lie-Gruppen und Lie-Algebren.* Vieweg, Braunschweig, 1991.

[52] Hilgert, J., K.-H. Neeb: *Lie Semigroups and Their Applications.* Lect. Notes Math. **1552**, Springer, 1993.

[53] Hilgert, J., K.-H. Neeb: A general setting for Wiener-Hopf operators. In: *75 Years of Radon Transform*, S. Gindikin and P. Michor, Eds., Wien, 1994.

[54] Hilgert, J., K.-H. Neeb: Wiener Hopf operators on ordered homogeneous spaces I. *J. Funct. Anal.* 132 (1995), 86–116.

[55] Hilgert, J., K.-H. Neeb: Groupoid C^*-algebras of order compactified symmetric spaces. *Jpn. J. Math.* **21** (1995), 117–188.

[56] Hilgert, J., K.-H. Neeb: Compression semigroups of open orbits on complex manifolds. *Arkif Mat.* **33** (1995), 293–322.

[57] Hilgert, J., K.-H. Neeb: Compression semigroups of open orbits on real flag manifolds. *Monat. Math.* **119** (1995), 187–214.

[58] Hilgert, J., K.-H. Neeb: Maximality of compression semigroups. *Semigroup Forum* **50** (1995), 205–222.

[59] Hilgert, J., K.-H. Neeb, B. Ørsted: The geometry of nilpotent coadjoint orbits of convex type in Hermitian Lie algebras. *J. Lie Theory* **4** (1994), 185–235.

[60] Hilgert, J., K.-H. Neeb, B. Ørsted: Conal Heisenberg algebras and associated Hilbert spaces. *J. reine und angew. Math.*, in press.

[61] Hilgert, J., K.-H. Neeb, B. Ørsted: Unitary highest weight representations via the orbit method I: The scalar case. Proceedings of "Representations of Lie Groups, Lie Algebras and Their Quantum Analogues," Enschede, Netherlands, 1994, in press.

[62] Hilgert, J., K.-H. Neeb, W. Plank: Symplectic convexity theorems. *Comp. Math.* **94** (1994), 129–180.

[63] Hilgert, J., G. Ólafsson, B. Ørsted: Hardy spaces on affine symmetric spaces. *J. reine und angew. Math.* **415** (1991), 189–218.

[64] Hilgert, J., G. Ólafsson: Analytic extensions of representations, the solvable case. *Jpn. J. Math.* **18** (1992) 213–290.

[65] Hirzebruch, U.: Über Jordan-Algebren und beschränkte symmetrische Räume von Rang 1. *Math. Z.* **90** (1965), 339–354.

[66] Hochschild, G.: *The Structure of Lie Groups.* Holden Day, San Francisco, 1965.

[67] Hofmann, K.H., J. D. Lawson: Foundations of Lie semigroups. Lect. Notes Math. **998** (1983), 128–201.

[68] Humphreys, J.E.: *Introduction to Lie Algebras and Representation Theory.* Springer, New York-Heidelberg-Berlin, 1972.

[69] Jaffee, H.: Real forms of Hermitian symmetric spaces. *Bull. Amer. Math. Soc.* **81** (1975), 456–458.

[70] Jaffee, H.: Anti-holomorphic automorphisms of the exceptional symmetric domains. *J. Diff. Geom.* **13** (1978), 79–86.

[71] Jakobsen, H.P.: Hermitean symmetric spaces and their unitary highest weight modules. *J. Funct. Anal.* **52** (1983) 385–412.

[72] Jakobsen, H.P., M. Vergne: Restriction and expansions of holomorphic representations, *J. Funct. Anal.* **34** (1979), 29–53.

[73] Jorgensen, P.E.T.: Analytic continuation of local representations of Lie groups. *Pac. J. Math* **125** (1986), 397–408.

[74] Jorgensen, P.E.T.: Analytic continuation of local representations of symmetric spaces. *J. Funct. Anal.* **70** (1987), 304–322.

[75] Kaneyuki, S.: On classification of parahermitian symmetric spaces. *Tokyo J. Math.* **8** (1985), 473–482.

[76] Kaneyuki, S.: On orbit structure of compactifications of parahermitian symmetric spaces. *Jpn. J. Math.* **13** (1987), 333–370.

[77] Kaneyuki, S., H. Asano: Graded Lie algebras and generalized Jordan triple systems. *Nagoya Math. J.* **112** (1988), 81–115.

[78] Kaneyuki, S., M. Kozai: Paracomplex structures and affine symmetric spaces. *Tokyo J. Math.* **8** (1985), 81–98.

[79] Knapp, A.W.: *Representation Theory of Semisimple Lie Groups.* Princeton University Press, Princeton, N. J., 1986.

[80] Kobayashi, S., T. Nagano: On filtered Lie algebras and geometric structures I, *J. Math. Mech.*, **13** (1964), 875-908.

[81] Kobayashi, S., K. Nomizu: *Foundation of Differential Geometry I, II.* Wiley (Interscience), New York, 1963 and 1969.

[82] Kockmann, J.: Vertauschungsoperatoren zur holomorphen diskreten Reihe spezieller symmetrischer Räume. Dissertation, Göttingen, 1994.

[83] Korányi, A., J.A. Wolf: Realization of Hermitian symmetric spaces as generalized half-planes. *Ann. Math.* **81** (1965), 265–288.

[84] Korányi, A., J. A. Wolf: Generalized Cayley transformations of bounded symmetric domains. *Amer. J. Math.* **87** (1965), 899–939.

[85] Kostant, B., S. Sahi: Jordan algebras and Capelli identities. *Invent. Math.* **122** (1993), 657–665.

[86] Koufany, Kh.: Semi-groupes de Lie associé à une algèbre de Jordan euclidienne. Thesis, Nancy, 1993.

[87] Koufany, Kh.: Réalisation des espaces symétriques de type Cayley. *C.R. Acad. Sci. Paris* (1994), 425–428.

[88] Koufany, Kh., B. Ørsted: Function spaces on the Ol'shanskii semigroup and the Gelfand-Gindikin program. Preprint, 1995.

[89] Koufany, Kh., B. Ørsted: Éspace de Hardy sur le semi-groupe metaplectique. Preprint, 1995.

[90] Koufany, Kh., B. Ørsted: Hardy spaces on the two sheeted covering semigroups. Preprint, 1995.

[91] Krötz, B.: Plancherel–Formel für Hardy–Räume. Diplomarbeit, Darmstadt, 1995.

[92] Kunze, R.A.: Positive definite operator-valued kernels and unitary representations. *Proceedings of the Conference on Functional Analysis at Irvine, California.* Thompson Book Company 1966, 235–247.

[93] Lawson, J.D.: Ordered manifolds, invariant cone fields, and semigroups, *Forum Math.* **1** (1989), 273–308.

[94] Lawson, J.D.: Polar and Olshanskii decompositions. *J. reine und angew. Math.* **448** (1994), 191–219.

[95] Libermann, P.: Sur les structures presque paracomplexes, *C.R. Acad. Sci. Paris* **234** (1952), 2517–2519.

[96] Libermann, P.: Sur le probleme d'equivalence de certaines structures infinitesimales. *Ann. Math. Pur. Appl.* **36** (1954), 27–120.

[97] Loos, O.: *Symmetric Spaces, I: General Theory.* W.A. Benjamin, Inc., New York, 1969.

[98] Makarevic, B.O.: Open symmetric orbits of reductive groups in symmetric R-spaces. *USSR Sbornik* **20** (1973), 406–418.

[99] Matsuki, T.: The orbits of affine symmetric spaces under the action of minimal parabolic subgroups. *J. Math. Soc. Jpn.* **31** (1979), 331–357.

[100] Matsuki, T.: Orbits on affine symmetric spaces under the action of parabolic subgroups. *Hiroshima Math. J.* **12** (1982), 307–320.

[101] Matsuki, T.: Closure relations for orbits on affine symmetric spaces under the action of minimal parabolic subgroups. *Hirsoshima Math. J.* **18** (1988), 59–67.

[102] Matsuki, T.: A description of discrete series for semisimple symmetric spaces II. *Adv. Studies Pure Math.* **14** (1988), 531–540.

[103] Matsumoto, S.: Discrete series for an affine symmetric space. *Hiroshima Math. J.* **11** (1981) 53–79.

[104] Mittenhuber, D., K.-H. Neeb: On the exponential function of an ordered manifold with affine connection. Preprint.

[105] Mizony, M.: Une transformation de Laplace-Jacobi, *SIAM J. Math.* **116** (1982), 987–1003.

[106] Mizony, M.: Algèbres de noyaux sur des éspaces symetriques de SL(2, $\mathbb{R}$) et SL(3, $\mathbb{R}$) et fonctions de Jacobi de premiere er deuxieme espéces. Preprint.

[107] Molchanov, V.F.: Holomorphic discrete series for hyperboloids of Hermitian type. Preprint, 1995.

[108] Muhly, P., J.Renault: C^*-algebras of multivariable Wiener Hopf operators. *Trans. Amer. Math. Soc.* **274** (1982), 1–44.

[109] Muhly, P., J. Renault, D. Williams: Equivalence and isomorphisms of groupoid C^*-algebras. *J. Operator Theory* **17** (1987), 3–22.

[110] Moore, C.C.: Compactification of symmetric spaces II. *Amer. J. Math.* **86** (1964), 358–378.

[111] Nagano, T.: Transformation groups on compact symmetric spaces. *Trans. Amer. Math. Soc.* **118** (1965), 428–453.

[112] Neeb, K.-H.: Globality in semisimple Lie groups, *Ann. Inst. Fourier* **40** (1991), 493–536.

[113] Neeb, K.-H.: Semigroups in the universal covering group of SL(2). *Semigroup Forum* **43** (1991), 33–43.

[114] Neeb, K.-H.: Conal orders on homogeneous spaces. *Invent. Math.* **134** (1991), 467–496.

[115] Neeb, K.-H.: Monotone functions on symmetric spaces. *Math. Annalen* **291** (1991), 261–273.

[116] Neeb, K.-H.: A convexity theorem for semisimple symmetric spaces. *Pac. J. Math.* **162** (1994), 305–349.

[117] Neeb, K.-H.: Holomorphic representation theory II. *Acta Math.* **173** (1994), 103–133.

[118] Neeb, K.-H.: The classification of Lie algebras with invariant cones. *J. Lie Theory* **4** (1994), 139–183.

[119] Neeb, K.-H.: Realization of general unitary highest weight representations. To appear in *Forum Math.*

[120] Neeb, K.-H.: Holomorphic Representation Theory I. *Math. Annalen* **301** (1995), 155–181.

[121] Neeb, K.-H.: Representations of involutive semigroups, *Semigroup Forum* **48** (1996), 197–218.

[122] Neeb, K.-H.: Invariant orders on Lie groups and coverings of ordered homogeneous spaces. Submitted.

[123] Neeb, K.-H.: Coherent states, holomorphic representations, and highest weight representations. *Pac. J. Math.*, in press.

[124] Neeb, K.-H.: A general non-linear convexity theorem. Submitted.

[125] Neidhardt, C.: Konvexität und projective Einbettungen in der Riemannschen Geometrie. Dissertation, Erlangen, 1996

[126] Nica, A.: Some remarks on the groupoid approach to Wiener Hopf operators. *J. Operator Theory* **18** (1987), 163–198.

[127] Nica, A.: Wiener-Hopf operators on the positive semigroup of a Heisenberg group. In: *Operator Theory: Advances and Applications* **43**. Birkhäuser, Basel, 1990.

[128] Ólafsson, G.: Fourier and Poisson transformation associated to a semisimple symmetric space, *Invent. math.* **90** (1987), 605–629.

[129] Ólafsson, G.: Causal Symmetric Spaces. *Math. Gotting.* **15** (1990), pp 91.

[130] Ólafsson, G.: Symmetric spaces of Hermitian type. *Diff. Geom. and Appl.* **1** (1991), 195–233.

[131] Ólafsson, G.: Spherical function on a causal symmetric space. In preparation.

[132] Ólafsson, G.: Fatou-type lemma for symmetric spaces and applications to the character formula. In preparation.

[133] Ólafsson, G., B. Ørsted: The holomorphic discrete series for affine symmetric spaces, I. *J. Funct. Anal.* **81** (1988), 126–159.

[134] Ólafsson, G., B. Ørsted: Is there an orbit method for affine symmetric spaces? In: M. Duflo, N.V. Pedersen, M. Vergne, Eds.: *The Orbit Method in Representation Theory*. Birkhäuser, Boston 1990.

[135] Ólafsson, G., B. Ørsted: The holomorphic discrete series of an affine symmetric space and representations with reproducing kernels, *Trans. Amer. Math. Soc.* **326** (1991), 385–405.

[136] Ólafsson, G., B. Ørsted: Causal compactification and Hardy spaces. Preprint, 1996.

[137] Ol'shanskii, G.I.: Invariant cones in Lie algebras, Lie semigroups, and the holomorphic discrete series. *Func. Anal. Appl.* **15** (1982), 275–285.

[138] Ol'shanskii, G.I.: Convex cones in symmetric Lie algebra, Lie semigroups and invariant causal (order) structures on pseudo-Riemannian symmetric spaces. *Sov. Math. Dokl.* **26** (1982), 97–101.

[139] Ol'shanskii, G.I.: Complex Lie semigroups, Hardy spaces and the Gelfand-Gindikin program. *Diff. Geom. Appl.* **1** (1991), 235–246.

[140] Oshima, T.: Poisson transformation on affine symmetric spaces. *Proc. Jpn. Acad. Ser. A* **55** (1979), 323–327.

[141] Oshima, T.: Fourier analysis on semisimple symmetric spaces. In: H. Carmona, M. Vergne, Eds. *Noncommutative Harmonic Analysis and Lie Groups*. Proceedings Luminy, 1980. Lect. Notes Math. **880** (1981), 357–369.

[142] Oshima, T.: Discrete series for semisimple symmetric spaces. *Proc. Int. Congr. Math.*, Warsaw, 1983.

[143] Oshima, T., T. Matsuki: Orbits on affine symmetric spaces under the action of the isotropic subgroups. *J. Math. Soc. Jpn.* **32**, (1980), 399–414.

[144] Oshima, T., T. Matsuki: A description of discrete series for semisimple symmetric spaces. *Adv. Stud. Pure Math.* **4** (1984), 331–390.

[145] Oshima, T., J. Sekiguchi: Eigenspaces of invariant differential operators on an affine symmetric space. *Invent. Math.* **57** (1980), 1–81.

[146] Oshima, T., J. Sekiguchi: The restricted root system of a semisimple symmetric pair. *Adv. Stud. Pure Math.* **4** (1984), 433–497.

[147] Paneitz, S.: Invariant convex cones and causality in semisimple Lie algebras and groups. *J. Funct. Anal.* **43** (1981), 313–359.

[148] Paneitz, S.: Determination of invariant convex cones in simple Lie algebras. *Arkiv Mat.* **21** (1984), 217–228.

[149] Penrose, R.: *Techniques of Differential Topology in Relativity.* Regional Conference Series in Applied Math. SIAM **7**, 1972.

[150] Renault, J.: A groupoid approach to C^*-algebra. *Lect. Notes Math.* **793**, Springer, Berlin, 1980.

[151] Renault, J.: Represéntation des produits croisés d'algèbres de groupoides. *J. Operator Theory* **18**(1987), 67–97.

[152] Rossi, H., M. Vergne: Analytic continuation of the holomorphic discrete series of a semi-simple Lie group. *Acta Math.* **136** (1976), 1–59.

[153] Rudin, W.: *Functional Analysis.* McGraw-Hill, New York, 1973.

[154] Ruppert, W.A.F.: A geometric approach to the Bohr compactification of cones. *Math. Z.* **199** (1988), 209–232.

[155] Satake, I.: *Algebraic Structures of Symmetric Domains.* Iwanami Shouten, Tokyo, and Princeton University Press, Princeton, N. J., 1980.

[156] Schlichtkrull, H.: *Hyperfunctions and Harmonic Analysis on Symmetric Spaces.* Prog. Math. **49**. Birkhäuser, Boston 1984.

[157] Segal, I.E.: *Mathematical Cosmology and Extragalactic Astronomy.* Academic Press, New York, 1976.

[158] Spindler K.: Invariante Kegel in Liealgebren. *Mitteilungen aus dem Math. Sem. Giessen*, **188** (1988), pp 159.

[159] Stanton, R.J.: Analytic extension of the holomorphic discrete series, *Amer. J. Math.* **108** (1986), 1411–1424.

[160] Stein, E.M., G. Weiss: *Introduction to Fourier Analysis on Euclidean Spaces.* Princeton University Press, Princeton, NJ, 1971.

[161] Thorleifsson, H.: Die verallgemeinerte Poisson Transformation für reduktive symmetrische Räume. Habilitationsschrift, Göttingen, 1994.

[162] Varadarajan, V.S.: *Lie Groups, Lie Algebras and Their Representations.* Prentice-Hall, Englewood Cliffs, N.J., 1974.

[163] Varadarajan, V.S.: *Harmonic Analysis on Real Reductive Groups.* Lect. Notes Math. **576**, Springer Verlag, Berlin, 1977.

[164] Viano, G.A.: On the harmonic analysis of the elastic scattering amplitude of two spinless particles at fixed momentum transfer, *Ann. Inst. H. Poincaré* **32** (1980), 109–123.

[165] Vinberg, E.B.: Homogeneous cones. *Sov. Math. Dokl.* **1** (1961), 787–790.

[166] Vinberg, E.B.: Invariant convex cones and ordering in Lie groups, *Func. Anal. Appl.* **15** (1982), 1–10.

[167] Wallach, N.: The analytic continuation of the discrete series, I, II. *Trans. Amer. Math. Soc.* **251** (1979), 1–17, 19–37.

[168] Warner, G.: *Harmonic Analysis on Semisimple Lie Groups I.* Springer Verlag, Berlin, 1972.

[169] Wolf, J.A.: The action of a real semisimple Lie group on a complex flag manifold, I: Orbit structure and holomorphic arc components. *Bull. Amer. Math. Soc.* **75** (1969), 1121–1237.

[170] Wolf, J.A.: The fine structure of hermitian symmetric spaces. In: W. Boothby and G. Weiss, Eds., *Symmetric Spaces,* Marcel Dekker, New York, 1972.

Index

Perspectives in Mathematics

Vol. 1 J.S. Milne, *Arithmetic Duality Theorems*
Vol. 2 A. Borel et al., *Algebraic D-Modules*
Vol. 3 Ehud de Shalit, *Iwasawa Theory of Elliptic Curves with Complex Multiplication*
Vol. 4 M. Rapoport, N. Schappacher, P. Schneider, editors, *Beilinson's Conjectures on Special Values of L-Functions*
Vol. 5 Paul Günther, *Huygens' Principle and Hyperbolic Equations*
Vol. 6 Stephen Gelbart and Freydoon Shahidi, *Analytic Properties of Automorphic L-Functions*
Vol. 7 Mark Ronan, *Lectures on Buildings*
Vol. 8 K. Shiohama, editor, *Geometry of Manifolds*
Vol. 9 F. Reese Harvey, *Spinors and Calibrations*
Vol. 10 L. Clozel and J.S. Milne, editors, *Automorphic Forms, Shimura Varieties, and L-functions, Volume I*
Vol. 11 L. Clozel and J.S. Milne, editors, *Automorphic Forms, Shimura Varieties, and L-functions, Volume II*
Vol. 12 Gabor Toth, *Harmonic Maps and Minimal Immersions Through Representation Theory*
Vol. 13 James W. Cogdell and Ilya Piatetski-Shapiro, *The Arithmetic and Spectral Analysis of Poincaré Series*
Vol. 14 D. Gieseker, H. Knorrer, and E. Trubowitz, *The Geometry of Algebraic Fermi Curves*
Vol. 15 Valentino Cristante and William Messing, editors, *Barsotti Symposium in Algebraic Geometry*
Vol. 16 Gerrit Heckman and Henrik Schlichtkrull, *Harmonic Analysis and Special Functions on Symmetric Spaces*
Vol. 17 Henrik Schlichtkrull and Bent Ørsted, editors, *Algebraic and Analytic Methods in Representation Theory*
Vol. 18 Joachim Hilgert and Gestur Ólafsson, *Causal Symmetric Spaces: Geometry and Harmonic Analysis*

Printed in Dunstable, United Kingdom

91342987R00178